# Intelligent Systems Reference Library

Volume 167

The aim of this series is to publish a Reference Library, including novel advances and developments in all aspects of Intelligent Systems in an easily accessible and well structured form. The series includes reference works, handbooks, compendia, textbooks, well-structured monographs, dictionaries, and encyclopedias. It contains well integrated knowledge and current information in the field of Intelligent Systems. The series covers the theory, applications, and design methods of Intelligent Systems. Virtually all disciplines such as engineering, computer science, avionics, business, e-commerce, environment, healthcare, physics and life science are included. The list of topics spans all the areas of modern intelligent systems such as: Ambient intelligence, Computational intelligence, Social intelligence, Computational neuroscience, Artificial life, Virtual society, Cognitive systems, DNA and immunity-based systems, e-Learning and teaching, Human-centred computing and Machine ethics, Intelligent control, Intelligent data analysis, Knowledge-based paradigms, Knowledge management, Intelligent agents, Intelligent decision making, Intelligent network security, Interactive entertainment, Learning paradigms, Recommender systems, Robotics and Mechatronics including human-machine teaming, Self-organizing and adaptive systems, Soft computing including Neural systems, Fuzzy systems, Evolutionary computing and the Fusion of these paradigms, Perception and Vision, Web intelligence and Multimedia.

** Indexing: The books of this series are submitted to ISI Web of Science, SCOPUS, DBLP and Springerlink.

More information about this series at http://www.springer.com/series/8578

Andree Woodcock · Louise Moody ·
Deana McDonagh · Ajita Jain ·
Lakhmi C. Jain
Editors

# Design of Assistive Technology for Ageing Populations

Springer

*Editors*
Andree Woodcock
Centre for Arts, Memory and Communities
Faculty of Arts and Humanities
Coventry University
Coventry, UK

Louise Moody
Centre for Arts, Memory and Communities
Faculty of Arts and Humanities
Coventry University
Coventry, UK

Deana McDonagh
School of Art and Design
University of Illinois at Urbana-Champaign
Champaign, IL, USA

Ajita Jain
Elizabeth Grove Surgery
Adelaide, SA, Australia

Lakhmi C. Jain
Faculty of Engineering and Information Technology
Centre for Artificial Intelligence
University of Technology
Sydney, NSW, Australia

Faculty of Science, Technology and Mathematics
University of Canberra
Canberra, ACT, Australia

KES International
Shoreham-by-Sea, UK

Liverpool Hope University
Liverpool, UK

ISSN 1868-4394 ISSN 1868-4408 (electronic)
Intelligent Systems Reference Library
ISBN 978-3-030-26294-5 ISBN 978-3-030-26292-1 (eBook)
https://doi.org/10.1007/978-3-030-26292-1



This Springer imprint is published by the registered company Springer Nature Switzerland AG
The registered company address is: Gewerbestrasse 11, 6330 Cham, Switzerland

# Preface

The world population is ageing. Longevity is typically associated with age-related illness disease and increased multi-morbidity. This places a significant and increasing burden on health and social care. However, the World Bank report,[1] on Golden Ageing argued that ageing societies are not pre-destined to experience stagnation or a decline in living standards.

As medical science and public health services seek to address this challenge at the individual and population level, the design community has an increasingly important role to play in the development of products, services and environments that not only maintain but also increase the quality of life. Designers, not only produce tools, systems and services that support independent and healthy ageing, but have also developed tools and methods, which give people the opportunity to create and shape their own future and the products that they will use. This book, featuring contributions from researchers across the world, provides examples of issues designers consider when they are designing, working with and for older people.

As well as the compelling moral and ethical arguments to ensure that older people are supported in enjoying a high quality of life, there is a commercial driver for design to be at the forefront of commercial enterprises. According to a new report for the European Commission by Technopolis Group and Oxford Economics, the total consumption by or on behalf of the 24m people over 50 years was close to £400 bn in 2015 and is expected to grow by 40% until 2025. The 80+ age group represents an expanding market, estimated at £21.4 billion a year, or 1.6% of GDP (Source Innovate UK/Technology Strategy Board 2014). The global connected home market is estimated to be $150b by 2020. As an example, '53% of American consumers predict a singular remote that controls everything in the home will be the norm in the next 10 years' (PWC).

The market is expected to grow strongly from 2020 onward, based on the assumption that a 'new' generation of older people, who are more tech-savvy than the previous generation, will be more inclined to invest in smart home solutions.

[1]http://www.worldbank.org/en/region/eca/publication/golden-aging.

In parallel, BCC Research predicts that the global market for ICT solutions for healthcare monitoring in private homes will triple from $11.3b in 2016 to roughly $33.1b by 2021. Predicted services change the focus from treatment to prevention or management.

The challenge for the design community is to ensure that across the life course, technology is being used to its best effect to help people solve challenges in living, to not let ageing and disabilities get in the way of their life. The approach shared here highlights the significant opportunity for the design community to ensure that assistive technology is not only people centred, but inclusive, adaptive and meets current and future needs.

This book has drawn together chapters from researchers and designers across the world[2] who are driven to define and develop solutions in this space. The approach is interdisciplinary, drawn from a range of related disciplines, e.g. product design, interaction design, engineering design, technology development, human factors, psychology.

Chapters range from describing methodological approaches for working with older adults to co-create and develop solutions, to explaining specific innovations and designs, to the demonstration and evaluation of products, systems and services. The book has been divided into six, overlapping part. Each chapter includes a full set of references to further knowledge and understanding in this area, and a set of recommendations for practitioners and future researchers, designers and healthcare practitioners.

Part I considers approaches to designing with and for older people. Chapter 1 by Joe Langley and colleagues outlines contemporary challenges to the design, use and uptake of assistive technology, before advocating the advantages and opportunities of co-design (a theme, which recurs in many of the chapters, and is a key innovation by designers in this area). Langley and colleagues present Lab4Living's approach to co-design, its principles and practical guidelines for its use. Hedley-Takhar and colleagues' chapter moves the discussion on to the use of technology in care homes in the UK, showing how a user-centred design approach has been used to design and evaluate digital tools for carers and residents, emphasizing the need to spend time and resources not only in development but in implementation. Chapter 3 by Spinelli et al. addresses the role of the under-researched of design, cultural and social factors in the uptake of assistive technology. It concludes with recommendations on how the perception of assistive devices may be shifted to facilitate the user's emotional investment in the devices, attachment to them, which, in return, may lead to better adherence and faster adoption. The last chapter, by Woodcock and colleagues, introduces the concept of empathy, its importance and how this can be introduced into the design curriculum as a means of giving young designers a greater awareness of what it is like to grow old.

[2]And as such you will find some variations in the terminology and spellings—most noticeably on the interchangeability of the terms seniors, elders and older people/persons, ageing/aging.

Part II addresses the usability and uptake of assistive technology. This has been a key area of concern for many years. Although assistive technology has been developed, it has not been widely adopted. This part looks at some of the reasons for this lack of uptake among different groups and leads on from some of the chapters in the previous section. Woodcock and colleagues look at the many reasons behind lack of uptake of technology by young older people. This group is classified as those in the md 50s, who might benefit from discrete aids, but who do not wish to be stigmatized as being old. A feature matrix is presented to help designers. This is extended further in McCarthy and Moody's chapter addressing usability and key device regulatory frameworks and standards to improve user acceptance, device usability and safety.

This theme resonates with chapters in Part III, related to dignity and ageing. In the first of these, McDonagh and Reardanz explore different approaches to expanding a person's understanding and felt experience of aging using a gerontology suit to experience ageing, and virtual reality simulation to experience the end of life. This chapter presents the researcher's firsthand accounts of the use and perceived value of these simulations for understanding ageing and the end of life. Jeays-Ward and colleagues from the National Institute for Health Research (NIHR) 'Devices for Dignity—D4D' describe the work of the National Institute for Health Research and their developmental model along with case studies of collaborative medical technology development, which seek to foster dignity in ageing through better, more thoughtful design.

Part IV has a more technological focus looking at the introduction of new technologies to assist in daily live and healthcare provision. Knight and colleagues report on the design of digital communication tools for e-consultation between healthcare professionals and patients. This service is designed not only to facilitate contact with hard to reach groups, but also to improve diagnosis, streamline operations and increase security of data collection. Moody and Cobley's chapter moves on to discuss advances in metallizing fabrics to embed assistive functionality in furniture, clothing and footwear which can respond to individual or environmental changes to provide the user with discrete support. The part concludes with a chapter by Wiczorek and colleagues which gives an overview of the design considerations for domestic robots to be used with older adults.

Part V focuses on mobility. Retaining independence and mobility is a key factor in maintaining quality of life in old age. This part takes a snapshot of some of the areas of research, such as the design of mobility scooters (Coxon and Oxley), improvements to the prescription and configuration of wheelchairs (Moody and colleagues), promotion of walking activities in older adults based on their hobbies and personal interests (den Haan and colleagues), mobile health applications to maintain healthy lifestyles (Şalgamcıoğlu), the information needs of older people using public transport.

The last part of the book includes research which is focusing on one specific, and at the time of writing, pressing area, that of dementia. Here, we have three chapters; Nayer and Coxon describe the design of touchscreen-based interface providing a variety of media experiences from music and photographs to family messages, all capable of being autonomously interacted with by elderly individuals. Pedell and colleagues again look at touchscreens, and the use of co-design methods to support creative and shared interactions. In the last chapter, Winton and Rodgers address the wicked problem of dementia and show how co-design projects have attempted to fully engage people living with dementia through the design process.

It is intended that the book is meaningful to not only the academic community but to the varied professions that are involved in the scoping, design, development, evaluation and implementation of technology and particularly assistive solutions to support older adults. Whilst exploring the challenges faced globally and in the development and delivery of solutions, it seeks to demonstrate innovation, good practice and pragmatic solutions through design.

Key issues which the book deals with includes:

- Methods—how to design with and for older people
- Design criteria and priorities—what people do and don't want
- Different care and support models—now and in the future
- Emotional needs—dignity, independence
- External drivers—demographic change, business models and regulation
- Types of technology and support—examples of a range of application areas, e.g. activities, services, products, hi-tech, robots, mobility aids, mobile apps, information
- Understanding interaction requirements—how ageing affects interaction with new and existing products
- Dementia—the increasing challenge of cognitive decline.

Coventry, UK — Andree Woodcock
Coventry, UK — Louise Moody
Champaign, USA — Deana McDonagh
Adelaide, Australia — Ajita Jain
Liverpool, UK — Lakhmi C. Jain

# Contents

## Part III Dignity and Ageing

# Part I
# Approaches to Designing With and for Older People

# Chapter 1
# Designing with and for Older People

**Joe Langley, Gemma Wheeler, Rebecca Partridge, Remi Bec, Dan Wolstenholme and Lise Sproson**

> *"I remember as a child we used to have one family computer on a special table in our lounge and we used it for specific tasks, deliberately having to disconnect the phone line to plug the computer into the internet, waiting for it to make a dialup connection and then having to be so patient with it to browse the web…these were mini events that sometimes brought the whole family together or, at the very least, required substantial, dedicated effort and time from (usually) my father."*
> Memory of Rebecca Partridge

## 1.1 Introduction

Assistive technologies are vital to enable older people to continue to live independent and fulfilling lives. Yet, current assistive technologies are predominately designed for the homogenous 'old' or 'disabled' and often don't fit the disease or disablement trajectory, requiring subsequent 'levels' of devices to cater to decreasing ability. As such, they often do not fit with the contexts of real people's lives and there are high rates of abandonment.

On the other hand, commercial technology, the people who use it and the ways in which it is used, changes rapidly. In the space of twenty years many families have moved from owning a few, large pieces of 'special' technology (a TV, a video recorder, a 'family computer') to multiple, small, embedded, networked devices.

---

J. Langley (✉) · G. Wheeler · R. Partridge · R. Bec · D. Wolstenholme
Lab4Living, Sheffield Hallam University, Sheffield, UK
e-mail: j.langley@shu.ac.uk

J. Langley · R. Bec · D. Wolstenholme
NIHR CLAHRC YH, Sheffield, UK

J. Langley · L. Sproson
NIHR Devices for Dignity MIC, Sheffield, UK



A. Woodcock et al. (eds.), *Design of Assistive Technology for Ageing Populations*, Intelligent Systems Reference Library 167,
https://doi.org/10.1007/978-3-030-26292-1_1

Moreover, young adults today have jobs in careers that did not exist even 10 years ago. Designing future technologies and future technology interactions for 'future old' generations, based on current experiences, is a challenge when so little is known about what their lifestyles might be. The very notion of 'technology,' and its assumed complexity, must also be challenged.

It is widely accepted nowadays that designing with the end user of a device or service is necessary; a valid means of addressing the challenges outlined above. Approaches with various labels such as participatory design, co-design and co-production have become increasingly prominent in both the Design and Healthcare literatures, but practical, nuanced guidance on how to conduct such approaches are sometimes incomplete or even conflicting.

In this chapter, we aim to highlight the contemporary challenges of Assistive Technology design, and how a co-design approach may help to address them. We will detail the basic principles that Lab4Living (a multi-disciplinary research group based at Sheffield Hallam University) follow within their co-design approach, illustrated by a key case study, and discuss the implications of such an approach in the future.

## 1.2 Literature Review

### *1.2.1 Assistive Technology*

Whilst a more thorough history of Assistive Technology (AT) is given elsewhere [1], several key issues are worth raising to support the main arguments of this chapter. Firstly, it is recognised that the AT market is limited, 'following the principle of "one-size-fits-all", [where] patients of different ages and socio-cultural backgrounds are often supplied with similar aids with little consideration for their personal preferences and socio-emotional needs,' or how these needs may evolve over time [2]. Although ATs are closely associated with ideas of identity and 'self,' 'narratives of medicine, decline, and functionality around ATs… remain predominant,' (ibid.), with research into the effectiveness of ATs often focusing on ergonomics or efficiency, rather than aesthetics or experience [3].

In light of this 'one size fits all' approach, ATs are often abandoned because people don't associate themselves with the stereotype of 'disabled' and don't want to be stigmatised [4]. This abandonment carries significant risk (particularly in case of mobility aids) including 'risk of falls, limited mobility and less engagement with social activities, that may compromise overall wellbeing,' [2, p. 7]. If barriers to adoption (through poor design) can be resolved, then it follows that patient outcomes may be improved, with potentially reduced demand on healthcare services.

### 1.2.2 *The Role of Design*

There is a growing understanding of the role of designers, and a call from the fields of both Design [5] and Healthcare for them to be involved in the area of AT [3]. Whilst some argue that adding some form of 'customisation' is vital to making AT 'work' [3, 6], many contemporary design practitioners (ourselves included) would argue that 'users' input (where the term 'user' encompasses multiple people involved in the context of use, such as the patient, family and healthcare professionals) should be incorporated from an early stage of design. This is known as a co-design[1] approach, known for its ability to move beyond mere functionality of a product/service, eliciting and incorporating the values, beliefs, contextual issues and emotional connotations of those who will be using it (all of which have been identified as barriers to adoption of ATs above).

Whilst an exploration of the rich and detailed history of co-design[2] is beyond the scope of this chapter, it is pertinent to highlight the key premise that those affected by a change *can* and *should* be involved in the design of AT, for practical reasons (i.e. they have vital experience and knowledge of the design context) and political reasons (i.e. they have a right to be involved). Over the decades the field of co-design has developed a range of tools, including prototyping, context mapping, visualisation, and storyboarding, to help bring multiple stakeholders together in the design process. These tools help participants to reflect on and express their understanding of current situations (including experiential, emotional and tacit knowledge) and to imagine future scenarios collaboratively and creatively.

### 1.2.3 *Co-design with Older Adults*

Rice and Carmichael [9] suggest that the current literature landscape paints a rather 'fragmented picture' of how best to involve older adults[3] in co-design. Perhaps unsurprisingly, many practitioners in this area have highlighted the physiological issues that may also hinder older people's involvement in co-design, such as 'sensory impairment, cognitive difficulties, mobility needs [and] fatigue' [11]. Since many of the published examples are concerned with the design of digital apps or interfaces,[4]

[1]For the purposes of this chapter, we will use Sanders and Stappers' definition of 'co-design' as referring to 'the creativity of designers and people not trained in design working together in the design development process' [7].

[2]Terms such as co-design, co-production, co-creation, and so on, have emerged from the field of Participatory Design which has its roots in the Scandinavian Workplace Democracy Movement of the 1970s—see Simonsen and Robertson [8] for a detailed history.

[3]Here the term is used to denote, roughly, the general adult population over the age of 55. It does not refer to adults with any specific health condition, although some researchers provide excellent guidance on involving older adults with dementia—see Branco et al. [10].

[4]However, some key examples of co-designing with older adults could be found in the contexts of hospital services and environments (see Wolstenholme et al. [12]; Macdonald et al. [13].

the challenges identified often also relate to technological literacy. Other challenges identified by authors are common across co-design work in general, regardless of age. These include a 'self-perception of inadequate drawing skills,' a tendency for participants to agree with the group rather than put forward their own, individual views [9] and 'difficulties in long-term projects with end-user motivation, invisibility of research results and instable [sic] prototypes' [14].

The guidance available in the literature on co-design with older adults can be roughly separated into four main categories:

- Individual considerations—participants may not consider themselves 'elderly.' Co-design practitioners must acknowledge that 'the older population is not a homogenous group' [15], and should involve participants 'without stigmatising them or making assumptions about them' [16].
- Empathic considerations—Xie et al. [17] recognise that older adults may not be accustomed to co-design practice, and as such encourage designers to validate their contributions, whilst also ensuring 'co-design partners feel physically and emotionally comfortable during the design session.'
- Practical considerations—generic guidelines include 'the use of simple and short questions, large type size when using written material, good lighting and appropriate seating arrangements for those with poor hearing' [18]. Many older people prefer face to face interviews within a home setting as the preferred method for participating [19].
- Methodological considerations—visual methods, such as visual aids, sketching, storyboards and vignettes have all been reported as successful in creating a common ground between participants, or to address assumptions [11, 20]. In addition, the use of both high- and low-definition prototypes is recommended for enhancing understanding and facilitating honest critique respectively [11, 21, 22].

### *1.2.4 Co-design for Assistive Technology*

The early 2000s saw a growing interest in 'Inclusive Design', or 'designing more accessible products and services for the widest possible range of users, regardless of age and capabilities' [23]. However, many self-declared Inclusive Design projects sought user input during initial research stages of the project only, to inspire designers rather than considering users equal partners in the design process, in line with the original values of participatory design [24].

Literature concerning co-design for AT in particular is less forthcoming. However, from the key examples available, it has been suggested that the involvement of older adults in co-design for AT can be a positive individual experience, as well as enhancing the experiences of a wider audience of AT users in the future [25, 26].

### 1.2.5 *Summary*

Whilst there are examples of engaging older adults in co-design, the resulting guidance consists mainly of 'tips' that work in a given context, or includes worthy but vague statements such as 'be respectful'. In fact, it is difficult to find clear, practical guidance on how to conduct co-design in healthcare contexts [27], or simple, robust frameworks for supporting the participation of older adults in co-design (for AT or otherwise).

In this chapter, we aim to give clear guidance and examples of how we approach involving distinct groups, such as older adults, in co-design for health and wellbeing. To help illustrate this, we present the key principles that Lab4Living follow within our co-design approach in the next section. To support this we first introduce a case study from the Lab4Living portfolio. This case study is then used as an example to illustrate how the principles have been employed in practice.

## 1.3 Case Study: The Parkinson's Service in the South West Peninsula, United Kingdom

This project sought to improve the patient and staff experiences of using and delivering Parkinson's Disease (PD) services in the South West Peninsula. It was recognised by the service that Parkinson's Disease is highly individualised and as such there are complexities in diagnosis and patient acceptance of the condition. There are also huge variations in symptoms and therefore ongoing management of PD is not a 'one size fits all' approach. Additionally, the service covers a large geographical area and has staffing pressures from unfulfilled roles and long-term sick leave.

The project consisted of five full day workshops that brought together people across the service. These included; six people living with PD, three of their partners, a PD nurse and therapy specialists, community care teams, specialist consultant neurologists, and a finance officer from a Parkinson's UK charity representative. There was high continuity and retention of participants throughout.

The five workshops were facilitated by designers from Lab4Living in the following format:

- Workshop One: What are the user experiences?
- Workshop Two: How does the service currently run?
- Workshop Three: What could be different? Coming up with new ideas
- Workshop Four: Idea development
- Workshop Five: Prototyping and testing.

Five concepts are currently in development as a result of this project. These include a Parkinson's patient passport, new service and local information, a media campaign and a card deck to support therapeutic sessions and patient determination of personal priorities.

## 1.4 Lab4Living's Approach to Co-design

The overall aim within our work is to bring more people into sharing, creating and using knowledge to create positive change, for both pragmatic and democratic reasons (in line with the origins of co-design). As such, we adapt our co-design processes according to the groups we work with to afford their *meaningful participation*[5] and to consider the *knowledge produced*[6] appropriately.

In this section, we describe the overarching principles that drive our approach with examples of how they have been employed in the Parkinson's Services case study. We then present a range of practical tips that emerged from our practice and that are useful when planning and conducting workshops.

### *1.4.1 Principle 1: Valuing Different Perspectives*

Within the 'democratic' roots of participatory design, we recognise the need for different perspectives to be shared and given a physical presence so that they cannot be dismissed or ignored. This is based on the idea that everyone has equally valuable experiences, including that of the Designer and their role in the process. The collective group experience that is shared and discussed is the most important.

*What Does This Look Like in Practice*?

The ways in which we employ this (seemingly abstract) principle will vary between projects, but the approach is the same; Give individuals time and space for reflection and response, record this in a visual way and allow participants to share without comment or conflict from others. Display these responses and, as a group, look for commonalities and themes through which people might come to a consensus (even if that consensus is accepting differences between experiences).

A method that valued different perspectives in the Parkinson's Services project was the LEGO® SERIOUS PLAY® method. This approach used the LEGO© medium, with which participants built models in response to a question (such as 'build a model that represents your experiences of PD services'). They then described this model to the group before selecting a crucial element of their model to negotiate into a shared model that represented the group experience. The subsequent shared model showed both commonalities and opposing experiences (such as good and bad care transitions) all visually represented together.

Another example of this principle is the use of the service design method 'personas' whereby the group created archetypal 'characters' of service providers or users. This was particularly useful since not all service providers and Parkinson's

[5] Whilst there are existing frameworks dedicated to exploring 'participation' (for example, [28–30]), participation for us is about an individual's ability to impart and share their knowledge.

[6] A discussion of this is available from Langley et al. [31].

disease patient groups could attend the workshops. The persona method required participants to take themselves outside their own experience to consider someone else's perspective. It also provided a chance for participants to present their own experiences anonymously. The personas were created in 3D form to be stood on tables and introduced to the wider group as though they were a person in the room. The service experiences of these 'persona participants' were considered and combined into the whole group experience.

### 1.4.2 Principle 2: Adapting Co-design Methods to the Needs of the Group and/or Individuals

Drawing on the 'pragmatic' roots of participatory design, we believe that time must be spent learning about the groups we work with so that we can adapt our process according to their particular functional, cognitive and emotional needs. This is important since there are huge variations between the groups of people we work with. Conducting research prior to co-design activities is important to understand and accommodate what might be needed to afford a group the maximum ability to participate.

*What Does This Look Like in Practice*?

The adaptations to methods in the Parkinson's Services project were necessary because of the variations of symptoms in Parkinson's Disease. The nature and severity of symptoms vary between individuals, but can include difficulties with writing such as dramatically reduced size of handwriting, difficulties in legibility or an inability to write due to a tremor. Other symptoms include rapid physical and mental fatigue. As such, we adapted our methods and delivery approach to ensure that those living with Parkinson's could fully participate. Written activities were kept to a minimum, and any written or recorded activity was done using large paper and pens with single word or short sentence answers. For some activities, participants could choose to respond using an image or have someone else record their response. All five workshops were very informal with refreshments available throughout, regular breaks scheduled, and the possibility for participants to leave the room if they felt fatigued.

### 1.4.3 Principle 3: 'Visibility and Transparency'

A key principle within our work is visibility, since it is a way to be:

1. Practical: keeping things visual ensures that everyone is able to capture and access information, providing a record of everyone's views and experiences;
2. Democratic: it is important to maintain visibility and transparency in the process to promote shared ownership of the project.

*What Does This Look Like in Practice?*

An important consideration in relation to this principle is sustained presence (visibility) of the participants' contributions during the workshops. This applies to the contributions of the current workshop as well as visual reminders of contributions from any that have happened previously.

For example, when participants created their personas in the Parkinson's Services project, the completed template that was provided could physically stand up independently on a table. This approach continued throughout the workshops where captured experiences and ideas were displayed around the room. This gave their contributions a sustained, physical presence and ongoing visibility making it harder for other participants to ignore, forget or dismiss what had been previously presented.

In the Parkinson's Services project, the time between the workshops was a few months. To keep momentum going and provide a record of the sessions, visual summaries (photographs, images and diagrams) were sent out between workshops, summarised at the beginning of the next workshop and displayed around the room. This recognised the high level of investment from both the patients and clinicians. Furthermore, due to the time between workshops it helped to encourage continued buy-in to the project from participants.

The same principle applies to the visibility and transparency of the project process itself. Due to the nature of the co-design workshops that we run, a degree of decision making and designing goes on outside of the sessions in the studio. As designers, we consider the outputs from a session and use these to guide design decisions on prototypes and the next interaction with the project participants. It is important that whatever these are, they clearly link back to a 'brief' derived from the preceding participatory event. A clear summary is provided of what happened, why, and how it links to any subsequent participatory event. This helps to maintain a sense of visibility and transparency for participants to take ownership and give critical feedback.

To ensure that participants can 'see' where we are in a project, a low resolution process map is displayed in the workshop space alongside a summary of what activities we intend to cover in the workshop time available. Keeping these displayed at all times allows everyone to see what they can expect.

### *1.4.4 Principle 4: Valuing Different Forms of Knowledge*

We recognise different forms of knowledge within each individual and we work to enable people to share their full 'repertoire' of knowledge (i.e. explicit, institutional, practical, tacit, experiential, emotional). This can be challenging within healthcare contexts, where scientific knowledge production dominates, but is nonetheless crucial to the success of co-design in this area.

*What Does This Look Like in Practice?*

Valuing different forms of knowledge is similar to principle one, valuing different perspectives. Therefore it can be hard to unpack the differences between them since perspectives are often based on knowledge. What is important about this principle is the recognition that someone's 'knowledge' might not be easy for them to surface or express, particularly if it is not 'trained' or 'institutional'. For example, someone with no institutional knowledge of how a service is organised might have strong emotional knowledge of how a service makes people feel. Another example is the tacit knowledge that practitioners have; they might find it hard to articulate exactly 'what' they know and 'why' because so much of their work has become unconscious.

It was important in the Parkinson's Services project to enable participants to recognise the range of their knowledge, to value its importance amongst all other 'knowledge' (for example a consultant's knowledge as equal to a patient's knowledge), to uncover it and to visualise it.

Throughout the Parkinson's Services project we provided a range of activities to draw out these knowledges from participants. Rather than simply asking people what they know, the range of making and recording methods allowed time and space for reflection and depth, to bring different forms of knowledge to the surface. For example the LEGO© method allowed those with emotional and experiential knowledge to share it. Similarly, a service mapping activity using six 'stages' of Parkinson's Disease uncovered much of the tacit knowledge of practitioners, as it forced them to break down their practice into clear, individual steps. All of our activities allow a time for personal reflection and sharing with the group before any wider discussion or consensus building.

### *1.4.5 Principle 5: Less Talking, More Doing*

We value the power of 'arts-based' (making) enquiry to draw out and express experiences, ideas and knowledge. Designers use making as a way of thinking, to reflect and consider new possibilities. When we enable others to 'make' and 'do', it encourages them to have the same self-reflective dialogue, to think in a different way and to move away from a reliance on words alone to express thoughts.

The use of making things to help imagine new possibilities is also particularly important. It helps reduce constraints and barriers, to allow participants to imagine new possibilities based on their experiences. The things that are made can serve many purposes: to make futuristic possibilities seem more real and achievable, to deliberately challenge preconceptions or to provoke people to see from a different perspective. This is important; considering we cannot predict or design technologies for a future based on the current information available, we must therefore take participants through a journey of reflective and creative thinking.

*What Does This Look Like in Practice?*

Something that is important to consider within this principle is that many participants might not have a lot of confidence in 'doing'. Therefore, asking them to approach a making task to share something from their experience that might be sensitive or emotive should not be rushed into. Instead, we invest time with participants in 'skills building', encouraging all parties to 'have a go' using the making/sharing medium on a trivial topic first to build their familiarity with it. These activities are fun, engaging and demonstrate the value of the approach. Furthermore, they help participants to realise that it is not about the 'quality' of the finished article; instead, it is about the thinking that the process facilitates and the enhanced communication that the 'thing' produced enables. This approach was used when any new making/sharing medium or technique was introduced in the Parkinson's Services Project. For example, before asking people to generate new ideas for the service we went through a warm up and familiarisation process. This involved using random image cards to combine, edit and generate ideas on a trivial subject; we encouraged 'wild' ideas and those that might not be technologically feasible currently.

Additionally, when set to specific tasks, 'doing' can help to focus information, reduce complexity and encourage participation from multiple, diverse stakeholders. For example, the complex process of the existing Parkinson's Service pathway was reduced by constraining people to use six boxes to share their experiences. This provided a focused area to consider what the 'six' phases might have been. As everyone took part and recorded, the final combined diagram included everyone's experiences, rather than the few who might have contributed if this was a spoken activity.

Recording and documenting the process and contributions of every participant is crucial as evidence of decision making within the co-design process. Yet it would be impossible for the designers to accurately do all of this *and* pay attention to the contributions being made. The use of templates and resources that enable the participants to record their own contributions succinctly are invaluable in enabling designers to capture evidence whilst also giving focused attention to what participants are saying and doing.

### *1.4.6 Practical Tips*

Based on our experiences in using the principles, we offer the following practical tips to the reader to support them in their own work designing with and for older people.

- Work with project partners to understand the needs of the population that will attend the workshop. If this is not possible, be open and honest; acknowledge that you don't know their needs and prepare an activity to understand what support or adaptations might be required.
- Build a sense of transparency; doing so is valuable in promoting shared ownership of a project. We endeavour to present the project timeline/aims concisely to each co-design group we work with, ensure workshop plans are visible and adhered to where possible, jointly establish ground rules about participation (i.e. 'respect each other's time'), and provide summaries of workshops soon after the session.

- Reduce the use of PowerPoint to keep information visible for longer and always use plain English summaries in short sections of text.
- Communicate a sense of value for the experiences of participants through well considered preparation, good quality templates/activity resources and even good quality stationary.
- Create the templates and resources so that the participants (creatively) record their contributions with text based summaries. This captures the evidence yet frees the designers up to really pay attention.
- Respect and appreciate the sharing of sensitive and vulnerable experiences. Within this, be prepared for personal stories and the emotions that come with them. Allow time, space and perhaps an activity (e.g. making a cup of tea) to help someone to compose themselves in their own way if this is helpful.
- Ensure that participants know that they are able to vacate the room or the session for however long they need, if they wish. Give them the opportunity to speak to someone 1:1 immediately, or to 'catch-up' later in the session, or even immediate transport home (with a check that they feel safe).
- Allow every participant to share their experiences, not just some. This involves providing appropriate tools and an element of skills building (see Sect. 1.4.5).
- Keep things as open as possible for as long as possible—suspend disbelief through creative making tasks (using a broad interpretation of 'making,' i.e. using means that are not solely written or verbal communication).

## 1.5 Discussion

It was presented earlier in this chapter that Assistive technologies (AT) can be vital in enabling older people to continue to live independent and fulfilling lives, and that designing with end users (in this case older adults) is necessary to ensure any technologies fit the contexts of people's lives. However, literature shows that there are limited examples of working and designing specifically with older adults.

The cases and research available produced four broad categories of advice that we would wholly agree with: individual considerations, empathic considerations, practical considerations and methodological considerations. However, for those who are looking to develop AT and overcome many of the contemporary challenges such as abandonment, a 'one-size-fits-all' culture, and rapid technological development [2, 4], there has been little support or guidance. We argue that this chapter adds detail to these considerations and, perhaps more importantly, demonstrates how putting them into practice can create an enhanced sense of engagement and equity, leading to better design outcomes.

Nicol et al. suggest that 'in the absence of a complete methodology for working with older users, researchers and designers are often left to improvise their own methods. This can result in co-design relationships being compromised and weak design insights emerging,' [32, p. 908]. We believe that our five principles and practical tips will help researchers to overcome this potential issue to achieve good insights *and* good designs.

Our case study illustrates how co-design is a structured, planned and considered approach through which the designer employs care, empathy and understanding with the expertise to adapt their methods accordingly. It is this adaptation of methods, and careful consideration for participant groups, that begins to move away from the historical notion of the homogenous 'old' or 'disabled'. For example, individual considerations are important, but in execution this involves research and careful planning, something clearly demonstrated when working with people living with Parkinson's.

We argue that one of the issues with the inclusive design movement outlined earlier in the literature review has been its 'tokenistic' user input to inspire designers, rather than considering users as equal partners. This partnership is crucial in co-design; it recognises the knowledge and experience of all participants, including the designers. If future assistive technologies are truly going to fit the contexts of people's lives, evolve with people's needs over time, and enable independent and fulfilling lives, then we argue it is not enough merely to *involve* end users in the design of these devices. It is crucial to ensure that user involvement is meaningful, by allowing users to access and share their experiences, insights and knowledge as equal partners.

There are two key elements to co-design that our principles demonstrate in this chapter that we believe enable this: 'doing' and 'making'.

'Doing' in co-design has two key benefits. It changes the way that participants think, share and express themselves by moving away from the sole use of words, to 'actions', 'images' and 'things' *supported* by words. As demonstrated by the case study, the act of 'doing' provides a much greater depth of knowledge and allows people to articulate potentially complex or forgotten things. This idea is supported by Information Theory which states that natural language is not an efficient code [33, p. 231]. Furthermore, from a participation point of view, 'doing' reduces the talking time of participants, flattens hierarchies and ensures equal participation where everyone has a chance to respond.

'Making' is equally valuable in co-design, but it is also important to describe what was made and why. This relates to Gaver and Bowers [34] who in their 'Annotated Portfolios' posit that a made artefact is a way to express a form of tacit knowledge which is then reinforced by written annotations. Therefore, using making that is shared through written or spoken words allows for exploring and reflecting on one's previous experiences. Practically, the Parkinson's Service case study demonstrates how doing this in a group setting, surrounded by other people who have related lived experiences, can provide a way to for participants to access and recognise their hidden, tacit forms of knowledge.

Finally, perhaps the challenges within the assistive technology market also indicate a need for a shift in the way we think about, and work with, older people and future older people, for assistive technologies. In the context of this chapter, we are talking about co-designing with the current older population. However, projects may take a long time to come to fruition and, as exemplified in the opening anecdote, the range of possible future technologies is expanding all the time. The danger then, is that in designing solely with today's current ageing population, experiences and technologies will be outdated or obsolete.

We therefore propose that perhaps we should not be designing with segregated age groups but instead in cross-generational initiatives (some suggestions of this have been made by Mitchell and Nørgaard [20], and Xie et al. [17]. Furthermore, rather than trying to anticipate what technological capability may be possible in the future, or drive new technologies, instead we could recognise an increasing need to be flexible, adaptable, and have the skills to try out, assimilate, discard and move onto new devices quickly.

Ageing is something that happens to everyone and developing assistive technologies to better fit the life course, and to adapt and evolve with people over time could be enhanced by breaking down these boundaries. Whilst our case study predominantly engaged people with PD over the age of 60, participants with early onset PD were also involved. The subsequent concepts from the project have been developed with all experiences and needs considered to support people at all stages of the disease. The principles outlined in this chapter are applicable to any age group, and as such go some way to facilitating a cross-generational approach that would also be reactive and adaptable to the future possibilities of Assistive Technology.

## 1.6 Conclusion

We have proposed a framework of principles to help guide the co-design of future AT, and illustrated how this framework and principles can be practically applied through a case study. We have demonstrated and discussed how user involvement is crucial to the success of future technologies and provided practical advice on how to go about meaningfully involving users. The landscape in this field is a fast-changing one of technologies, populations and contexts. Therefore we propose that rather than focusing on the current older generations to design and develop assistive technologies for future generations, there is a need to design with and for people, of all ages.

## Bibliography

1. McCreadie, C., Tinker, A.: The acceptability of assistive technology to older people. Ageing Soc. **25**, 91–110 (2005)
2. Spinelli, G., Micocci, M., Martin, W.: Objects of desire and of disgust: analysis and design of assistive technologies. In: Proceedings of the 5th Conference on Design4Health Sheffield, 4th–6th Sept 2018 (2018)

3. Renda, G., Jackson, S., Kuys, B., Whitfield, T.W.A.: The cutlery effect: do designed products for people with disabilities stigmatise them? Disabil. Rehabil. Assist. Technol. **11**(8), 661–667 (2016). https://doi.org/10.3109/17483107.2015.1042077
4. Bright, A.K., Coventry, L.: Assistive technology for older adults: psychological and socio- emotional design requirements. In: Proceedings of the 6th International Conference on Pervasive Technologies Related to Assistive Environments (p. 9). ACM (2013)
5. Pullin, G.: Design Meets Disability. MIT Press (2009)
6. Procter, R., Greenhalgh, T., Wherton, J., Sugarhood, P., Rouncefield, M., Hinder, S.: The day-to-day co-production of ageing in place. Comput. Supp. Coop. Work (CSCW) **23**, 245–267 (2014)
7. Sanders, E., Stappers, P.: Co-creation and the new landscapes of design. CoDesign **4**(1), 5–18 (2008)
8. Simonsen, J., Robertson, T.: Routledge International Handbook of Participatory Design. Routledge, New York (2013)
9. Rice, M., Carmichael, A.: Factors facilitating or impeding older adults' creative contributions in the collaborative design of a novel DTV-based application. Univ. Access Inf. Soc. 12(5), (2011). https://doi.org/10.1007/s10209-011-0262-8
10. Branco, R.M., Quental, J., Ribeiro, Ó.: I'm Always Well When We are Together: Observation of Institutional Activities With People With Dementia and Its Implications for a Co-Design Research Project. Design4Health, Sheffield, UK, 13–16 July 2015 (2015)
11. Wherton, J., Sugarhood, P., Procter, R., Hinder, S., Greenhalgh, T.: Co-production in practice: how people with assisted living needs can help design and evolve technologies and services. Implement. Sci. 10(75), (2015)
12. Wolstenholme, D., Cobb, M., Bowen, S., Wright, P., Dearden, A.: Design-Led service improvement for older people. Australasian Med. J. **3**(8), 465–470 (2010)
13. Macdonald, A., Teal, G., Bamford, C., Moynihan, P.: Hospitalfoodie: an interprofessional case study of the redesign of the nutritional management and monitoring system for vulnerable older hospital patients. Qual. Primary Care **20**, 169–177 (2012)
14. Müller, C., Hornung, D., Hamm, T., Wulf, V.: Practice-based design of a neighborhood portal: focusing on elderly tenants in a city quarter living lab. In: Proceedings of CHI 2015, Seoul, Republic of Korea, 18–23 Apr 2015 (2015)
15. Lim, C., Frohlich, D., Ahmed, A.: The challenge of designing for diversity in older users. Gerontechnology **11**(2), 297 (2012). https://doi.org/10.4017/gt.2012.11.02.483.00
16. Göllner, S., Lindenberg, J., Conradie, P., Le, J., Sametinger, F.: The enchanted neighborhood: using metaphorical devices for the inclusion of seniors in the co-designing process. In: Proceedings of the 3rd International Conference for Universal Design, 30th Oct–3rd Nov (2010)
17. Xie, B., Druin, A., Fails, J., Massey, S., Golub, E., Franckel, S., Schneider, K.: Connecting generations: developing co-design methods for older adults and children. Behav. Inf. Technol. **31**(4), 413–423 (2012)
18. Barrett, J., Kirk, S.: Running focus groups with elderly and disabled elderly participants. Appl. Ergon. **31**, 621–629 (2000)
19. Patmore, C., Qureshi, H., Nicholas, E.: Consulting older community care clients about their services: some lessons for researchers and service managers. Res. Policy Plann. J. Soc. Serv. Res. **18**(1), 4–11 (2000)
20. Mitchell, R., Nørgaard, M.: Using DIY cartoon storyboard, live sketching and co-sketching to involve young and older users in participatory design. In: Proceedings of IASDR2011, the 4th World Conference on Design Research, 31 Oct–4 Nov, Delft, The Netherlands (2011)
21. Bec, R., Langley, J., Wolstenholme, D.: Designing in health: developing shared knowledge through (un-)prototyping. Design4Health, Sheffield, UK, 4–7 Sept 2018 (2018)
22. McGee-Lennon, M., Smeaton, A., Brewster, S.: Designing home care reminder systems: lessons learned through co-design with older users. In: 6th International Conference on Pervasive Computing Technologies for Healthcare (PervasiveHealth) and Workshops (2012)
23. Dong, H., Clarkson, P.J., Cassim, J., Keates, S.: Critical user forums—an effective user research method for inclusive design. Des. J. **8**(2), 49–59 (2005)

24. Herriot, R.: Are inclusive designers designing inclusively? an analysis of 66 design cases. Des. J. **16**(2), 138–158 (2013). https://doi.org/10.2752/175630613X13584367984820
25. Bush, P., ten Hompel, S.: An integrated craft and design approach for wearable orthoses. Des. Health **1**(1), 86–104 (2017)
26. De Couvreur, L., Dejonghe, W., Detand, J., Goossens, R.: The role of subjective well-being in co-designing open-design assistive devices. Int. J. Des. **7**(3), 57–70 (2013)
27. Partridge, R.: Exploring design mindsets: Developing design skills in adolescents with long-term conditions. Ph.D. Thesis (2019)
28. Arnstein, S.: A ladder of citizen participation. J. Am. Inst. Plann. **35**(4), 216–224 (1969)
29. Slay, J., Stephens, L.: Co-production in mental health: A literature review. Lond. New Econ. Found. (2013)
30. Kraff, H.: A tool for reflection—on participant diversity and changeability over time in participatory design. CoDesign **14**(1), 60–73 (2018)
31. Langley, et al.: BMC Health Serv. Res. **18**, 585 (2018). https://doi.org/10.1186/s12913-018-3397-y
32. Nicol, E., Dunlop, M., Komninos, A., McGee-Lennon, M., Baillie, L., Eslambolchilar, P., Foong, P., Gault, P., Hakobyan, L., Lumsden, J., Velàsquez, F., Horcher, A.: 2nd workshop on designing with older adults: towards a complete methodology. In: Proceedings of MobileHCI '15 Adjunct, Copenhagen, Denmark, 24–27 Aug 2015 (2015)
33. Hutchins, E.: Cognition in the Wild. MIT Press (1995)
34. Gaver, Bill, Bowers, John: Annotated portfolios. Interactions **19**(4), 40 (2012). https://doi.org/10.1145/2212877.2212889
35. Chamberlain, P., Yoxall, A.: 'Of Mice and Men': the role of interactive exhibitions as research tools for inclusive design. Des. J. **15**(1), 57–78 (2012)
36. Collerton, J., Jagger, C., Yadegarfar, M.E., Davies, K., Parker, S.G., Robinson, L., Kirkwood, T.B.L.: deconstructing complex multimorbidity in the very old: findings from the newcastle 85+ study. BioMed Res. Int. **2016**, 15, Article ID 8745670 (2016). http://dx.doi.org/10.1155/2016/8745670
37. Dawe, M.: Desperately seeking simplicity: how young adults with cognitive disabilities and their families adopt assistive technologies. In: Proceedings of the SIGCHI Conference on Human Factors in Computing Systems (pp. 1143–1152). ACM (2006)
38. Evans, S., Corley, M., Corrie, M., Costley, K., Donald, C.: Evaluating services in partnership with older people: exploring the role of 'community researchers'. Work. Older People **15**(1), 26–33 (2011)
39. Fernández, M.A., Delchevalerie, P., Van Herp, M.: Accuracy of MUAC in the detection of severe wasting with the new WHO growth standards. Pediatrics **126**(1), 195–201 (2010). https://doi.org/10.1542/peds.2009-2175
40. Grant A., Njiru, J., Okoth, E., et al.: Comparing performance of mothers using simplified mid-upper arm circumference (MUAC) classification devices with an improved MUAC insertion tape in Isiolo County, Kenya. Arch Public Health. **76**, 11. Published 22 Feb 2018. https://doi.org/10.1186/s13690-018-0260-x
41. Levin, B.: Mobilising research knowledge in education. London Rev. Educ. **9**(1), 15–26 (2011)
42. Moss, G.: Research, policy and knowledge flows in education: what counts in knowledge mobilisation? Contemp. Soc. Sci. **8**(3), 237–248 (2013). https://doi.org/10.1080/21582041.2013.767466
43. Rycroft-Malone, J., Seers, K., Titchen, A., Harvey, G., Kitson, A., McCormack, B.: What counts as evidence in evidence-based practice? J. Adv. Nurs. **47**(1), 81–90 (2004)
44. Sanders, E.: Scaffolds for building everyday creativity. In: Frascara, Jorge (ed.) Design for Effective Communications: Creating Contexts for Clarity and Meaning. Allworth Press, New York (2006)
45. Salive, M.E.: Multimorbidity in older adults. Epidemiol. Rev. **35**(1), 75–83 (2013)
46. Almqvist, F.: The fuzzy front-end and the forgotten back-end: User involvement in later development phases. Des. J. **20**(1), 2524–2533 (2017)

47. Barnett, K., Mercer, S.W., Norbury, M., Watt, G., Wyke, S., Guthrie, B.: Epidemiology of multimorbidity and implications for health care, research, and medical education: a cross-sectional study. Lancet **380**(9836), 37–43 (2012)

**Joe Langley** studied a Masters and Ph.D. in Mechanical Engineering at The University of Sheffield, starting his research career developing the use of medical imaging data to build biomechanical engineering simulations.

Later, Joe moved to the Art and Design Research Centre at Sheffield Hallam University and helped establish Lab4Living; a trans-disciplinary group of researchers from design, nursing, visual communication, fine art, and media, applied to enhancing health and wellbeing.

Joe's work focuses on coproduction; involving service users and providers in health technology and service innovation, particularly on adoption and implementation—bridging the gap between research and impact.

Joe has over £5 M of grant capture including an NIHR Research Fellowship.

**Gemma Wheeler** is a Design Researcher at Lab4Living, an interdisciplinary research cluster with a focus on bringing Design-led research to fields of health and wellbeing. Her background is in Product Design, and has developed an interest in Participatory Design, Service Design and Design Research.

Gemma has a broad experience of using participatory methods in complex healthcare contexts, and enjoys adapting her approach to the particular needs and sensitivities to each community she collaborates with. She is interested in the role of creative, empathic, design-led methods in creating a 'common language' between patients, family members and healthcare professionals for mutual learning and real-world innovation.

**Rebecca Partridge** is a Design Researcher for the interdisciplinary research cluster Lab4Living. She has a broad range of research experience in the field of healthcare. Her background is in Product design, Design Research and Service design. She often works on projects that have diverse stakeholders with hierarchical boundaries based in complex medical systems. Her approach is participatory, visual, creative, experiential and champions the role of Design to include these diverse stakeholder voices in research, develop interventions, improve services and create lasting change. She is passionate about the role of the Design professional in the facilitation of Design methods, and core to her work is understanding the role that these design methods and approaches play and the impact and benefits they bring to healthcare.

**Remi Bec** is a design researcher in Lab4Living. He studied model making and interior design in France before following this up with BA, MA and PhD in Product Design at Sheffield Hallam University. Since then he has worked in Lab4Living on a number of projects as a design researcher in residence for the NIHR CLAHRC YH as part of the Translating Knowledge to Action theme. His research interest focus on the use of games to promote engagement and behaviour change, applying this predominantly, but not exclusively, in the field of public health.

**Dan Wolstenholme** is a visiting researcher at Lab4Living, in his day job he is the Director of the Centre for Quality Improvement and Clinical Audit at the Royal College of Obstetricians and Gynaecologists for the Royal College of Obstetricians and Gynaecologists. He is general nurse who has worked in health services research for the last 12 years as part of the NIHR CLAHRC Yorkshire and Humber. Dan has eclectic research interests and writes on knowledge mobilisation, specifically creative coproduction, or how using creative methods we can get people to work together to deliver meaningful, successful change in health and social care.

**Lise Sproson** works for NIHR Devices for Dignity MIC as NHS Innovations Manager, Diabetes Theme Manager and PPI Lead.

Devices for Dignity is a national MedTech and In vitro diagnostic Co-operative (MIC) funded by the NIHR to act as a catalyst within the NHS for the development of new medical devices and healthcare technologies for people with one or more chronic health conditions.

Lise is also a Speech and Language Therapist with specialist expertise in the assessment and management of dysphagia (swallowing) and communication difficulties. She is a committee member of the Council for Allied Health Professions Research (CAHPR) and the UK Swallow Research Group (UKSRG) and Specialist Advisor on dysphagia for the Royal College of Speech and Language Therapists (RCSLT) and the National Institute for Health and Care Excellence (NICE).

# Chapter 2
# A User-Centred Approach to Digitalising Care Homes

**Philippa Hedley-Takhar, Angel Jimenez-Aranda, Vitaveska Lanfranchi, Sue Pownall, Lise Sproson and Wendy B. Tindale**

**Abstract** The care sector requires support and new ways of designing and delivering services to respond to the needs of an increasing ageing population, characterised by significant numbers of people who encounter difficulties in living independently in their own home due to the collective effect of ageing, frailty and overlapping complications of long term conditions. The use of technology in care homes has a role to play as an enabler for new care models. In this chapter we review a selection of digital technologies for integrated care, present two case studies evaluating innovative digital solutions in care homes, and consider the user-centred design principles to aid successful and sustainable implementation.

**Keywords** Care home · Ageing population · Digital · Co-design · Technology evaluation · Monitoring · Dysphagia · Preventive interventions

## 2.1 Introduction

The difficulty of the health care system to respond appropriately to the growing demands of the ageing population is widely recognised. A large proportion of elderly people currently live in care homes (approximately 416,000 people in the UK,[1] mostly managed by independent care providers). The demand for care home place is likely to increase in the next year. Kingston et al. [19] predicts a need for extra 189,043 care home places by 2035, increasing the market size by 86%.

---

[1] www.laingbuisson.com/wp-content/uploads/2016/06/Care_OlderPeople_27ed_Bro_WEB.pdf.

---

P. Hedley-Takhar (✉) · A. Jimenez-Aranda · L. Sproson · W. B. Tindale
NIHR Devices for Dignity MedTech Co-Operative, Sheffield, UK
e-mail: Philippa.Hedley-Takhar@nihr.ac.uk

P. Hedley-Takhar · A. Jimenez-Aranda · V. Lanfranchi · S. Pownall · L. Sproson · W. B. Tindale
Sheffield Teaching Hospitals NHS Foundation Trust, Sheffield, UK

V. Lanfranchi
University of Sheffield, Sheffield, UK

 
A. Woodcock et al. (eds.), *Design of Assistive Technology for Ageing Populations*, Intelligent Systems Reference Library 167,
https://doi.org/10.1007/978-3-030-26292-1_2

Most of the care home residents are affected by co-morbidities and suffer from issues such as poor mental health, mobility problems, incontinence and pain [10].

Care home residents have 40–50% more hospital admissions and Accident and Emergency attendances than the general population age 75 and over. In Sheffield in 2016/17 there were around 3000 attendances to Accident & Emergency (A&E) from care home residents of which 40% were subsequently admitted to the hospital's Frailty Unit.

It is widely recognised that older people in care homes have such a variety of interconnected needs that co-ordination and integration of care between health, social care and independent care providers is essential [3]. Providing integrated care has been proven to improve patient's quality of life and care efficiency [6, 11]; Ham et al. (2013) and whilst a common model is hard to apply due to differences in contexts [15], digital technologies and innovation can be fundamental to support integrated care [1].

The National Health Service (NHS) Five Year Forward View set a target of 'enhanced care in care homes' informed by the learning of the Vanguards. The Enhanced Health in Care Homes Vanguards were six collaborations established as part of the NHS England New Models of Care programme to offer older people better, joined up health, care and rehabilitation services.[2] The Vanguards cooperated to create model of evidence-based interventions, designed to be delivered within and around a care home in a coordinated manner in order to make the biggest difference to its residents.

This chapter focuses on understanding the challenges and opportunities of innovative digital technologies combined with integrated working of healthcare professionals to provide enhanced clinical interventions to care home residents, to reduce and prevent the risk of an emergency admission to hospital.

We start from the principle that embracing new digital technologies does not necessarily guarantee positive outcomes if other factors are not taken into account, such as end-user engagement, usability of the system, and integration with existing services.

The paper is organized as follows. The first section illustrates how new digital technologies have been adopted in healthcare, with a special attention for their adoption in the care homes domain. We then present two case studies carried out with care homes in UK that are undergoing digital transformation. The first one has been carried out as part of the Perfect Patient Pathway NHS England Test Bed programme and the second one as a collaboration to develop a Care Homes Guide for Dysphagia.

Both studies aimed at understanding the impact of introducing new technologies in an existing service with vulnerable user. We identify key themes pertaining to technological innovation across care homes to understand how people and technological innovation have influenced and impacted each other.

In the final section, we outline some findings and reflections from both case studies and reflect on the implications that they have for going forward with introduction of technologies in care homes.

---

[2]www.england.nhs.uk/publication/the-framework-for-enhanced-health-in-care-homes/.

## 2.2 Literature Review

This literature review focuses on a set of digital technologies for integrated care that have been recommended as key by the South East Health Technologies Alliance (SEHTA—a health technology net-work that support businesses, health professionals, care providers etc.) in their review of technology and innovation in care homes.[3] We highlight the technologies that have been piloted in care homes, with a focus on understanding impact and barriers to adoption.

### *2.2.1 Electronic Patient Records (EPR)*

An Electronic Patient Record (EPR) "is an electronic set of information about a single patient" [12] used to collect and share information with the aims of facilitating care provision [16].

The introduction of electronic patient's records has been proven to have a substantial impact on quality and efficiency of care in healthcare settings [21], as it reduces the time required to manage documents and information and facilitates integration between different care providers via easy information sharing. These effects on the processes have freed the healthcare professionals' time and improved their quality of care, thus ultimately improving to patient outcomes (e.g. reducing the occurrence of infections, high-risk pressure sores, neurolepsis and improving activities of daily living).

However a number of barriers and challenges to the adoption of EPRs have also been recognised in literature, especially regarding organisational challenges, such as to the need to alter existing processes and to provide training to staff [20].

Electronic Nursing Records (ENR) are a specialised type of EPRs that contains data about nursing assessment, diagnoses, interventions, and outcomes. Nurses use electronic records, often integrated in Nursing Information Systems (NIS), to design and manage care plans, record interventions and outcomes, evaluate the quality of services [2]. ENR have been introduced success-fully in care homes [23], resulting in improvements in precision and quality of information recorded about the residents, better quality of care and quicker decision-making.

ENR are very similar in nature to the technology we are presenting in the first case study, although in our case we have investigated how ENRs can be successfully linked with other existing services to provide remote monitoring and early detection of changes in patients' health.

---

[3]www.sehta.co.uk/cms-data/depot/sehta/Technology-Innovation-in-Care-Homes-The-SEHTA-Review.pdf.

### 2.2.2 *Wearable Health Devices and Remote Sensing*

Wearable Health Devices and Remote Sensing are emerging technologies to support constant and often real-time monitoring of patients vital signs and can be integrated with telehealth applications. They can take the form of specialised devices or they can be embedded in mainstream products such as smartphones or watches.

Wearable Health Devices can be classified into three categories ac-cording to their type of use (home/remote or clinical), type of monitoring (online or offline), and user type (patient or healthy individual) [4, 7]. A review of the main type of wearable devices for health such as self-monitoring bracelets and home sensors is presented in Wang et al. [30].

Their usage in the context of care homes has been explored in a study carried out by Hall et al. [14] in three care homes in North-West England. The study focused on the acceptability of the technology by patients and staff and on the challenges faced. The participants felt that this type of technologies would be most helpful to preserve the safety of the patients, for example in the case of fall detection devices. The main organisational barriers were due to the lack of training.

### 2.2.3 *Telehealth*

Telepresence in the field of Health is used to deliver remote care, with a variety of technologies such as videoconferencing, telephone systems, video systems etc.

Jenkins and White [18] carried out a study on the potential benefits of telehealth and concluded how telehealth is particular useful as a complement to traditional services, to increase accessibility to healthcare and range of available options and the use of telehealth in nursing home resulted in reduced hospital admissions. A systematic review of studies of telehealth in care homes [9] has found that the most used mode of service was real-time videoconferencing to improve communication between care providers and this had a great impact in specialities such as allied health, dermatology, general practice, neurology, psychiatry, geriatrics. The care delivered using telehealth services was well received by all stakeholders (patients, clinicians, family and carers) and in some cases resulted in cost efficiencies. Driessen et al. [8] carried out a survey with physicians and practice providers in nursing homes in Kentucky that revealed how staff highly rated the role that telehealth can play in reducing possible hospital readmissions as long as technical specifications are of the highest quality (high quality sound and video).

A study carried out in Australia [13] focused on the use of telehealth in care homes to support patients in getting better and increased access to health professionals and health services to understand which services would be most useful to match the resident needs. The study concluded how, over a period the most used consultations were with general practice specialists and the most useful services were consultation appointments either before or after a procedure.

Following the NHS Five Year Forward View recommendations an number of care homes in UK have adopted telehealth services to better manage care. For example, in Croydon two pilots[4] have been carried out in Care Homes using a tablet to monitor the health condition of the residents and providing communication with the Croydon's Community Matrons technical triage team that can access the tele-health systems, receive alerts in case of abnormal parameters and can take appropriate action. These pilots resulted in positive outcomes for both patients and clinical staff. Patients were more relaxed with their context of care and were able to be proactive and self-monitor and also had increased social interaction with members of staff. Staff reported the ability to better manage the care in a proactive manner.

## 2.3 Case Study—Digital Care Home

The Perfect Patient Pathway NHS England Wave 1 Test Bed programme ran from January 2016 to December 2018, and comprised multiple projects in collaboration with stakeholders from health and social care providers, academia, industry and the voluntary sector, with the aim of benefitting people with multiple long-term conditions through combining and integrating innovative digital technologies and service redesign to keep people well and independent.

The Digital Care Home was a service evaluation project as part of the Perfect Patient Pathway Test Bed in Sheffield, which sought to not only test and evaluate the design of the technology in terms of usability, acceptability and efficacy, but also the re-design of the care pathway and the impact of this new way of working on behaviours and culture within a busy care home environment.

### *2.3.1 The Design Process*

The project involved engagement and active participation from a number of stakeholder groups. Primary stakeholders are defined here as stakeholders that are directly concerned and/or affected by the daily care that happens in care homes:

- Residents
- Care Home managers
- Care Home clinical staff
- Care Home admin staff
- Family, friends and carers
- Primary and Community Care
- Secondary Care

[4] www.kingsfund.org.uk/sites/default/files/media/T3(10)_Using%20telehealth%20to%20avoid%20hospital%20-%20Croydon%20Tunstall.pdf.

Secondary stakeholders are those who are more distant from the day-to-day care but still can have an input in it/be affected, e.g.:

- Care Quality Commission (CQC)
- Local authorities.

Project design was initiated through consultation with clinical networks (representing primary, secondary and community care) where it was recognised that telehealth (with care home staff as the primary users) could potentially play a role in co-developing an intervention or series of interventions to reduce the number of *avoidable* A&E attendances, hospital admissions and readmissions from care home residents.

Introduction to the care home community across Sheffield was supported by the Clinical Commissioning Group (CCG) and local authority. The timing of this engagement coincided with city-wide activity to build and strengthen relationships between care homes and the local health and care services in a coordinated manner using learning from the Enhanced Health in Care Homes framework (2016).

Engagement with the 10 participating care homes was led by the Digital Care Home Project Management Office hosted by Sheffield Teaching Hospitals NHS Foundation Trust and led within the homes by the care home manager (in some instances the regional manager or turnaround manager) and their senior clinical leads.

To ensure the experience of people recruited to participate was considered in the project design, Healthwatch Sheffield supported the Test Bed Advisory Group (TAG). This group comprised volunteers that either had long term health conditions or were the relative or carer of someone with such conditions, who actively identified issues relating to the design of the Digital Care Home that could matter from a patient, relative and carers perspective and suggested ways in which the project could be developed to better suit individuals and encourage them to take part (e.g. content, language and presentation of information materials).

Decision-making support in response to the remote monitoring was provided by the nurses in the Single Point of Access (SPA) team (hosted by Sheffield Teaching Hospitals NHS Foundation Trust), and the 11 GP practices participating in the project (through pre-existing commissioned service arrangements to each of the care homes).

### *2.3.2 The Technology*

The Digital Care Home service was designed to test and evaluate the theory that digital technology could be used by care home teams in a range of roles (Managers, Registered General Nurses, Care Assistants) as an ‘early warning system’ to highlight changes in resident’s health which could have otherwise been undetected and were early signs for further health deterioration, and potentially reduce avoidable attendance to A&E or admission to hospital. The service, through this early identification of health changes, sought to facilitate earlier preventative interventions through coordinated decision making, by the care home and local NHS services, using resident's health data monitored over a period of time to aid those decisions.

As with the Pathway of Care work stream of the Newcastle Gateshead Care Home Vanguard [17], the project sought to test whether early detection, and competency of clinical response supported by the use of (National Early Warning Score) NEWS in care homes would have an impact on clinical outcomes in the resident population.

Care teams were asked to use a portable recording tool that removed the need to record potentially sensitive personal information on paper that could be misplaced or misinterpreted, which was introduced into their typical care home monitoring routines.

Participating care homes were provided with a tablet device with the Digital Care Home app pre-installed that allowed them to submit a resident's vital signs (National Early Warning Signs (NEWS) observations) - respiration rate, oxygen saturation, temperature, blood pressure, heart rate and level of consciousness. This information was then stored within a web-hosted portal that could be viewed and shared securely with a team of nurses, located at the Single Point of Access (SPA) at Sheffield Teaching Hospitals NHS Foundation Trust.

If early deterioration of health was identified and an 'alert' generated (based on the NEWS), the SPA team would contact the care home for a joint assessment of the residents' health, support with clinical decision making; and subsequently make referrals to GPs and community nursing teams (see Fig. 2.1 for the process).

Examples of potential responses as a result of the intervention could include prioritising residents for immediate attention during GP ward rounds or additional GP visits; increased resident monitoring; medication reviews; or community nurse home visits.

The service integrated directly into clinical systems including SystmOne and EMIS Web, for shared information to be uploaded into patient records and accessed by NHS teams following referrals for a joined-up response between local healthcare services.

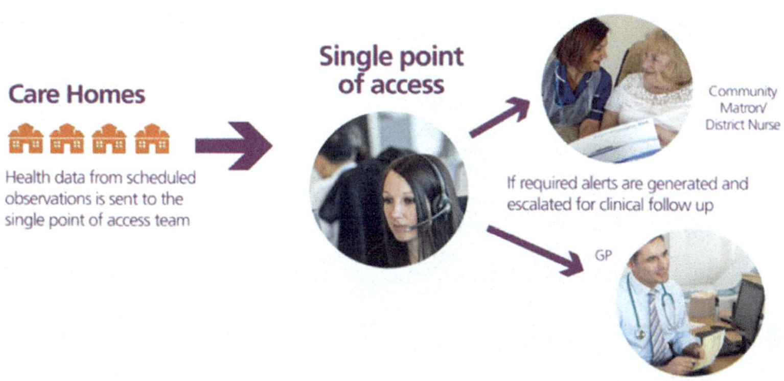


**Fig. 2.1** Digital Care Home support system

The residents monitoring information was also stored and accessible to the care home as a longitudinal record for more knowledge for the care home, GP, family and the resident themselves about their 'normal' health state (or baseline) and to contribute to care planning conversations.

### *2.3.3 The User Interface*

The design of the user interface to the Digital Care Home application was developed initially with a number of clinicians and care homes in the North East of England when deploying nutritional monitoring programmes within care homes. At the time the user interface was initially designed to enable care home staff to collect nutritional information from residents and was adapted to other clinical areas to not only to collect information but as a two way messaging between health and social care teams.

The design of the application has followed an iterative design process and was improved numerous times to be able to calculate scores and provide advice and guidance to the care homes based upon the information that they enter into the application, with continuous improvements to the application built into the service based on user feedback, predominantly based on improving visual clarity and usability of the interface (for example size and colour of the font, use of different colours for different sections, and use of symbols) (Fig. 2.2).

**Fig. 2.2** Digital Care Home user interface—©Inhealthcare

### 2.3.4 *Evaluation*

The Digital Care Home project was evaluated during a series of focus groups with stakeholders. The aim was to carry out a qualitative evaluation of the technology and of the user acceptance, focusing on lessons learnt and understanding barriers to adoption and possible mitigation strategies.

A number of key themes and barriers emerged from the focus groups.

**Theme 1: Personalisation of Adoption Process**
During the focus groups it clearly emerged how participating homes took varied approaches to using the technology such as different timings in morning routines, sharing information with colleagues at flash meetings or huddles, submitting data using the tablet whilst with the resident or with the care home manager. These ways of working were all developed by the care home teams themselves based on their knowledge of 'what works' within their home – using the service had to *work for them.*

**Theme 2: Usability of the Technology**
Usability and responsiveness of technology is particularly important in busy care home environments due to demands on carer time, high staff turnover or short-term agency staff placement.

Care home teams involved in the Digital Care Home were generally positive about the design simplicity of the user interface of the smartphone app which was used to share data, stating it was intuitive to use. It was noted that expectations of user interfaces for work-based tools were very different from the expectations of interfaces for apps for personal use—with less concern about use of colour or features as long as the technology was easy to use, secure, and posed no risk to residents' safety. In case of technology not working as anticipated (and if the way of fixing technological issues were not evident) staff were quicker to become disenchanted with using it, therefore higher attention should be paid to the interface design itself, to avoid disengagement from users.

**Theme 3: Need of Continuous Engagement**
The complex nature of the care homes and their daily routines, often with the need to respond to emergency situations amongst the frailest residents or challenging behaviour in dementia residents, impacted on the ability to adhere to monitoring at set times and the tablets were sometimes kept securely in cupboards until the next monitoring days. Embedding digital into the heart of a care home requires designing technologies that people actively reach for regularly and see the value in using rather than intermittent use where hardware is misplaced or batteries are not charged.

The care homes that maintained consistency in submitting vital signs observations during the project were the ones that as a consequence obtained more value in the subsequent data/information captured (by creating a comprehensive longitudinal record) and hence continued to engage and develop their own monitoring routines.

**Theme 4: Striking a Balance Between Monitoring and Causing Distress**
The act of regular monitoring showed the potential to increase the knowledge and understanding of non-nursing care home staff, and increase vigilance and attention. This was corroborated by responses to 'alerts' (where the NEWS score identified potential ill-health), where care home staff had taken actions before contact from SPA, indicating increased autonomy and confidence in clinical handovers and escalation of any concerns.

It was recognised during the project that regular monitoring by carers could cause distress in some residents (especially those with dementia) and medicalise their experience of living in a care home (residential settings). This could be addressed in future service designs by using wearable devices for monitoring (removing the need for a carer to take readings) however, taking away vital carer interaction with residents could potentially adversely impact on the purpose of the monitoring by losing that extra human contact and insight by relying on a data only approach.

**Theme 5: Communication Between Care Homes and Other Care Givers**
The care homes did not in general use the information stored against each resident in wider care planning conversations or with visiting health professionals (although it was recognised during the evaluation phase that this would have been useful to them), suggesting more time committed between introducing a technology for one application would eventually support adaption and adoption over the longer term to integrate or replace other existing ways of working, and that co-design of how data is presented is essential for digital platforms to add value and change the dynamics from '*this is the information you need to use*' to '*what data/information would you/I like to have?*'

**Theme 6: Wi-Fi and Connectivity Issues**
One of the main barriers to adoption was the variance of Wi-Fi connection quality across a home and between units and rooms, which caused problems submitting information using the Digital Care Home service. This negatively impacted on workforce confidence that information had been shared and the ability to maintain monitoring routines. For technological issues to not detract from positive experiences and outcomes, care homes and local authorities will need to identify innovative ways to boost Wi-Fi connectivity [17].

**Theme 7: Training**
The introduction of new technology requires a period of staff training to ensure that every member of staff is familiar with the new technology and confident in the usage. The focus group highlighted the need of providing guidance in-built in the application to support staff during the initial period in gaining confidence during everyday usage. Over the project timeline, it was identified that building more 'self-help' features into the app (such as motion graphic film content to explain the monitoring process, and 'how-to' guides) contributed to the autonomy of the home to resolve any technology issues or become accustomed to the digital service, and were a valuable resource for new members of staff where time for training was limited.

**Theme 8: De-Skilling**

The value in the project, as seen from the care home managers perspective, was in increased resources and investment in preventative interventions (i.e. identifying and supporting residents before their health deteriorated), and improving the quality of information available to them for clinical handovers. However, there was some concern from the nursing teams about being ‘deskilled’ by the introduction of technology and vital signs monitoring taking the place of carers noticing the softer signs of health deterioration.

**Theme 9: Workload Increase**

During the initial consultations, a fear about increase workloads for participating care homes emerged, as staff perceived the usage of the app as an addition to their everyday routine.

However during the project, care home teams reported using the time spent taking observations to talk to their residents and learn more about how they were feeling, using the monitoring data captured to supplement detection of the softer signs of health deterioration. This provided an interesting perspective from initial fears about increased workloads—viewing the Digital Care Home service as an ‘enabler’ to complement human interactions.

**Theme 10: Data Visualisation**

How the data was presented and whether that information was seen as ‘adding value’ to care impacted significantly on whether it was used to its full potential (e.g. used to understand trends, shared with GPs, ECPs, family etc.) to change behaviours and encourage innovation in how data/information could be used to improve quality of resident care.

## 2.4 Case Study—Care Homes Guide for Dysphagia

Dysphagia (difficulty swallowing) is common in frail older people particularly in those with long term conditions. Studies suggest as many as between 50% and 75% of residents in nursing homes, and up to 68% of people in EMI care homes have swallowing difficulties [26, 27].

When individuals have difficulty swallowing the challenge to optimise nutrition and hydration, in a way that is safe for the individual, is exacerbated. In frail older adults dysphagia can be life threatening and is thus a patient safety concern because of the increased risk of malnutrition, dehydration and aspiration pneumonia which can lead to death [24, 25]. Additionally, it is one of the major health conditions that can result in aspiration pneumonia which is the second most common infection found in Nursing Homes [22, 28].

Effective management of dysphagia in care homes is complex and multi-dimensional. Provision of food and drink which is of both good quality and which is safe for the person with dysphagia to swallow is known to result in better health and well-being of the person and reduces feelings of social isolation as the individual is able to join in social activities many of which revolve around eating and drinking. The Essential Standard of Quality and Safety [5] requires care homes to identify and take action on swallowing difficulties, including appropriate diet provision and support.

Without proper understanding of what is required for good dysphagia management within a care home environment, resident's health and well-being are compromised. Good management requires an integrated whole systems approach across both catering and clinical services, including policy development, quality monitoring, communication systems, staffing, specialist equipment needed, and specialist training for cooks, nurses, and support workers. Lack of guidance as to what is required is leading to suboptimal care.

Funding was secured from Abbeyfield Research Foundation to complete a two year project to address some of these barriers. In this project a digital tool was developed and evaluated for the effective management of dysphagia in a care home setting. The four objectives in this project were:

1. To identify and examine the evidence-base and views of residents, staff and company representatives to better understand management and shared decision-making in dysphagia care.
2. To co-design a framework that promotes adoption of dysphagia knowledge and evidence-based practice in a care home setting.
3. To implement the framework via a digital application for tablets.
4. To conduct an acceptability and validation process of the framework.

### *2.4.1 Methodology*

Stage 1: Literature Review

Evidence for dysphagia management in care homes was synthesized from a review of current literature. Findings were used to inform the semi-structured interviews and focus groups developed in next stage.

Stage 2: Qualitative Interviews and Focus Groups

Ethical approval was secured from London—West London and GTAC Research Ethics Committee to carry out the interviews and focus groups. Eight focus groups were conducted with members of the nursing, care assistant team and catering staff across 4 participating care homes. Four semi-structured interviews were carried out with the care home managers and quality managers from the organisations. In addition, semi-structured interviews were completed with six residents. The information from the interviews and focus groups was analysed and coded to determine themes around good practice patterns and to determine barriers to implementing good practice into daily routines. Recurring themes emerged in the interviews and focus groups:

training, food, workforce and quality and safety, with the first two ranked as the highest priority. The information in each theme was structured in three levels: individual (mealtime experience), care home (systems in place to support resident care) and organisational (policies and procedures). Themes and levels were digitated into an initial set of graphics, subsequently undertaken in the next stage.

Stage 3: Co-design of Dysphagia Management Protocol

The co-design process for our study expanded to include the expertise of designers and software developers from a digital health care company called Elaros who had secured funding from Innovate UK to explore the management of health and wellbeing in care homes with digital solutions.

The framework we were producing as part of our study lent itself to digital formatting thus via the collaboration of the two studies we were able to develop an app for tablets with the digital version of the resource ready for piloting.

The digital solution was initially tested for usability before the pilot was commenced in the care homes, thus ensuring the structure was a workable format for people with a range of technical and digital experience. Nine participants answered questions mainly centred around the layout, colours used, content of the application, navigation around the application. The implementation of usability testing during the development phases highlighted important considerations for final users of the application to allow them to navigate successfully through the information and use the features of the application in a successful manner. Participants' suggestions received for improvements were considered and implemented in the final version of the digital guide.

### 2.4.2 Technology

This collaboration resulted in the development of the Digital Care Home Guide to Dysphagia. The application was developed with the open source development framework Apache Cordova for iOS and Android devices. This scalable solution for use in care homes included text, images, links and videos embedded to support the workforce with gaining the skills and knowledge acquisition for managing residents experiencing swallowing difficulties.

The Guide was structured around 5 key topic areas identified during the focus groups with final users. The structure enabled rapid access to the most relevant information with easy navigation. The 5 topic areas were:

- Dysphagia Essentials
- Food
- Quality and Safety
- Workforce
- Training

The Guide included a total of 52 pages. A top navigation bar allowed direct access to the topic areas, each one easily represented with a different colour. The Guide allowed users to send feedback directly to the development team via a 'Like'/'Dislike' button included in all the pages that opened a form to send comments, and suggestions (Fig. 2.3).

**Fig. 2.3** Screenshots of the care home guide to dysphagia

### *2.4.3 Evaluation*

The Care Home Guide to Dysphagia was been piloted in 4 care homes. Each home was provided with two tablets with the Guide preinstalled to be used over a 12 week period.

The evaluation included the analysis of digital analytics, feedback provided via the Guide and a short questionnaire for final users. By integrating an analytics service in the software code, we were able to monitor remotely the activity of all users in each page. This included the frequency each page was visited, the time spent on each page and the navigation pathway. Although all pages were visited results demonstrated a major interest of the users in two topic areas: Food and Essentials. Pages with videos discussing Dysphagia and including recipes for residents were the most visited.

The Dysphagia Guide included a box at the bottom of all pages for users to provide direct feedback to the development team. This feedback was in the form of a question "Was this article helpful?" with two options "Yes" or "Not". If a user decided to answer the question, then a new box appeared to expand the answer and include a text message. Almost half of the pages included in the Dysphagia Guide received feedback from at least one user. Aligned with the results of the digital analytics, the topic areas Essentials and Food received 8 of each 10 feedback messages sent by final users.

At the end of the 12-week pilot period participants were asked to complete a short questionnaire about the Guide. The findings indicated that final users reported the Guide to be helpful to carry-out their job, to be easy to use and with clear organisation of the information.

Following the success of this study and the positive findings from the pilot evaluation of the Guide we are seeking further funding to scale up and spread usage of the Guide to support more care homes with the management of their residents who present with dysphagia.

## 2.5 Conclusions and Next Steps

One of the key challenges in designing technology solutions to complement how care is delivered in care homes is creating solutions that respond to the needs of a range of stakeholders (within each care home and across the health and social care system) when the perceived value of a technology within a service varies depending on the viewpoint of the stakeholder.

Without meaningful co-production with all stakeholder groups when 'digitalising' the care home, there is a risk of creating solutions which fail to respond to the subtleties of different environments, users and situations.

Adopting a 'one-size' implementation approach to using digital in care homes is unlikely to be successful—adaption of the application of technology to the unique environment of each care home is more likely to result in systems that are used and welcomed by the teams involved and that fit within the routines and rhythms of a typical day.

Through reviewing the learning from the Digital Care Home and the Care Homes Guide for Dysphagia (and similar projects in the north east involving the Vanguards), it is suggested that investing time and effort into creating ownership within care home teams for using digital tools could nurture and grow the enthusiasm of the workforce to use (and keep using) the technology, even amongst users who are more fearful using technology at work and have low confidence in their own digital skills.

In addition, this investment could encourage creativity in potential further applications and benefits of the technology, which in the case of Digital Care Home would be using information collected to not only to provide an immediate intervention on occasions where health deterioration was identified, but also over a longer timeframe to influence advanced care planning or decisions regarding end-of-life care.

The challenge for the long term sustainability of the digitised care home will be designing tools and resources that carers and residents' enjoy using—rather than using them because they have to—by contributing positively about how they feel about themselves, such as feeling more confident, empowered, or reassured about their health of the health of people cared for. Reassurance, empowered communication and enhanced feelings of professionalism were also identified as outcomes in Sunderland [29] and the Pathway of Care work stream of the Newcastle Gateshead Care Home Vanguard [17].

Effective implementation requires investment in education and training; engagement with care home teams, residents, and their family and friends; and resources to help teams trouble-shoot any problems they have using technology to encourage independence and increase confidence. It takes time to change behaviours and work in a new way – and stakeholders designing new care systems will need to take the time to understand the multitude of factors involved in running a care home in the present day but will also respond to a potentially very different and more 'activated' care home population of the future.

**Acknowledgements** The Digital Care Home study was supported by NHS England Wave 1 Test Bed programme and the National Institute of Health Research (NIHR) Devices for Dignity MedTech and In vitro diagnostics Co-operative (D4D). The views expressed are those of the author and not necessarily those of the NHS, the NIHR or the Department of Health. We thank our colleagues from Inhealthcare, Sheffield Teaching Hospital NHS Foundation Trust, University of Sheffield, Healthwatch Sheffield and from the participating Care Homes.
The Care Homes Guide for Dysphagia project was supported by Abbeyfield Research Foundation and InnovateUK. The authors gratefully acknowledge the insight and expertise of the research team Elizabeth Barnett, Julie Skilbeck, Davlyn Jones, Sally Fowler-Davis, Melanie Gee and Paul O'Brien.

## References


1. Addicott, R., Ham, C.: Commissioning and Funding General Practice: Making The Case For Family Care Networks. The King's Fund, London (2014). Available at https://www.kingsfund.org.uk/publications/commissioning-and-funding-general-practice
2. Alkouri, O., AlKhatib, A.J., Kawafhah, M.: Importance and implementation of nursing documentation: review study. Eur. Sci. J. **12**(3), 101–106 (2016)
3. Baylis, A., Perks-Baker, S.: Enhanced Health in Care Homes: Learning From Experiences So Far. The Kings Fund, London (2017)
4. Banaee, H., Ahmed, M.U., Loutfi, A.: Data mining for wearable sensors in health monitoring systems: a review of recent trends and challenges. Sensors **13**, 17472–17500 (2013). https://doi.org/10.3390/s131217472
5. Care Quality Commission.: Essential Standards Of Quality and Safety. Care Quality Commission, London (2010)
6. Curry, N., Ham, C.: Clinical and Service Integration: The Route To Improved Outcomes. The King's Fund, London (2010). Available at https://www.kingsfund.org.uk/publications/clinical-and-service-integration
7. Dias, D., Cunha, J.P.S.: Wearable health devices—vital sign monitoring, systems and technologies. Sensors **18**(8), (2018). https://doi.org/10.3390/s18082414
8. Driessen, J.R., Bonhomme, A., Chang, W., Nace, D.A., Kavalieratos, D., Perera, S., Handler, S.M.: Nursing home provider perceptions of telemedicine for reducing potentially avoidable hospitalizations. J. Am. Med. Dir. Assoc. **17**(6), 519–524 (2016)
9. Edirippulige, S., Martin-Khan, M., Beattie, E., Smith, A.C., Gray, L.C.: A systematic review of telemedicine services for residents in long term care facilities. J. Telemed. Telecare **19**(3), 127–132 (2013). https://doi.org/10.1177/1357633X13483256
10. Goodman, C., Davies, S.L., Gordon, A.L., Dening, T., Gage, H., Meyer, J., et al.: Optimal NHS service delivery to care homes: a realist evaluation of the features and mechanisms that support effective working for the continuing care of older people in residential settings. Health Serv Deliv Res. **5**(29) (2017)
11. Goodwin, N., Sonola, L., Thiel, V., Kodner, D.: Co-ordinated Care For People With Complex Chronic Conditions: Key Lessons and Markers For Success. The King's Fund, London (2013). Available at https://www.kingsfund.org.uk/publications/co-ordinated-care-people-complex-chronic-conditions
12. Gordon, D., Geiger, G., Lowe, N., Jickling, J.: What is an electronic patient record? Proceedings of the AMIA Symposium, pp. 240–244 (1998)
13. Gray, L.C., Edirippulige, S., Smith, A., Beattie, E., Theodoros, D., Russell, T., Martin-Khan, M.: Telehealth for nursing homes: The utilization of specialist services for residential care. J. Telemed. Telecare **18**, 142–146 (2012). https://doi.org/10.1258/jtt.2012.SFT105
14. Hall, A., et al.: Implementing monitoring technologies in care homes for people with dementia: a qualitative exploration using Normalization Process Theory. Int. J. Nurs. Stud. **72**, 60–70 (2017)
15. Ham, C., Walsh, N.: Making integrated care happen at scale and pace: lessons from experience. The King's Fund, London (2013). Available at https://www.kingsfund.org.uk/publications/making-integrated-care-happen-scale-and-pace
16. Health IT: Benefits of EHRs: improved diagnostics and patient outcomes (2017). https://www.healthit.gov/providers-professionals/improved-diagnostics-patient-outcomes
17. Hodgson, P., Cook, G., Thompson, J., Abbott-Brailey, H.: Assessment and clinical decision making of the acutely ill older care home resident: implementation of NEWS in Gateshead Care Homes. Final report. Northumbria University (2017). Accessible at https://static1.squarespace.com/static/5893239037c581b39142e013/t/59dddeccd2b8572ff2c11bd5/1507712721712/NEWS+REPORT+FINAL_2.pdf
18. Jenkins, R.L., White, P.: Telehealth advancing nursing practice. Nurs. Outlook **49**, 100–105 (2001)

19. Kingston, A., Wohland, P., Wittenberg, R., Robinson, L., Brayne, C., Matthews, F., Jagger, C.: Is late-life dependency increasing or not? a comparison of the Cognitive Function and Ageing Studies (CFAS) (2017). https://doi.org/10.17863/cam.21762
20. Kruse, C.S., Mileski, M., Alaytsev, V., Carol, E., Williams, A.: Adoption factors associated with electronic health record among long-term care facilities: a systematic review. BMJ Open **5**(1), (2015)
21. Kruse, C.S., Mileski, M., Vijaykumar, A.G., Viswanathan, S.V., Suskandla, U., Chidambaram, Y.: Impact of electronic health records on long-term care facilities: systematic review. JMIR Med. Inf. (2017)
22. Marik, P.E., Kaplan, D.: Aspiration pneumonia and dysphagia in the elderly. Chest **124**(1), 328–336 (2003)
23. Meissner, A., Schnepp, W.: Staff experiences within the implementation of computer-based nursing records in residential aged care facilities: a systematic review and synthesis of qualitative research. BMC Med. Inf. Dec. Making **14**, 54 (2014)
24. National Patient Safety Agency: Ensuring safer practice for adults with learning disabilities who have dysphagia (2007). http://www.nrls.npsa.nhs.uk/resources/clinical-specialty/learning-disabilities/?entryid45=59823&p=2. Accessed 10 Mar 2015
25. National Patient Safety Alert: Risk of death from asphyxiation by accidental ingestion of fluid/food thickening powder (2015). https://www.cas.dh.gov.uk/ViewandAcknowledgment/ViewAlert.aspx?AlertID=102291. Accessed 10 Mar 2015
26. O'Loughlin, G., Shanley, C.: Swallowing problems in the nursing home: a novel training response. Dysphagia **13**(3), 172–183 (1998)
27. Steele, C.M., Greenwood, C., Ens, I., Robertson, C., Seidman-Carlson, R.: Mealtime difficulties in a home for aged. Dysphagia **12**(1), 43–50 (1997)
28. Tanner, D.C.: Lessons from nursing home malpractice litigation. J. Gerontol. Nurs. **36**(3), 41–46 (2010)
29. Unsworth, S., Bell, T.: Interim Evaluation Report Digital Care Home Tablet. South Tynesides NHS Foundation Trust (2017). Accessible at: http://www.atbsunderland.org.uk/wp-content/uploads/2017/10/Ditigal-Care-Home-Tablet-Evaluation-March-2017.pdf
30. Wang, Z., Yang, Z., Dong, T.: A review of wearable technologies for elderly care that can accurately track indoor position, recognize physical activities and monitor vital signs in real time. Sensors (Basel, Switzerland) **17**(2), 341 (2017). https://doi.org/10.3390/s17020341

**Philippa Hedley-Takhar** is Business Development Manager for the NIHR Devices for Dignity MedTech and In vitro diagnostic Co-operative (MIC), a catalyst organisation for the development of new technology-dependent interventions to support people with one or more chronic health conditions typically associated with loss of dignity and independence and poor patient outcomes.

Philippa previously worked on the NHS England Perfect Patient Pathway Test Bed programme to test and evaluate novel combinations of technology and service re-design supporting the care home community; and develop training and resources for health professionals to support people with long-term conditions to use technology for self-management.

**Angel Jimenez-Aranda** is the Digital Technology Lead at NIHR Devices for Dignity MedTech Co-operative (D4D). He has prime responsibility for promoting the development and adoption of digital healthcare technologies in the D4D consortium and maximising the operational efficiency in the selection and execution of high impact research and innovation projects. He has an extensive experience developing and evaluating digital solutions for clinical decision support and patient empowerment. Dr Angel Jimenez-Aranda has been an active member of the delivery team in the Perfect Patient Pathway Test Bed Programme (2016–2018). This programme, funded by NHS England and led by Sheffield Teaching Hospitals NHS Foundation Trust, aimed to build infrastructure and test innovative combinations of technology and care pathway development to improve patient outcomes and experience.

**Vitaveska Lanfranchi** is a Senior Lecturer in Medical Computing at The University of Sheffield in the OAK Group and in Sheffield Teaching Hospital NHS Foundation Trust as part of the Medical Imaging and Medical Physics Group and affiliated with NIHR Devices for Dignity MedTech Co-operative with the mission of enhancing dignity and independence for people with prolonged ill-health and addressing health technology needs of children and adults. Her work has a strong applied focus, designing and implementing applications for real-world scenarios, at all technology readiness levels. She has a long experience in stakeholder analysis and translation of stakeholders' needs into technology, providing a useful link between technical innovation and user perspective. She has published over 50 papers in international conferences and journals.

**Sue Pownall** is Head of Sheffield Adult SLT Service, Honorary Professor at Sheffield Hallam University and the Academic Director for the Therapeutics and Palliative Care Directorate in Sheffield Teaching Hospitals NHS Foundation Trust. Sue is co-lead for the Yorkshire and Humber CRN AHP Specialty Group and also co-lead of the South Yorkshire CAHPR Hub.

Sue's clinical specialism is in the area of dysphagia and she has extensive experience of working with patients who present with dysphagia resulting from a range of medical conditions, across clinical pathways and from both acute and community settings. Sue completed her Ph.D. at the University of Sheffield in 2009 after investigating the management of patients presenting with dysphagia following stroke. Sue is currently evaluating the implementation of electrical stimulation into routine clinical practice for the treatment of dysphagia and has recently completed a two year study exploring ways to optimize nutrition and hydration for residents with dysphagia in care homes by the use of a digital Dysphagia Guide.

**Lise Sproson** works for NIHR Devices for Dignity MIC as NHS Innovations Manager, Diabetes Theme Manager and PPI Lead. Lise is a Speech and Language Therapist with specialist expertise in the assessment and management of dysphagia (swallowing) and communication difficulties. She is a committee member of the Council for Allied Health Professions Research (CAHPR) and the UK Swallow Research Group (UKSRG) and Specialist Advisor on dysphagia for the Royal College of Speech and Language Therapists (RCSLT) and the National Institute for Health and Care Excellence (NICE).

**Wendy B. Tindale** (OBE, M.Sc., Ph.D., CSci, FIPEM, ARCP) is Consultant Clinical Scientist and Scientific and Innovation Director at Sheffield Teaching Hospitals. She holds a Professorship in Medical Physics at the University of Sheffield and nationally is Director of the NIHR 'Devices for Dignity' MedTech Co-operative. She has an interest in combining clinical academic excellence and patient benefit with business opportunities through the translation of innovative medical technologies and has worked with both public and private sectors to facilitate successful collaborations. Wendy has contributed to numerous national and international committees. She received a Beacon Award for outstanding contribution to healthcare and was honoured in the Health Service Journal's inaugural Inspirational Women awards for outstanding leadership in her field, becoming Healthcare Scientist of the Year in 2016.

# Chapter 3
# Contemporary Themes in the Design of AT for the Ageing Population: Materiality, Co-design and Cultural Influences

**Gabriella Spinelli, Massimo Micocci, Emmanuel Tsekleves, Yu-Han Wang, Wendy Martin, Yonghun Lim and Umber Shamim**

**Abstract** Products we purchase are much more than artefacts that fulfil functional needs in our life. We have grown to enact our consumer choices, even those regarding fast moving consumable goods, with careful considerations informed by numerous trials, recommendations and, growingly, environmental concerns in mind. Advanced manufacturing and progress in research and development are providing more choices for consumers even in quite specific and complex product markets. An exemption to this market trend is represented by assistive technologies (ATs). This is a rela-

G. Spinelli (✉) · U. Shamim
Brunel University London, College of Engineering, Design and Physical Science, Kingston Ln, London, Uxbridge UB8 3PH, UK
e-mail: gabriella.spinelli@brunel.ac.uk

U. Shamim
e-mail: umber.shamim@brunel.ac.uk

M. Micocci
Department of Surgery and Cancer, Imperial College London, NIHR London In Vitro Diagnostics Co-op., QEQM Building, St. Mary's Campus, Praed Street, London W2 1PE, UK
e-mail: m.micocci@imperial.ac.uk

E. Tsekleves
ImaginationLancaster, Lancaster University, LICA Building, Bailrigg, Lancaster LA1 4YW, UK
e-mail: e.tsekleves@lancaster.ac.uk

Y.-H. Wang
Department of Product Innovation and Entrepreneurship, College of Innovation Management, National Taipei University of Business, No. 100, Sec. 1, Fulong Rd., Pingzhen Dist., Taoyuan City 324, Taiwan (R.O.C.)
e-mail: wangyuhan@ntub.edu.tw

W. Martin
Brunel University London, College of Health and Life Sciences, Kingston Ln, London, Uxbridge UB8 3PH, UK
e-mail: wendy.martin@brunel.ac.uk

Y. Lim
Bournemouth University, P301, Poole House, Talbot Campus, Fern Barrow, Poole BH12 5BB, UK
e-mail: ylim@bournemouth.ac.uk

 
A. Woodcock et al. (eds.), *Design of Assistive Technology for Ageing Populations*, Intelligent Systems Reference Library 167,
https://doi.org/10.1007/978-3-030-26292-1_3

tively underdeveloped context despite the growing demands for assistive devices by those in later life who need either support in accomplishing everyday life to stay independent or have complex co-occurring conditions. In this chapter, we explore why ATs, especially for older adults, are underdeveloped by exploring issues related to design approaches and cultural and social perceptions that have contributed to making consumers more or less sensitive and demanding towards the role of ATs in their lives. The chapter will conclude with recommendations that may be able to shift the perception of assistive devices so as to facilitate the user's emotional investment in the devices, attachment to them, which, in return, may lead to better adherence and faster adoption.

**Keywords** Assistive technology · Adoption · Identity · Desire · Lifestyle · Older adults

## 3.1 The Design of Assistive Technologies (ATs) Beyond Functional Requirements

Products we purchase are much more than artefacts that fulfil functional needs in our life. We have grown to enact our identity through the consumer choices we make and in return to be defined by the goods we surround ourselves by Escalas et al. [1]. In making consumer choices we follow friends' suggestions, online reviews, experts' advice, price and brand credibility. Personal taste and peers' acceptance are also factors that have great impact in our choice. Despite a growing number of older adults, extended longevity and, consequently, a longer period of comorbidity, the ATs market has not kept pace with what is required for a diverse group of consumers who need support in everyday tasks. ATs are still mostly designed with functional support in mind, little personalisation in the support of different physical requirements and almost no devices catering for aesthetic, emotional and cognitive differences. After an introduction about the long-standing debate on what requirements ATs are meant to support, the chapter reviews the causes that lead to the abandonment of ATs and the heighten challenges represented by designing for people with severe disabilities. The chapter then explores three trends in the design of AT for an ageing population: co-design, materiality and the cultural framework that surrounds their design and use.

'Assistive Technologies' (ATs) is used to describe devices aimed to increase or maintain the functional capabilities of individuals with injuries or declining abilities and to enhance overall well-being [2]. The range of ATs has been populated with subcategories differentiating between 'replacement parts' such as artificial limbs; 'orthopaedic products' for external applications; off-the-shelf products to help people coping with declining abilities in daily life settings [3]. Despite these categorisations, ATs often fail to move beyond the mere functional support they offer and to embrace the holistic role of support they could offer.

With the need to consider how ATs may support people holistically, the concept of Inclusive Design has been widely accepted in the design arena by considering extreme users so as to expand both the type of support and the user groups of ATs [4]. Inclusive design is defined as "The design of mainstream products and/or services that are accessible to, and usable by, as many people as reasonably possible … without the need for special adaptation or specialised design" by the British Standards Institute [5]. Inclusive design is also known under several different terminologies, i.e. universal design, design for all, barrier-free design and accessible design in different countries, which are used in a similar way [6]. Inclusive design has been recognised internationally as a well-known design approach in mainstream products, services, and environments offering a clear business benefit; that of targeting people with differing abilities and needs, hence a wider market audience [7].

Hocking [8], in the US, established that 56% of ATs was abandoned, and 15% were never used in the first place. This may be attributed to a lack of empathy designers imbue in the design of ATs, and the resulting neglect of non-functional requirements such as aesthetics (ibid). The importance of non-physical aspects in design has already been stressed, and design movements concerned with psychosocial design attributes have been captured in approaches such as *meaning-centred design* [9]; *human-centred design* [10, 11]; *experience design* [12–14] and *emotional design* [15]. However, a growing number of standardised devices that compensate for people's missing abilities continue to be developed with little consideration for the individual preferences, changing physical needs [16] and the positive/negative connotations encapsulated in the device [17]. As a result, the development of dedicated products that fall under the definition of ATs often promotes isolation rather than inclusion, being apart from the aesthetic and social standards of more appealing and versatile consumer products. The consequence is that more than one-third of the ATs that are purchased are abandoned when they are still needed [17, 18].

Further difficulties emerge when older adults require life-enhancing devices that are not routinely used by the general population and that convey messages of vulnerability and stigmatisation. This, in return, affects the self-esteem and overall mental wellbeing, treatment adherence and social relations [19–22]. This picture is exacerbated by the normal decline of cognitive abilities that occurs with ageing, resulting in hindrance of novel technologies' usage and limitations to their adoption [23]. Learning difficulties affect information processing and the consequent decision-making older adults exercise when making informed decisions; rather than embarking on products that require a superior cognitive workload, older adults significantly rely on more spontaneous and effortless experiential-based knowledge [24, 25] that eventually leads towards the purchase of standardised and popular products. This per se creates obstacles to innovation in the market of ATs, as mass-market products are better recognised and for the reasons aforementioned, they continue to be preferred, even when they may only partially address user needs.

The resignation to 'functional' rather than 'pleasing' products, has resulted in an increased perception of older adults as disabled [8] with their growing fear of being stigmatised [26]. This process is intensified by the role that ATs assume in a social context and how in return, social contexts may influence personal preferences

[27, 28]. In a study conducted by Pape et al. [29], it was demonstrated that social factors, beyond usability and functionality, impact the choice of ATs. The use of an assistive device often piggybacks the acceptance of the disability and the device is more likely to be abandoned if it makes the users feel excluded from their social context, e.g. it doesn't match with the values of the community they are part of. In those cases, the dissonance between the individual self-representation and the expected social norms may cause deviation from others and the social interaction may be compromised [28, 30].

## 3.2 Adoption and Abandonment of ATs

The review conducted by Kraskowsky and Finlayson [31], and Wessels et al. [32] about non-use of ATs, identifies four main factors that affect ATs use:

1. Factors related to the person: this set of factors revolves around the expectations that a person has of himself/herself as well as the expectations that people in his or her social circle have of them using the device. The list includes items such as age, gender, diagnosis, acceptance of the disability, emotional maturity, inner motivation, progression of disability, severity of disability, change in disability and use of multiple devices;
2. Factors related to the device: e.g. quality and appearance;
3. Factors related to the user's environment: social circle support, physical barriers, the presence of engagement opportunities for those who use ATs, access and availability of ATs on the market;
4. Factors related to the professional assessment of the users and nature of the intervention planned for the same: participation of the users and of their views in the assessment, instruction and training provided re the ATs, correct provision and installation process, length of the ATs delivery period and follow-up service.

The definition of 'non-use' in ATs entails a complex interconnection of several elements that go beyond the mere usability and functionality of the device. As stressed in studies conducted by Federici and Borsci [33] with healthcare professionals and end users, the user experience of an assistive solution is affected not only by the quality of the interaction between the user and the solution itself but also by the perceived quality of the professional service provided. The interplay between personal related factors and factors related to the user's environment are explored in a comparative study [34] between young and older users demonstrating that while products for children may imbue a higher level of enjoyment, devices for older adults are perceived exclusively in relation to function, e.g. they predominantly convey the physical support they provide to a person with health decline.

The finding of this study suggest the need for a double design intervention: (1) a functional customisation by means of a personalisation that allow product changes for multiple purposes and functions to reflect the complex disability of the users and its evolution; and (2) an effort to include technical and futuristic features in ATs so as to empower the users and excite them about the deployment of technology for smart ATs, in line with technologically advanced mainstream products, that were more likely accepted with enthusiasm by the study participants. In a further study, Shinohara and Wobbrock [28] conducted an interview with 20 people with disabilities to understand the use of ATs in social and professional contexts. They found two common perceived misperceptions when using ATs in a social context:

1. the ATs functionally eliminate a disability;
2. people who normally use ATs, are helpless without them.

An example described in the study is that of two participants with visual impairments who used the iPhone, a mainstream technology, due to its accessibility features. As ATs appear different from mainstream products, a common misperception would be that the two people involved would be unable to use mainstream technologies without help. Therefore, design interventions should also be aimed at providing support in the form of ATs that are indistinguishable by mainstream products to alleviate the common misperception that disabilities dominate and control those who are affected by them.

## 3.3 ATs Design for Mobility Impairments Among Older People

According to the statistics released by the UK Office for Disability Issues, the prevalence of disability rises with age. Around 6% of children are disabled, compared to 16% of working age adults and 45% of adults over State Pension age [35]. The scale of age-associated disability can vary in the future depending upon the health status of the older people. However, the current indicators point towards an increase in the number of impaired mobility issues due to old age [36].

In extreme cases of mobility impairments, the ATs prescribed resemble medical technology devices even more [37]. Hence what mainly characterises the approach to their design, is a problem-solving oriented process [38]. In such problem-solving approach patients feel that AT products are designed "for them and not with them" [39].

Similarly, due to the lack of participation in professional assessment and prescription of ATs, patients are also deprived of the chance to choose an AT device, which may be purchased or prescribed for them by a third party [4]. Increasingly, efforts have been made to adapt, standard design techniques like Inclusive Design (ID) and co-design for ATs, to encourage end-user evaluation of prototypes [40] and to improve the design process itself [41]. The design of ATs in such cases may be seen as extremely challenging due to the abilities of the users, the involvement of multiple stakeholders and input of cross-disciplinary expertise [42].

People with severe mobility impairments are prescribed Environmental Control (EC) devices. An EC device allows the patients to control several peripheral devices, for example, TV, lights, radio, telephone etc. [43]. These devices may help to provide some independence, to decrease social isolation, and finally to open up venues for education and employment. However, there is little research in support of the effectiveness of EC devices [44].

EC devices can operate through numerous methods for control and input. The most widely used method of input is a simple switch-based (single click) control device. Customised input options for patients have emerged recently, according to the type and severity of their mobility impairment. However, as the severity of the impairment increases, the chances of operating these devices gracefully continues to decrease.

The availability of off-the-shelf, voice-controlled technology has influenced patients' requests and aspirations. The ease of use of speech recognition has led to its inclusion into many types of EC integrated systems [45–47] which mobility impaired users seem to prefer [48] due to its speed and relatively effortless interaction [49].

However, the voice-controlled technology embedded in EC devices are not very reliable and limited to a set of commands [50]. On the other hand, speech recognition technology used nowadays in Interactive Smart Agents (ISA) like Amazon Echo, Google Home etc., have the ability to understand natural language. These ISA devices also offer comparable functionalities in terms of controlling smart devices around the home, potentially contributing more significantly to the independence and dignity of disabled patients. Whilst supporting patients in controlling their surroundings more effectively, voice control devices are also beneficial as they afford interactions that are more natural and socially and emotionally acceptable [51].

## 3.4 Co-design Approaches for ATs: Designing with Older Adults

Designers have a social responsibility to design products and services for the common good. Design should, therefore, promote healthy behaviours and enhance the wellbeing of everyone, including older adults. However, there is a significant disparity in cases of design to improve life, health and wellbeing products and who it is actually benefiting from them [52, 53]. Older adults form a minority when it comes to new product development.

Part of the problem lies in the traditional user-centred methodologies employed for designing assisted technologies. Although such approaches are effective in developing empathy for mainstream adult users [54], they fall short when it comes to older adults. This is because it is not always clear at the time of design, to know what constitutes actual needs for the intended older adult users, and how ATs may affect these. This knowledge may be difficult to gather because design typically concerns products or services that do not yet exist [55]. This is where co-design can have a real impact, by designing "with" as opposed to designing "for" people. Therefore, one approach that has been proposed to design more effective technology for older adults is to include them in the design process from the requirement stage throughout the development phase [56].

As co-design [57] enables a wide range of people, including older adults, to make a creative contribution in the solution but critically also in the formulation of a problem (a task predominantly led previously by designers), it creates new opportunities for using materiality for eliciting aspirations. A key element of co-design is that users are seen as 'domain experts' of their own needs and experiences [58] providing ways for people to engage with each other as well as providing ways to communicate, be creative, share insights and test out new ideas [59]. Therefore, through this, we can gain a better understanding of the everyday relationship of older users with existing technologies but also material objects in their homes as well as other aspects of social life. Within the context of ATs, co-design enables going *beyond 'obvious health technologies' to explore more mundane aspects of materiality, such as the built environment* [60], which form key aspects within the sociology of health.

Involving older adults in co-design comes with challenges, these normally are: general decline in sensory perception, cognitive difficulties, mobility needs, fatigue, and lack of technical knowledge [61]. However, several visual aid tools and story-telling techniques have been developed to help engage older adults in the co-design process [62, 63]. Nevertheless, the design of products for older consumers tends to focus on specific chronic health conditions, such as for stroke survivors [64, 65] people living with dementia [66, 67] and other health-related conditions [68, 69], rather than the mainstream older adult market.

However, co-design-based research conducted with older consumers across different health and ATs reveals several benefits. More precisely, these benefits are related to improving the creative process, the service or product, project management, and longer-term effects [70], whilst also impacting significantly on the design of assistive technology devices, to make them more fit for purpose [71]. The experience of the 'Ageing Together' project also shows that co-design can succeed in introducing helpful technology into the lives of older adults [72].

Furthermore, the work of De Couvreur et al. [73] shows how the process of collaborative co-designing, making and using artifacts fosters several elements of subject well-being in itself. Besides engaging in a productive approach that empowers older people in the process of co-designing and evaluating technologies for themselves, the findings from Leong and Johnston [74] reveal that co-design is capable of enhancing older adults' independence, social agency and well-being.

In the current UK and European political climate, as the government passes more control to communities and individuals, co-design might have a significant role to play in the transformation of public services [75]. This represents an opportunity for the wider adoption of co-design and co-production approaches to develop ATs for growing ageing communities.

## 3.5 ATs as Objects: Materiality, Emotions and Everyday Life

Narratives of medicine, decline, and functionality around ATs remain predominant, with minimal changes towards narratives of consumerism, flexibility, and style. Borgerson [76] has argued the significance of the relationship between social and self-identity and material objects in consumption—and how a focus on materiality may enhance our knowledge and understanding of consumer relationships, processes and practices. Objects are symbols of consumption and are significant to how people develop, portray and enhance our narratives of self as part of constructing our identities and lifestyles. In this context, objects such as ATs become extensions of the embodied self—and co-constitute our sense of identity. Material objects are therefore central to our social identities, for example, in relation to gender, age, ethnicity and social class [77] and presentation of our embodied self in everyday life. It is therefore notable that there are limited choices and styles associated with many ATs. The negative connotations around a sense of stigma, dependence and decline can moreover limit the opportunities for self-expression. ATs are rarely viewed as objects of desire and the people who use them as consumers. This is where the design for flexibility and consumption in ATs can challenge and change the milieu and opportunities for self-expression in people's everyday lives. A focus on material culture may, therefore, open up the possibilities of enhancing a shift in narratives and language around assistive technologies that may result in improvements in their adoption and effective use, as well as the emotional connection to ATs.

There has been increasing recognition of the significance of material culture in health and social care [60]. This can include objects that are central to our mundane and habitual everyday lives and are thereby taken for granted and unnoticed as the objects are embodied and embedded in our tacit and daily routines, this may include, for example, bowls, glasses and clothing [60, 78]. For Miller, '[O]bjects are important because they are evident, and they physically constrain or enable, but often because we do not 'see' them' [79: 5]. ATs are objects—either very small (glasses and hearing aids) or larger in size (walking sticks, wheelchairs)—that become central to people's daily routines, the everyday care of the body, and may be experienced as mundane, taken for granted and invisible. Alternatively, assistive technologies can feel ever present, imposing and highly visible to personal and social worlds.

For Buse et al. [60], 'materialities of care' can be a means to make visible mundane, frequently unnoticed aspects of material culture within health and social care contexts. In particular, Buse et al. [60] identify three distinct but interconnected analytical ways in which material culture can be explored: namely, *spatialities of care*, *temporalities of care* and *practices of care*. With *spatialities of care*, the researchers refer to the way in which space influences the possible embodied actions, social interactions and care practices that ATs can enable. The ways ATs are used in different spaces from the hospital, home and public areas in this context would be significant. Second, time is central to care, and the *temporalities of care* highlight the multiple and intersections of time and routines associated with care, from transitory moments to everyday routines, and institutional regimes of care. People who use ATs have usually been assessed as needing an aid and do therefore have journeys and transitions with their objects across and within time. Third, *care practices* highlight the dynamic relations between objects, meanings and the body in which practices are attuned to tactile and 'sensory ways of knowing' [60: 249]. Material objects therefore have an active role in the ways health and social care interactions are created and in the ordering of care relationships, in which the AT as an object can be central to regulating bodies in health and social care through the predominance of assessments of mobility and safety, the monitoring of everyday activities of life, and underlie discourses of discharge, safety and risk assessments of the user. In this context, the materiality of ATs may shape and facilitate caring relationships, relationships that often denote hierarchical structure and power, in which, for example, the current very limited options of design of ATs in health care systems constrain and limit practices of care, and result in fewer choices and self-expression for the user. However, through the process of 'material imagining' [80] in which more personalised and flexible ATs as objects can be reimagined and designed, novel and reimagined 'possibilities for care' and care relationships can also be opened up (cf. Buse and Twigg [81]). Materialities of ATs, therefore, focus the attention on relationships within the care journey as well as on interconnections between bodies, objects and spaces.

Everyday practices can reveal the ways in which bodies, materials and identities are constituted, and how the use for ATs may disrupt everyday routines that then need to be reformed in new contexts. A focus on the materiality of ATs as everyday objects provides an original lens to look at the interplay between body, object and the self, which renegotiate their respective roles in order to adjust to the introduction of ATs [82]. As objects in everyday life, ATs are therefore imbued with sociocultural and emotional meanings and are invested with of social and emotional significance. There is a wide range of emotions associated with objects that Ahmed [82] explores, including pleasure, pain, shame, fear and hate. ATs are permeated with emotions which enhance or limit the way the objects are seen, experienced and utilised in everyday life. For example, emotions and meanings associated with ATs can range from desire to disgust, independence to stigma, and feelings of shame to enhanced well-being [83, 84]. The emotions connected to the ATs can not only influence the persons' sense of identity, their relationships with the materiality of the objects, but also the extent to which they are responsible for emotions that hinder or enable the users' presence and participation in public and private spaces. The meanings around ATs as objects are continually performed, negotiated and, at times, resisted. ATs therefore not only highlight the significance of the aesthetics of care but how the use of ATs as objects in everyday life is lived and felt [78].

A methodology in which the materiality of ATs as objects is used as ways to elicit data can also be a means to understand more about the ATs in everyday life. A focus on materiality may be a novel lens through which the doing of health and social care practices and relationships can be re-examined with a focus on the way objects may shape health and social care encounters and moments [60]. In this context, novel and reimagined designs of ATs may result in a move away from a predominant focus on function in care practices that surround older people, to more personalised pathways of approaching care relationships that can facilitate a more sensate and flexible approach by 'caring through things' [85]. A material approach can moreover make visible the unseen, the mundane and the taken for granted nature of ATs in health and social care, making it more feasible to tease out attributes, values and design language elements that may scaffold very different interactions with ATs as companions and objects of desire. The desire to facilitate the user to have more power and control in the design, choice and consumption of ATs has, therefore, the potential to reconfigure care relationships, care journeys and imaginaries of need.

## 3.6 Culture as an Interpretative Framework for the Use of ATs

The ageing phenomenon is global and cuts across cultural frameworks. However different cultures have experienced this ageing transformation at varied speed. Researchers globally believe that at least a partial solution to this societal issue resides in technology, its design and application [86]. However cultural frameworks are likely to impact how technological products are perceived and consequently used. The term 'culture' is all-encompassing and has found definitions based on discipline-based research [71, 87–91]. Its meaning shifts from time to time, though in the current literature there is an agreement in considering culture as the body of social programs, economic systems, political ideologies, and technological systems [92]. Culture also tends to be defined within descriptive and symbolic contexts; in the former, culture is viewed as race or ethnicity where a group of people share the same beliefs, ideas, values and artifacts; in the latter, culture refers to the meanings and experiences represented in members' actions and behaviours [93]. Culture is dynamic, and it enables its members to evolve and adapt in the same ecosystem they contribute to shaping [94]. Various cultural groups reflect their attitudinal, spiritual and emotional explanations of health behaviours in very different ways, and this reflects the role that culture plays in the mental and physical health of human beings. Kagawa-Singer et al. [95] propose a cultural framework for health, which integrates culture into health research. Burke et al. [96] also stress how the impact of culture in health research affects the language, meanings and interpretation of terminology used in relation to health and social care.

The trend of product design has evolved from one-size fits all to personalization that fits diverse consumers' needs and aspirations. The Usability and functionality of a product used to be key factors when people chose to buy an item, but they are now the basic attributes that consumers are looking for. Gradually, as the needs for usability and functions of a product have been fulfilled, consumers have started to pay more attention to the aesthetic value and the meaning that products can bring to their lifestyle.

In modern society, a product is considered an extension of the owner, representing the owner's personal or social identity [8, 97, 98]. Whilst the appearance of products may express the owner/user personal identity [99], the complex meaning of products has been recognised almost a century ago and defined in the space populated by the users, the socio-cultural groups, the availability of other products and the changing nature of the interplay between these parts over time [100].

The importance of cultural sensitivity in interpreting the space in which the meaning of product is identified and evolves has been highlighted by Wang [101] and in practical terms, such understanding of the cultural framework that surrounds products has been found to increase sales and reach out to larger market shares [102, 103]. As for any other products' category, it is reasonable to assume that culture would inform the way people belonging to such culture would interpret and use ATs; specifically, how a culture perceive disability and ageing would pervade how people would relate to those ATs that become the embodiment of either disability and fragility or empowerment and self-determination. When studying a set of ATs with participants from different cultural settings [84], feedback demonstrated some interesting differences, especially concerning the appreciation of technological intervention, imbued with divergent cultural connotations across countries. For example, while an electric scooter is associated with decline and laziness in the UK study, a sense of empowerment and independence was attributed to the same device by Taiwanese participants.

The most popular ATs purchased in Taiwan in 2017 from a major online catalogue where ATs for basic individual needs: walking frames, wheelchairs and bath chairs [104]. Digital ATs, able to detect and monitor the geographical location and physical exercise are also becoming more popular and this trend follows the increase of smartphone in the young-old population [105]. This is generally contributing to a wider acceptance of ATs [106] and in a study we conducted it was clear that, even within a culture of homogenisation such as that of Asia, older adults felt relatively comfortable to make use of ATs that would set them apart. We found that Asian participants were more conforming to social etiquette and acceptance and whilst they identified ATs as tools for living, they yet wished to somehow comply with what their peers would consider normalized appearance. For example, we discovered that ATs in proximity to the face were considered stigmatizing more than other ATs as the face is the most observed and noticed part of the human body during social interaction. Boldness, on the other hand, was considered an important attribute in ATs' design by the English participants who had accepted the gradual physical fragility that the ageing process may cause [83, 84].

The enablement of core values such as personal freedom, independence and feeling in control caused users to emotionally invest in their ATs and to consider them as precious possessions. By elevating ATs to *enabling objects*, ATs participate in redefining the meaning of the assistance they provide. It is this gradual shift in adoption, emotional investment and cultural definition that may enable designers, manufacturers and health professionals to reconsider the participation of users in the creation and prescription of novel and appropriate ATs.

## 3.7 Reflections on the Design of ATs

Humans are the intricate stratification of needs, aspirations, expectations and values. Our identity is also construed by the physical ability and appearance that our bodies enable. In considering the complex interplay between social, physical, emotional and economic factors that determine the circumstances of our lives we often neglect to consider how much of our outlook is manifested in the products and experiences we chose to own, and we decide to take part in. The freedom to select and use what we like and can afford is not often an option for individuals who are ageing or with a disability. Their world includes devices that they must use if they wish to retain some independence, mobility and dignity. However, the limited choice of ATs available to them creates a vicious cycle where reduced choice leads to either abandonment of the prescribed devices, or to resignation to use what is perceived as fundamentally hostile because it lacks thoughtfulness and has no merit or place in enabling lives.

Why the ATs market is still so behind in an era where equality and inclusion are at the forefront of the media debate is rather puzzling, especially when we have developed design approaches, manufacturing processing and culturally sensitive tools that could allow a courageous reconsideration of the bland language of AT's design. Costs are often offered as the excuse that jeopardises and limits innovation in this domain. However, the invisible costs of falls, lack of social participation, spiralling comorbidity and isolation are very tangible, albeit felt at individual rather than at the systemic economic level.

In this chapter, we have discussed co-design as a valid and practical design process to engage in the meaningful understanding of the users and their needs as they are experts. We have also looked at how cultures can frame differently the role of technology and technology inspired devices for the ageing population and we have also discussed materiality as an approach that can enhance our consideration of ATs as objects playing a normal, albeit very specialised, role in the life of their users. The three trends described are tools to imagine how the ATs market could accelerate and diversify and could provide emotionally and socially invested artefacts having impact that goes beyond the mere functional supports we have considered them to offer thus far.

## References

1. Escalas, J., White, K., Argo, J.J., Sengupta, J., Townsend, C., Sood, S., Van Boven, L.: Self-identity and consumer behavior. J. Consum. Res. **39**(5), xv–xviii (2013)
2. Assistive Technology: Retrieved 17 Mar 2019 from https://www.who.int/en/news-room/fact-sheets/detail/assistive-technology (2018)
3. Consumer Focus: Equipment for Older and Disabled People: An Analysis of the Market. Consumer Focus, London (2010)
4. Newell, A.: Inclusive design or assistive technology. In: Inclusive Design, pp. 172–181. Springer, London (2003)

5. British Standards Institute: 7000-6: 2005 Design Management Systems—Managing Inclusive Design—Guide. Author, London (2005)
6. Dong, H.: Global perspectives and reflections. In: Trends in Universal Design, p. 38 (2013)
7. Clarkson, P.J., Coleman, R.: History of inclusive design in the UK. Appl. Ergon. **46**, 235–247 (2015)
8. Hocking, C.: Function or feelings: factors in abandonment of assistive devices. Technol. Disabil. **11**(1, 2), 3–11 (1999)
9. Verganti, R.: Design Driven Innovation: Changing the Rules of Competition by Radically Innovating What Things Mean. Harvard Business Press (2009)
10. Giacomin, J.: What is human centred design? Des. J. **17**(4), 606–623 (2014)
11. Brown, T., Katz, B.: Change by design. J. Prod. Innov. Manag. **28**(3), 381–383 (2011)
12. Hassenzahl, M., Diefenbach, S., Göritz, A.: Needs, affect, and interactive products—facets of user experience. Interact. Comput. **22**(5), 353–362 (2010)
13. Pullman, M.E., Gross, M.A.: Ability of experience design elements to elicit emotions and loyalty behaviors. Decis. Sci. **35**(3), 551–578 (2004)
14. Hekkert, P., Mostert, M., Stompff, G.: Dancing with a machine: a case of experience-driven design. In: Proceedings of the 2003 International Conference on Designing Pleasurable Products and Interfaces, pp. 114–119. ACM (2003)
15. Norman, D.A.: Emotional Design: Why We Love (or Hate) Everyday Things. Basic Civitas Books (2004)
16. Phillips, B., Zhao, H.: Predictors of assistive technology abandonment. Assist. Technol. **5**(1), 36–45 (1993)
17. Dawe, M.: Desperately seeking simplicity: how young adults with cognitive disabilities and their families adopt assistive technologies. In: Proceedings of the SIGCHI Conference on Human Factors in Computing Systems, pp. 1143–1152. ACM (2006)
18. Kintsch, A., DePaula, R.: A framework for the adoption of assistive technology. In: SWAAAC 2002: Supporting Learning Through Assistive Technology, pp. 1–10 (2002)
19. Lebbon, C., Boess, S.: Wellbathing-attitudes and well-being of older consumers in relation to bathing. In: 3rd International TIDE Conference Quality of Life for the European Citizen, Helsinki, Finland (1998)
20. Luborsky, M.R.: Sociocultural factors shaping technology usage: fulfilling the promise. Technol. Disabil. **2**(1), 71 (1993)
21. Polgar, J.M.: Using Technology to Enable Occupation. Retrieved 12 Mar 2003 (2002)
22. Parette, J.M., Scherer, M.: Assistive technology use and stigma. Educ. Train. Dev. Disabil. **39**(3), 217–226 (2004)
23. Barnard, Y., Bradley, M.D., Hodgson, F., Lloyd, A.D.: Learning to use new technologies by older adults: perceived difficulties, experimentation behaviour and usability. Comput. Hum. Behav. **29**(4), 1715–1724 (2013)
24. Epstein, S.: Integration of the cognitive and the psychodynamic unconscious. Am. Psychol. **49**(8), 709 (1994)
25. Reyna, V.F.: How people make decisions that involve risk: a dual-processes approach. Curr. Dir. Psychol. Sci. **13**(2), 60–66 (2004)
26. Bright, A.K., Coventry, L.: Assistive technology for older adults: psychological and socio-emotional design requirements. In: Proceedings of the 6th International Conference on Pervasive Technologies Related to Assistive Environments, p. 9. ACM (2013)
27. Shinohara, K., Tenenberg, J.: A blind person's interactions with technology. Commun. ACM **52**(8), 58–66 (2009)
28. Shinohara, K., Wobbrock, J.O.: In the shadow of misperception: assistive technology use and social interactions. In: Proceedings of the SIGCHI Conference on Human Factors in Computing Systems, pp. 705–714. ACM (2011)
29. Pape, T.L.B., Kim, J., Weiner, B.: The shaping of individual meanings assigned to assistive technology: a review of personal factors. Disabil. Rehabil. **24**(1–3), 5–20 (2002)
30. Elliott, G.C., Ziegler, H.L., Altman, B.M., Scott, D.R.: Understanding stigma: dimensions of deviance and coping. Deviant Behav. **3**(3), 275–300 (1982)

31. Kraskowsky, L.H., Finlayson, M.: Factors affecting older adults' use of adaptive equipment: review of the literature. Am. J. Occup. Ther. **55**(3), 303–310 (2001)
32. Wessels, R., Dijcks, B., Soede, M., Gelderblom, G.J., De Witte, L.: Non-use of provided assistive technology devices, a literature overview. Technol. Disabil. **15**(4), 231–238 (2003)
33. Federici, S., Borsci, S.: Providing assistive technology in Italy: the perceived delivery process quality as affecting abandonment. Disabil. Rehabil. Assist. Technol. **11**(1), 22–31 (2016)
34. Spinelli, G., Massimo, M., Martin, W.: Objects of desire and of disgust: analysis and design of assistive technologies. In: Design4Health 2018, Sheffield, 4–6 Sept 2018
35. Department of Work and Pensions. Office for Disability Issues. Official Statistics: Disability Facts and Figures. https://www.gov.uk/government/publications/disability-facts-and-figures/disability-facts-and-figures (2014)
36. Metz, D.H.: Mobility of older people and their quality of life. Transp. Policy **7**(2), 149–152 (2000)
37. Cook, A.M., Hussey, S.: Assistive Technologies: Principles and Practice (2001)
38. Pullin, G.: Design Meets Disability. MIT press (2009)
39. Ravneberg, B.: Usability and abandonment of assistive technology. J. Assist. Technol. **6**(4), 259–269 (2012)
40. Herriott, R., Cook, S.: Inclusive design for assistive technology. In: Universal Design 2014: Three Days of Creativity and Diversity. Proceedings of the International Conference on Universal Design, UD 2014 Lund, Sweden, 16–18 June 2014, vol. 35, p. 175. IOS Press (2014)
41. Williamson, T., Kenney, L., Barker, A.T., Cooper, G., Good, T., Healey, J., Heller, B., Howard, D., Matthews, M., Prenton, S., Ryan, J., Smith, C.: Enhancing public involvement in assistive technology design research. Disabil. Rehabil. Assist. Technol. **10**(3), 258–265 (2015)
42. Blanco, T., Berbegal, A., Blasco, R., Casas, R.: Xassess: crossdisciplinary framework in user-centred design of assistive products. J. Eng. Des. **27**(9), 636–664 (2016)
43. Dickey, R., Shealey, S.H.: Using technology to control the environment. Am J Occup. Ther. **41**(11), 717–721 (1987)
44. Craig, A., Tran, Y., McIsaac, P., Boord, P.: The efficacy and benefits of environmental control systems for the severely disabled. Med. Sci. Monit. **11**(1), RA32–RA39 (2005)
45. Noyes, J.M., Haigh, R., Starr, A.F.: Automatic speech recognition for disabled people. Appl. Ergon. **20**(4), 293–298 (1989)
46. Jiang, H., Han, Z., Scucces, P., Robidoux, S., Sun, Y.: Voice-activated environmental control system for persons with disabilities. In: Proceedings of the IEEE 26th Annual Northeast Bioengineering Conference (Cat. No. 00CH37114), pp. 167–168. IEEE (2000)
47. Aguilera, F., Ataeefar, A., Brothers, R., Castellano, M., Ginart, A., Grangeia, G., Noel, Y.: A personal computer based environmental control system for the disabled. In: 1992 14th Annual International Conference of the IEEE Engineering in Medicine and Biology Society, vol. 4, pp. 1533–1534. IEEE (1992)
48. Geggie, C.: Voice control of environmental control systems. ACNR **3**(4) (2003)
49. Hawley, M.S.: Speech recognition as an input to electronic assistive technology. Br. J. Occup. Ther. **65**(1), 15–20 (2002)
50. Judge, S., Robertson, Z., Hawley, M., Enderby, P.: Speech-driven environmental control systems—a qualitative analysis of users' perceptions. Disabil. Rehabil. Assist. Technol. **4**(3), 151–157 (2009)
51. Greenhalgh, T., Wherton, J., Sugarhood, P., Hinder, S., Procter, R., Stones, R.: What matters to older people with assisted living needs? A phenomenological analysis of the use and non-use of telehealth and telecare. Soc. Sci. Med. **93**, 86–94 (2013)
52. Polak, P.: Design for the other ninety percent. In: Kim, C.R. (ed.) Design for the Other Ninety Percent, pp. 19–25. Cooper-Hewitt, New York (2007)
53. Chamberlain, P., Wolstenholme, D., Dexter, M., Seals, E.: The State of the Art of Design in Health: An Expert-Led Review of the Extant of the Art of Design Theory and Practice in Health and Social Care. Sheffield Hallam University, Sheffield (2015)
54. Kouprie, M., Visser, F.S.: A framework for empathy in design: stepping into and out of the user's life. J. Eng. Des. **20**(5), 437–448 (2009)

55. Van de Poel, I.: Can we design for well-being? In: Brey, P., Briggle, A., Spence, E. (eds.) The Good Life in a Technological Age, pp. 295–306. Routledge, New York (2012)
56. Spinelli, G., Micocci, M., Ajovalasit, M.: Behavioural strategies of older adults in the adoption of new technology-based products: the effects of ageing and the promising application of smart materials for the design of future products. In: Tsekleves, E., Cooper, R. (eds.) Design for Health. Design for Social Responsibility Series, pp. 358–374. Taylor & Francis (2017)
57. Sanders, E.B.N., Stappers, P.J.: Co-creation and the new landscapes of design. Co-design **4**(1), 5–18 (2008)
58. Visser, F.S., Stappers, P.J., Van der Lugt, R., Sanders, E.B.: Contextmapping: experiences from practice. CoDesign **1**(2), 119–149 (2005)
59. White, S., Pettit, J.: Participatory approaches and the measurement of human well-being. Human Well-Being, pp. 240–267. Palgrave Macmillan, UK (2007)
60. Buse, C., Martin, D., Nettleon, S.: Conceptualising 'materialities of care: making visible mundane material culture in health and social care contexts'. Sociol. Health Illn. **40**(2), 243–255 (2018)
61. Lindsay, S., Jackson, D., Schofield, G., Olivier, P.: Engaging older people using participatory design. In: Proceedings of the SIGCHI Conference on Human Factors in Computing Systems, pp. 1199–1208. ACM, New York (2012)
62. Xie, B., Druin, A., Fails, J., Massey, S., Golub, E., Franckel, S., Schneider, K.: Connecting generations: developing co-design methods for older adults and children. Behav. Inf. Technol. **31**(4), 413–423 (2012)
63. Davidson, J.L., Jensen, C.: Participatory design with older adults: an analysis of creativity in the design of mobile healthcare applications. In: Proceedings of the 9th ACM Conference on Creativity & Cognition, pp. 114–123. ACM (2013)
64. Mawson, S., Nasr, N., Parker, J., Davies, R., Mountain, G.: Developing a personalised self-management system for post stroke rehabilitation: utilising a user-centred design methodology. Disabil. Rehabil. Assist. Technol. **9**(6), 521–528 (2013)
65. Nasr, N., Leon, B., Mountain, G., Nijenhuis, S.M., Prange, G., Sale, P., Amirabdollahian, F.: The experience of living with stroke and using technology: opportunities to engage and co-design with end users. Disabil. Rehabil. Assist. Technol. **11**(8), 653–660 (2016)
66. Craven, M.P., De Filippis, M.L., Dening, T.: Quality of life tools to inform co-design in the development of assistive technologies for people with dementia and their carers. In: International Workshop on Ambient Assisted Living, pp. 394–397. Springer (2014)
67. Hwang, A.S., Truong, K.N., Cameron, J.I., Lindqvist, E., Nygård, L., Mihailidis, A.: Co-designing ambient assisted living (AAL) environments: unravelling the situated context of informal dementia care. BioMed Res. Int. **2015**(Article ID 720483), 12 (2015)
68. Craig, C., Gwilt, I., Langley, J., Partridge, R.: Thinking through design and rehabilitation. In: Encarnacao, P., et al. (eds.) AssistiveTechnology: From Research to Practice, pp. 798–803. IOS Press, Amsterdam (2013)
69. Reed, H., Langley, J., Stanton, A., Heron, N., Clarke, Z., Judge, S., Tindale, W.: Head-up; an interdisciplinary, participatory and co-design process informing the development of a novel head and neck support for people living with progressive neck muscle weakness. J. Med. Eng. Technol. **39**(7), 404–410 (2015)
70. Steen, M., Manschot, M., De Koning, N.: Benefits of co-design in service design projects. Int. J. Des. **5**(2), 53–60 (2011)
71. Williams, R.: A Vocabulary of Culture and Society. Fontana, London (1988)
72. Botero, A., Hyysalo, S.: Ageing together: steps towards evolutionary co-design in everyday practices. CoDesign **9**(1), 37–54 (2013)
73. De Couvreur, L., Dejonghe, W., Detand, J., Goossens, R.: The role of subjective well-being in co-designing open-design assistive devices. Int. J. Des. **7**(3), 57–70 (2013)
74. Leong, T.W., Johnston, B.: Co-design and robots: a case study of a robot dog for aging people. In: International Conference on Social Robotics, pp. 702–711. Springer, Cham (2016)
75. Szebeko, D., Tan, L.: Co-designing for society. Australas. Med. J. **3**(9) (2010)

76. Borgerson, J.: Materiality, agency, and the constitution of consuming subjects: insights for consumer research. Adv. Consum. Res. **32**, 439–443 (2005)
77. Hamlett, J., Hoskins, L.: Comfort in small things? Clothing, control and agency in county lunatic asylums in nineteenth- and early twentieth-century England. J. Victorian Cult. **18**(1), 93–114 (2013)
78. Latimer, J.: Afterword: materialities, care, 'ordinary affects', power and politics. In: Buse, C., Martin, D., Nettleton, S. (eds.) Materialities of Care. Encountering Health and Illness Through Artefacts and Architecture, pp. 136–147. Wiley Blackwell, Oxford (2018)
79. Miller, D. (ed.): Materiality. Duke University Press, Durham (2005)
80. Woodward, S.: Objects, interviews, material imaginings and 'unsettling' methods: interdisciplinary approaches to understanding materials and material culture. Qual. Res. **16**(4), 359–374 (2015)
81. Buse, C., Twigg, J.: Dressing disrupted: negotiating care through materiality of dress in the context of dementia. In: Buse, C., Martin, D., Nettleton, S. (eds.) Materialities of Care. Encountering Health and Illness Through Artefacts and Architecture, pp. 97–109. Wiley Blackwell, Oxford (2018)
82. Ahmed, S.: The Cultural Politics of Emotion, 2nd edn. Edinburgh University Press, Edinburgh (2014)
83. Spinelli, G., Micocci, M., Martin, W.: Objects of desire and of disgust: analysis and design of assistive technologies. In: Design4Health 2018, Sheffield, 4–6 Sept 2018
84. Spinelli, G., Micocci, M., Martin, W.: Objects of desire, objects of disgust. In: British Society of Gerontology 47th Annual Conference—Ageing in an Unequal World: Shaping Environments for the 21st Century, Manchester, UK, 4–6 July 2018
85. Puig de la Bellascasa, M.: Matters of care in technoscience: assembling neglected things. Soc. Stud. Sci. **41**(1), 85–106 (2011)
86. Mynatt, E., Melenhorst, A.-S., Fisk, A.D, Rogers, W.A.: Aware technologies for aging in place: understanding user needs and attitudes. Pervasive Comput. **3**(2) (2004)
87. Throsby, D.: Economics and Culture. Cambridge University Press, Cambridge (2001)
88. Storey, J.: Cultural Theory and Popular Culture, 3rd edn. Pearson Education Limited, England (2001)
89. Eagleton, T.: The Idea of Culture. Blackwell, Oxford (2000)
90. Gay, P.D., Hall, S., Janes, L., Mackay, H., Negus, K.: Doing Cultural Studies: The Story of the Sony Walkman. Sage, London (1997)
91. Kroeber, A.L., Kluckhohn, C.: Culture: A Critical Review of Concepts and Definitions. Museum of Cambridge Massachusetts, USA (1952)
92. Statistics Canada: Conceptual Framework for Culture Statistics 2011. Statistics Canada. Retrieved from https://www150.statcan.gc.ca/n1/en/pub/87–542-x/87-542-x2011001-eng.pdf?st=10sbfza5 (2011)
93. Kasnitz, D., Shuttleworth, R.P.: Introduction: anthropology in disability studies. Disabil. Stud. **21**(3), 2–17 (2001)
94. McElroy, A.; Townsend, P.K.: Medical Anthropology in Ecological Perspective, 5th edn. Westview Press (2009)
95. Kagawa-Singer, M., Dressler, W.M., George, S.M., Elwood, W.N.: The Cultural Framework for Health: An Integrative Approach for Research and Program Design and Evaluation. National Institutes of Health (2014)
96. Burke, N.J., Villero, O., Guerra, C.: Passing through meanings of survivorship and support among Filipinas with breast cancer. Qual. Health Res. **22**(2), 189–198 (2012)
97. Belk, R.W.: Possessions and the extended self. J. Consum. Res. **15**(2), 139–168 (1988)
98. Holt, D.B.: How Brands Become Icons: The Principles of Cultural Branding. Harvard Business School, Cambridge, MA (2004)
99. Krippendorff, K.: On the essential contexts of artifacts on the proposition that "design is making sense (of things)". Des. Issues **5**(2), 9–39 (1989)
100. Dewey, J.: Art as Experience. Capricorn Books, New York (1934)

101. Wang, Y.: Chinese cultural features for new product design development. Ph.D. Thesis, Brunel University London. Retrieved from https://ethos.bl.uk/OrderDetails.do?did=1&uin=uk.bl.ethos.681203 (2016)
102. Marcus, A., Aykin, N.M., Chavan, A.L., Prabhu, G.V., Kurosu, M.: SIG on one size fits all? Cultural diversity in user interface design. In: CHI Extended Abstracts, pp. 342 (1999)
103. Sparke, P.: An Introduction to Design and Culture: 1900 to the Present, 2nd edn. Routledge, Oxon (2004)
104. Appledaily: Sales of aging products increased by 1.5 times. AppleDaily News. Retrieved from https://tw.lifestyle.appledaily.com/daily/20170618/37686865/ (2017)
105. FIND team: A Report of Senior Citizens' Shopping Behaviour in Taiwan. Institute for Information Industry. Available at https://www.iii.org.tw/Press/NewsDtl.aspx?nsp_sqno=1917&fm_sqno=14 (2017)
106. Lin, L.: A study on intention of adopting living aids for aged people. Master Dissertation. Available from National Digital Library of Theses and Dissertations in Taiwan (2017)

**Dr. Gabriella Spinelli** is a Reader in Design in the College on Engineering, Design and Physical Science and a membere of the Research Institute for the Environment, Healthand Societies at Brunel University London, interested in research that explores the relationship between, health and wellbeing, social circular economy and co-designed interventions. In the last ten years her research interests have developed in the area of innovation and ageing as this is an unprecedented challenge that poses interesting societal questions and provides opportunity for social innovation and development.

**Dr. Massimo Micocci** is a Research Associate in Human Factors at the NIHR London IVD Coop., Imperial College London. His areas of interest and contribution include Inclusive Design and Human-Centred technology development. In the past years he has investigated cognitive differences in the adoption of technological devices between younger and older generations and advanced strategies to design intuitive and immediate interactions for health and wellbeing products.

**Dr. Emmanuel Tsekleves** is a senior lecturer at ImaginationLancaster, at Lancaster University. He leads international research in design interactions for health, which lies at the nexus of *public health, design for behaviour change* and *technology*. His research focuses in the design of chronic and public health interventions aimed at improving the quality of life and wellbeing of people into old age. He is the editor of the Routledge book on Design for Health.

**Dr. Yu-Han Wang** is a lecturer in the College of Innovation Management, Department of Product Innovation and Entrepreneurship at National Taipei University of Business. She has developed a range of design research interests which include, but are not limited to, innovation in product development, cultural sensitivity in products, aesthetics and emotional engagement of products, and design for aging.

**Dr. Wendy Martin** is a Senior Lecturer in the College of Health and Life Sciences, Brunel University London, UK. Her research focuses on ageing, embodiment and daily life and the use of visual and material methods in ageing research. She was Principal Investigator for the Economic and Social Research Council (ESRC, UK) research project Photographing Everyday Life: Ageing, Lived Experiences, Time and Space. Wendy is co-editor of the Routledge Handbook of Cultural Gerontology published in 2015.

**Dr. Yonghun Lim** is a lecturer in Product design in the Department of Design & Engineering in the Faculty of Science & Technology at Bournemouth University. He is specialised in design for inclusion and human-centred innovation. His research interest is psychosocial (i.e. non-physical aspects such as emotional, social, cognitive factors) inclusivity in the design of products, services and built environments through the human-centred approach. He also teaches both undergraduate and postgraduate units with an emphasis on the design process and human-centred approach. He is a member of the international Inclusive Design Research Group (IDRG) and Human-Centred Design Institute (HCDI).

**Umber Shamim** is an MRes student in College of Engineering, Design and Physical Sciences, Brunel University London, UK. Her research project focuses on the Environmental Control (EC) devices issued to people with severe mobility impairments. Her research interest includes, Assistive Technology Design and Human Centred Design. In the past, she has worked on projects related to adoption of technology in older adults and computer supported collaborative learning tools.

# Chapter 4
# The Value and Place for Empathy in Designing for Older People

**Andree Woodcock, Deana McDonagh, Paul Magee and Mike Tovey**

**Abstract** It may be argued that developing empathy for the people we design for is a prerequisite to more effective design outcomes. Although we can intellectually appreciate aging and disability, only when we 'experience another person's experience' can we begin to appreciate the lived experience we are trying to enhance. It is difficult for student designers to gain a perspective of the lived experience of 'the other'. However, this is the very skill that is needed when they move into professional practice. With an ageing and more discerning population, future designers need to understand both the needs and aspirations of older populations who may experience age related decline in mobility, and life changing illnesses and diseased if they are to design for the whole person. This chapter describes the development of Discrete Learning Interventions (DLIs) to increase the empathic horizon of student designers and engineers, who may not have access to elderly populations. The DLIs used low-level simulations and the GERT (gerontology) aging suit, to complete activities of daily living while experiencing a range of physical impairments. While a relatively short immersion into disability and aging could be considered superficial, the authors believe that low level experiential simulations, together with the GERT suit could be valuable in training Industrial Design students and lead to more age sensitive, inclusive design.

**Keywords** Empathy · Aging · Gerontology · Disability · Industrial design

A. Woodcock (✉) · M. Tovey
Centre for Arts, Memory and Communities, Faculty of Arts and Humanities, Coventry University, Coventry, UK
e-mail: a.woodcock@coventry.ac.uk

M. Tovey
e-mail: m.tovey@coventry.ac.uk

P. Magee
Faculty of Health and Life Sciences, Coventry University, Coventry, UK
e-mail: p.magee@coventry.ac.uk

D. McDonagh
Beckman Institute of Advanced Science and Technology, University of Illinois at Urbana-Champaign, Urbana, USA
e-mail: mcdonagh@illinois.edu



A. Woodcock et al. (eds.), *Design of Assistive Technology for Ageing Populations*, Intelligent Systems Reference Library 167,
https://doi.org/10.1007/978-3-030-26292-1_4

## 4.1 Introduction

We are moving towards an empathy economy that prizes the user experience. It has long been acknowledged that at the key responsibility of designers is to represent the needs and wants of users [1]. Products and services should fulfill user expectations, make them feel valued and understood. Empathic design was born out of a realization that consumers wanted more from their products. Increasingly, consumers search for deeper meaning from their material possessions, the services they receive and the environments with which they interact. Function needs to be enhanced by meeting the ephemeral, emotional needs of users.

The meaning of disability and aging is changing, as it does not necessarily mean dependent living, fragility or cognitive decline [2]. Saga's strategy director, Pethick [3] said: 'People are living longer and have more active social and economic lives, which is changing how we think about 'retirement'. As a consequence, the over 50s have become, through their hard work, big earners and even bigger spenders – today they account for some 76 per cent of the UK's financial wealth.'

Unfortunately, despite a changing world, consumer landscape and legislative requirements, companies have been slow to design for those with different abilities and older users. Ageing consumers are knowledgeable, discerning and wish to enjoy life to the full and they have disposable income. For example, in the UK 11.9 million people have disabilities with a disposable income of £80 bn ($112 bn) a year.

Most undergraduate design students tend to be able-bodied 18–24 year olds with 'limited life experience' [4]. They begin by basing their design decision-making on their personal perspective (e.g. designing for themselves), but may continue to want to design exciting, innovative products for people like themselves. Experience changes the way that designers' 'function' in this initial period. Sanders and Stappers [5] illustrate how chaotic and early this can and should be. Fear of this stage means that younger designers will seek something to work from, rather than be comfortable in having no new knowledge, but should have the confidence to know that the knowledge is coming.

Following graduation, they will be required to design for 'the other' and hence need to be provided with the skills to enable them to do this in an expert way. For this, they need to develop a range of research approaches, techniques and tools to help them elicit specific information to support their design-decision making. They need to be agile, resourceful and skilled at recognising patterns in behaviour, anticipating future behaviour and be mindful of peoples' needs, whilst not allowing the data/information to dampen their creativity [6].

The design community has an opportunity and responsibility to contribute to the quality of life for all. Looking at the need to encourage independence in old age, there are issues in the degree to which user centred design approaches have been used successfully to inform the design and understanding of assistive technology products [7, 8]. For example, a product defined by an OT to lift a user's chair and thereby better enable sitting movement, may not match to a designer's or a manufacturers interpretation of the same requirement. The OT solution is often taken from a basic mechanical understanding of in this instance, lifting the seat. But in doing so, the aesthetic qualities of the seat (for which it was originally purchased by the user) are compromised, immediately ‘visually’ changing the purpose of the chair from luxurious statement of personal taste to a functional item with a practical focus. This is not to say that one method is better than another, but rather sensitivity to the meaning of both is important

Many products are not conceived so as to recognize a population’s particular needs. For example AT products are often designed for just one disability or just one instance of disability. What is needed is for the whole evolving person and the context of use to be taken into account. Currently higher level, ‘belongingness’ needs is neglected [9]; or designers do not have the right tools to understand users with specific, critical and additional needs [8]. Often a designer will be given a design brief which has not truly examined or challenged the problem at hand and instead provides a manufacturers’ assumption of an issue, within their existing constraints of production. More problematically what is often seen is that what is addressed is only an issue that exists for the specific individual, not for a commercially viable population.

Key questions relate to how we can help students to design for the ‘other’, so that they have the most appropriate tools and mindset to deal with requirements of older consumers. Student design briefs typically require them to design products for the ‘mid future’. This effectively invites them to project their current selves into the future, and design ‘technocentric’ products without considering how usable such products may be in futures in which they might be disabled or disempowered. The current, tangible, everyday problems of older users are ignored and complacency is created. Problems such as not being able to read prescriptions, open water bottles or medical packages, or walk downstairs are ignored. In her 2012 paper addressing the needs of laggards and late adopters, who form over 48% of potential users, Woodcock [10], concluded that there was a need for student designers to have:

> ‘greater empathy with end users, more challenging design briefs and topics and courses which are structured and allow time for in depth study, and research led design briefs’ to ‘increase the awareness, ability, understanding and interest of future designers in not just shaping products, but the world’. (p. 58)

## 4.2 Empathy

Empathy is defined as ‘the intuitive ability to identify with other people’s thoughts and feelings – their motivations, emotional and mental models, values, priorities, preferences, and inner conflicts’ [11]. It is distinguished as feeling *with* someone, rather than feeling *for* someone. McDonagh and Denton [12] introduced the term ‘empathic horizon’ to indicate a designer’s ability to empathise beyond the characteristics of their own group. For example, an ageing population requires designers to reach further into their understanding of what it means to be in a different age group and to look through the disabilities and frailties associated with old age.

Empathy's strength lies in raising awareness of 'what makes life rich, personal, and meaningful' [13] as such it can be advantageous by enabling designers to realise the aspirations, preferences and needs that consumers may be unaware of or be unable to articulate. An association has been found between intuitive thinking and affective empathic reactions. Products need to satisfy both functional and emotional needs whilst also intuitive to use. Expanding our understanding of users is supported by both affective (feeling) and cognitive (knowing) empathy. Seeking authentic human behaviour will help to identify disconnects between the user and their material landscape, which leads to real problem identification. Reducing bias and avoiding assumption further supports more relevant design outcomes [14] in that a deeper level of *design integrity* (deeper understanding of users' needs) is achieved that can be reflected in more meaningful products, services or environments [15].

Empathy is not simply 'walking in someone else's shoes' or using a wheelchair for a day to simulate wheelchair use. It is a process by which deeper understanding is elicited that can impact on both the skill and mind-sets of the designer.

Scott et al. [8] concluded that designers need empathy and that this requires making an emotional connection with the user, understanding their situation and why certain experiences are meaningful to them. Research suggests that effective human-centred design requires empathic qualitative approaches to inform and inspire designers to help them understand the personal experience and private context of the 'other' [16, 17]. Designers needed empathy and to make an emotional connection with the user, understanding both their situation and why certain experiences are meaningful to them [18]. Empathy in design requires deliberate practice. Battarbee et al. [19] stressed that the willingness of designers to engage in empathic experiences is key e.g. through immersion in the life of the users, design probes, and imaginative projection and experiential simulations.

Figure 4.1 illustrates from right to left (top row) how our perception of others and the wider world is directly impacted by our experiences in life, such as our education, gender, sex, culture, religion and family. Our early life experiences are typically strong cultural imprints that can impact how we view the world for the rest of our lives. Being awareness of our personal viewpoints enables us to acknowledge when we need to seek further understanding of others that may hold different values and offer different viewpoints.

If our empathic horizon is symbolic of the limit to our understanding beyond our own personal experience, one can begin to acknowledge such limitations to knowledge. A key question is what activities can be used to expand this capability? Designing 'for the other' poses a problem for young designers, who may lack the experience, knowledge, confidence and (even) interest in designing for those unlike themselves. Therefore, there is a potential gap between the skills, which graduates possess, and the needs of industry.

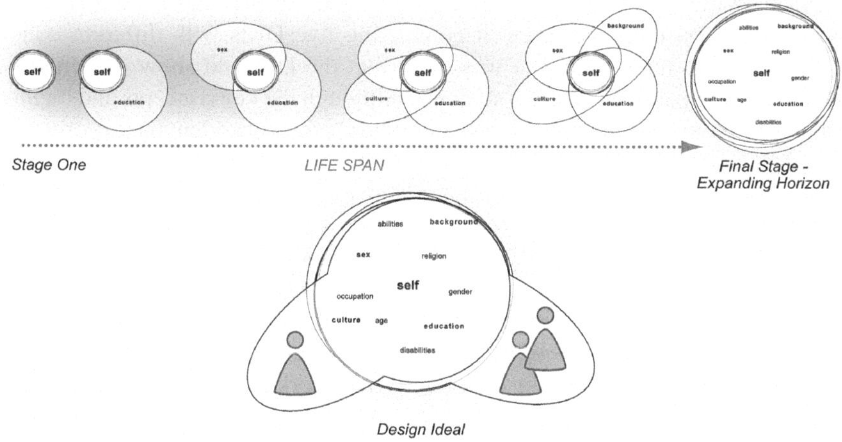


**Fig. 4.1** Empathic Horizon: expanding one's person understanding of the world beyond their personal (and biased) viewpoint. The bottom row illustrates how one needs to acknowledge a boundary to their knowledge when designing for others and therefore expand their empathic understanding (horizon) [15]

## 4.3 Discrete Learning Interventions (DLIs)

It is often lamented that there are few opportunities within industry to have 'continual informal encounters with users' [13], and this lack is reflected within design education. To address this gap, we have used DLIs both in the form of short workshops and more extended activities. These are directed to developing in students the confidence to look at more experiential ways of understanding 'the other' (in this case older people and those with disability) and showing them a set of techniques they can apply in different design situations. Design students typically tend to perceive user engagement experiences/exercises as valuable but are ultimately keen to get back to core design activities such as sketching, concept generation and model making. Simply providing an empathic design experience is not enough and does not lead to any transformative change. It is argued that coupling lectures with design tasks, experiential simulations and opportunities for reflection and sharing will help students empathise with users and act upon this improved understanding. Consequently, experiential simulations need to be scaffolded within a reflective cycle [20] which not only enables knowledge and meaning to be extracted from encounters and influence design, but which also leads to improved designerly practice and self-awareness.

For the purposes of this discussion a Discrete Learning Intervention is one, which is delivered by expert tutors external to the course for a fixed amount of time. Any design exercise in which the students engage is bounded in time and may not necessarily feed directly into the curriculum. Participation in the DLIs was voluntary, and provided students with experiences which they would not normally be offered on their courses. Where a DLI was offered above and beyond normal teaching and learning activities, students were offered financial incentives to take part. The research draws on the experience of conducting five DLIs with different cohorts over 2 years. The main objective was to perfect the DLI and show its benefits in increasing the empathic horizon of students. Although the key concepts and methods remained, implementation varied between groups being dependent on time, resource and student availability (Table 4.1).

The approach used evoked the designer's own experiences in the context of the user, with simulations selected to represent a user's experience. Any type of representation designed to understand, explore or communicate what it might be like to engage with the product, space or system has been described as 'experience prototyping' [21]. Focusing on experiential knowledge allows the designer new insights; rather than external observation, they can focus on behavioural aspects. This technique seems most appropriate for building a framework that could extend student's empathic horizons when they are unlikely to have extensive opportunity to engage and study others.

**Table 4.1** Summary of cohorts

| Ref. | Cohort details | Discrete learning interventions experience |
|---|---|---|
| DLI1 | 5 transport design undergrads undertaking final assessment | Students self-selected following a lecture on design for the elderly and empathy and were supported with tutorials, classroom based and real-world experiential simulations during the concept design phase of their project involving design of transport for older users. This ran over 2 terms |
| DLI2 | 30 undergraduate engineering design students enrolled on a humanitarian engineering module | 2-h low fidelity classroom based experiential simulation workshop to encourage them to develop a more empathic understanding of older transport users |
| DLI3 | 20 Serbian undergraduate engineers attending the 6th Humane Cities Conference | 2-h low fidelity classroom based experiential simulation workshop to encourage them to develop a more empathic understanding of older transport users |
| DLI4 | 10 transport planners/professionals, delegates enrolled for an interactive workshops at the IPATH conference in US | 90-min low fidelity classroom at the IPATH conference in US based experiential simulation workshop to encourage them to develop a more empathic understanding of older transport users |
| DLI5 | 5 MA transport design students participating in challenge to design products to support companionship | This involved low and high-fidelity simulation in and out of the classroom, group tasks and design critique over a 3-day period |

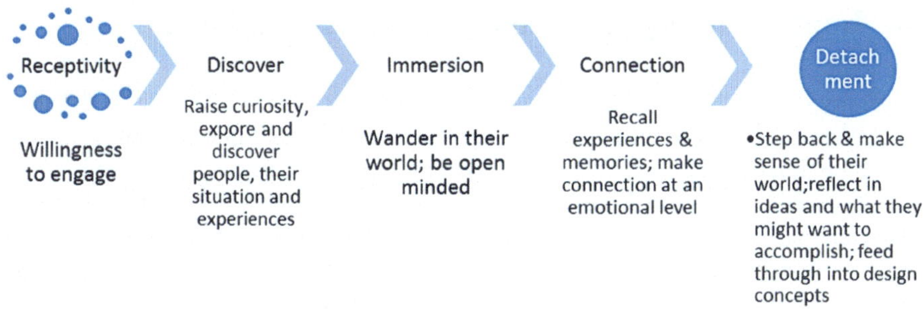


**Fig. 4.2** Empathy framework after [22]

The framework underpinning all DLIs was based loosely around that of [22] including stages: (1) discovery, (2) immersion, (3) connection and (4) detachment to which we have added the pre-stage of receptivity and our own evolving practice. As shown in Fig. 4.2.

The students took part in individual and group activities related to mobility and travel—in a classroom setting (e.g. route finding, using a phone, eating and drinking, playing games (cards and word search) and removing outer garments) and out of the classroom (in studies 1 and 5) where they used different forms of transport while having restricted mobility and sensation. In relation to this, a technique, which is gaining attention, is the use of whole-body simulation suits in design and transport research to provide designers with an immersive empathic experience [23, 24]. A gerontology suit was used with DLI5.

The DLI workshops attempted to take all students as far as connection, using reflection in, on and through action [20] to encourage them to think more deeply before, during and after their experiential simulations. Time was allowed after each experience for students to verbalise their personal reflection to the group: what they experienced, how they felt, what was different, and how this could relate to their current and future design activities.

Given the short length of time available for DLIs 2, 3 and 4, participants were encouraged to brainstorm ideas about how transport could be made elder friendly, based on their empathic experiences and the low fidelity experiential simulations. DLIs 1 and 5 allowed for more protracted simulations, leading to design solutions. The following section provides an account of the way in which the framework was implemented and evolved over the 5 cohorts, following an action research cycle [24], with participants reflecting on the activities they were engaged in, thereby informing the design of later DLIs.

## 4.4 Implementation, Participant Activities and Buy in

University students tend to have excelled academically to ensure a place on a degree programme. The DLIs deliberately reduced their abilities and impaired them, possibly for the first time as young adults. For young student designers, diminishing their abilities even slightly can have a dramatic impact upon them. The realisation that others may struggle at activities of everyday living becomes a 'felt' realisation. At the start of each DLI the participants were introduced to the concept of 'empathy' with a motivational lecture and an overview of the human factors associated with an ageing transport user.

**Stage 1 Receptivity**
The motivational lecture and financial incentives were not always sufficient in encouraging participation in our research, even if increasing empathy towards older people through experiential simulation may have been beneficial to the work. This was discouraging, but perhaps understandable for students with busy timetables, who may have to juggle many tasks.

However, willingness to engage was also noted in other DLIs. In terms of the short workshops, some transport professionals in DLI4, were both reluctant and skeptical about trying low fidelity simulations (where fingers are taped together to reduce dexterity) or use vision impairing goggles/tinnitus headsets. A shift in attitude occurred when they struggled to complete tasks considered childish and trivial when once their senses and mobility were restricted as part of the simulation. Resistance was also noted from one student in DLI5, whose first written comment was that he doubted the value of low levels experiential simulations. Although he participated in the exercises, these had little impact on his initial ideas or his practice, where he persisted in using the Internet and developing technological solutions with no consideration of the needs of the older person.

Therefore, we concur with Buchenau and Fulton Suri [21]. Based on our experiences we have included 'receptivity', defined as 'willingness to engage' as the first stage in our framework. Some students may be unable to engage because of training or preconceived ideas. In terms of the development of the DLI it is important to encourage the students to try things out, design activities which are fun, engaging and relevant, and to encourage the student to join in and actively listen to the experience of the 'other'.

**Stage 2 Discovery**
In this stage participants were provided with classroom-based immersive experiences to start to expand their empathic horizons. Materials provided included:

- Glasses: simulating a variety of conditions, such as macular degeneration and cataracts
- Ear plugs: to reduce hearing in one or both ears
- Masking tape used to tape thumb and forefinger together on dominant hand, and bind three fingers together on non-dominant hand

- Thin gloves to reduce sensation in hands
- Bandages to restrict movement

Figure 4.3 illustrates how tape can be used to restrict mobility and dexterity. Though relatively unsophisticated, utilising basic, inexpensive and easily accessible materials encourages empathic modelling throughout the design process.

Initial tasks included trying to read labels/open packages and eat with reduced vision, hearing, mobility and tactile impairments. Over the course of the DLIs the tasks became more transport oriented relating to journey planning, reading maps, understanding bus ticket and activities normally undertaken on transport, taking shoes on and off (or someone else's shoes when impaired), walking around and going to the toilet. Participants worked in groups, and all experienced a range of disabilities. They were encouraged to share reflections on their experience with and between groups and relate this to their typical journeys. Participants in DLI1 were encouraged to explore this in their home environment where they undertook everyday tasks, comparing length of time and difficulties in undertaking tasks. Uploading photos and sharing experiences, reinforced group cohesion and added new insights.

**Stage 3 Immersion**

Typically at this stage the designer moves out of the office and explores the user's world. DLI1 and DLI5 were able to do this using simulation suits. DLI1 students took part in a 'walkabout' where they were required to travel from the university to the main rail station, board a train to a local station and return. To support this 'experience prototyping', low fidelity simulations were used, including a range of visual impairment glasses (e.g. glaucoma, cataracts), mobility impairments (e.g. crutches, wheelchair) and hearing impairments (e.g. tinnitus and hearing loss).

Figure 4.4 captures a student wearing glasses that simulate Retinosa Pigmentosa (e.g. tunnel vision) while making a journey by train. The first image illustrates the

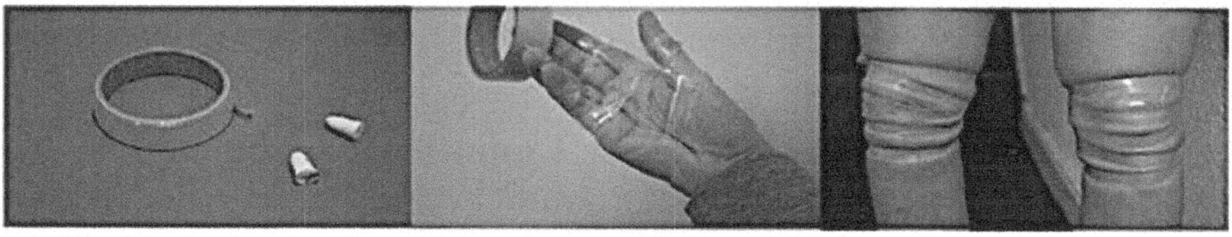

**Fig. 4.3** Using everyday materials to restrict mobility and hearing

**Fig. 4.4** Student wearing empathic glasses that simulate Retinosa Pigmentosa

glasses. The second shows her engaging with the ticket machine interface, which she finds too demanding and resorts to seeking help from a member of the station staff (third image). The final image illustrates how close she needs to place the ticket to her eyes in order to read it.

A companion looked after the students and videoed significant moments. On arrival at the destination, students took on another simulation, so they experienced different or multiple disabilities. DLI5 participants had a one-hour walk about in the Gerontology suit, and a conducted walk around the city and to the market to buy food wearing tinnitus headsets and vision restricting goggles.

**Stage 4 Connection with Users**
Combining the information provided on empathy and ageing, the student's own experiences and relating this to real world situations and people was supported in debrief sessions where students were given 'quick note' sheets to record thoughts before, during and after the experience. These thoughts were shared with their peers. Although they have a felt experience, they still may not be able to transfer this feeling into design activities.

**Stage 5 Detachment**
This involves the student or participant stepping back into the role of designer, to deploy the new insights into the current design task. Our participants were a convenience sample, all except the engineers in DLI3 had elected to attend our DLIs and were aware that we were trying to improve transport for older people; by increasing their empathy towards elders in whatever capacity they encountered them. As it was not possible to include a design activity in all groups, at the end of the session, participants were asked to step back and think how they would employ what they had learnt either through brainstorming in groups (DLIs 2–4), creating individual pieces of work (DLI5) or showing how their broader empathic horizons might affect their attitude to older users.

## 4.5 Helping Students Address the Empathy Horizon

We evolved a number of strategies to help students make the leap from personal experience to relating this to real people. For example:

- DLIs 1 and 3 were given semi-structured interviews and prompts to help them capture mobility experiences of older people.
- DLI1 created videos for a shared repository, some participants in DLI3 produced posters to share with their groups, see Fig. 4.5.
- DLI4 consisted of transport planners and academics that were attending a transport conference which had required the use of 3 or more modes of transport (including a ferry and a horse drawn cart). They were asked to draw on this experience, relate how they had felt during the simulated exercises and apply this to problems elders might have experienced if they had to undertake a similar journey.

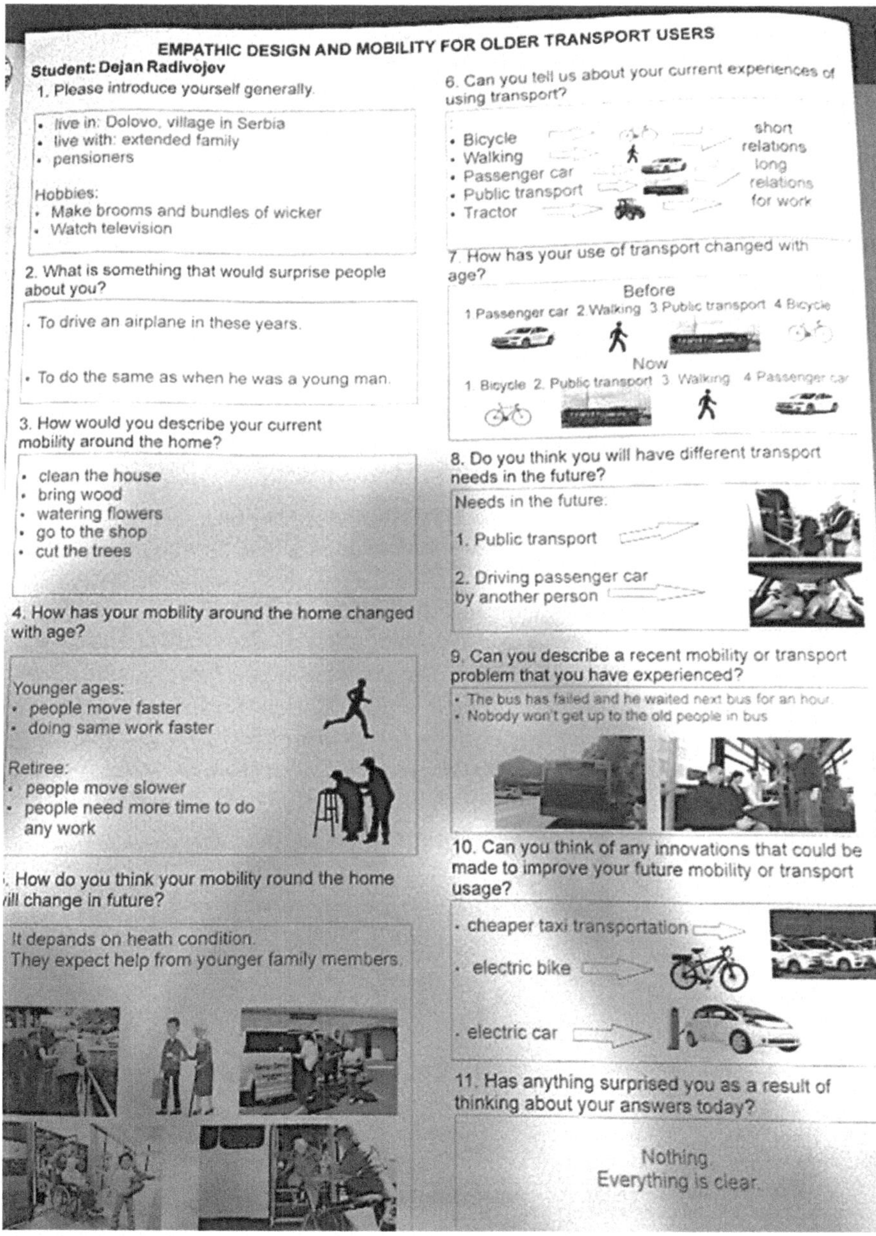


**Fig. 4.5** Poster created by a student in DLI4 from their interview with an elder

- DLI5 participants were asked to create mind maps of grandparents, on Day 1 and revisit these/consider their grandparents as end users of their designs. Asking them to draw their grandparents on Day 2, to refocus their minds, reinforced this. On Day 3 they were asked to produce mood boards, which captured how they felt when wearing the gerontology suit.

## 4.6 Results

The overall aim of the research has been to look at developing a DLI that could be used to increase student empathy towards older transport users. If designers, engineers and transport planners apply empathic thinking to urban and transport planning, design of vehicles and transport infrastructure then it is hypothesized that more age friendly transport will start to emerge. Participant feedback suggested that there was a change in the worldview of some students, and they backed this up by suggesting how they would apply this new understanding in their work and everyday experiences (Fig. 4.6).

For example, the Serbian engineering students in DLI3 had never experienced such a method and were keen to use it in their classes. They felt that they were too much interested in the mathematics of construction and forget about people. They wanted to champion the users in their activities. Their feedback included '*the workshop is a great practice for all people to accept disabled person and much more to understand these people. Our responsibility is to help them to have as much normal world (normal means accessible, like for non-disabled people, Thank you for this opportunity*'.

**Fig. 4.6** Students experiencing reduce abilities through empathic modelling

### *4.6.1 Transference of Learning Out of the Classroom*

DLI3 students reflected on how they would be more thoughtful and tolerant of older people using public transport. This transference out of the classroom was also noticeable in one of the participant's comments in DLI1. He started to interpret things in a new way. For example he commented that he found himself watching an old lady struggling down some steps in Coventry, and for the first time understood why she was moving so slowly. It was because it was painful for her. This could be directly attributed to the binding up of limbs to restrict mobility and walking with stones in the shoes to simulate diabetic nerve pain.

Engineering students in DLI2 considered how they might change their everyday behaviour by being more patient and helpful, show more understanding towards people in traffic, and by educating others about the effects of disability.

### *4.6.2 The* Felt *Experience*

Typically one is able to appreciate intellectually other peoples' contexts, challenges, and needs. As a design researcher this is how we begin to design with sensitivity especially if the designer has not experienced the activity being designed. Acknowledging there is a gap in the understanding is the first stage. Reading statistics on ageing populations and user requirements does not necessarily mean that the students understand the felt sense, but it will certainly provide more context that if they were to design within a vacuum. Empathic modelling transforms the felt sense (intellectually understanding) to a felt experience. This subtle but significant change repositions the platform on which decisions are based. We would argue that simulations can have profound and unexpected emotional effects, and that this can also be transformative if it is acknowledged and brought into play.

A student from DLI1 experienced anxiety when walking in a busy street with vision restricting goggles. He relied on vision to read people's faces to judge their moods and feelings. With restricted vision he could not do this. All students in this group reported that they felt vulnerable—which they had not expected. The student engineers in DLI2 had the shortest intervention. They were asked to reflect on their experience when working in groups—socialising, way finding or doing group work. Many quickly became isolated and left behind in tasks. This is reflected in comments such as:

> … disinclined to get involved when no one can understand me

> … don't want to ask question as I can't hear the answers.

> … other people lose attention quickly if you can't communicate with them.

They learnt that everyday tasks were much harder and more painful. They felt embarrassed that they could not do things, and quickly lost confidence. This was also reflected in the comments of DLI3 participants who expressed insecurity, loss of confidence during group work when they were wearing low fidelity simulations. This is important as it shows that even experiential simulations of short durations can have an effect.

### 4.6.3 Application to Design

Participant 1 in DLI1 embraced the method, being our most regular attendee experimenting with low-level simulations at home. Refer to Fig. 4.7 where he is conducting tasks of his daily living while experiencing restricted body movement and vision hand dexterity impairment. He saw the benefits of the method, asserted that he would use it as part of his future practice, concluding that:

> … It made multitasking very difficult and delayed every task—the whole thing took double the time, I became more frustrated as it went on and I even thought about not bothering. This is not an option if you're having to travel somewhere important. I tried to do the task how I normally do it and it was just not possible … In some cases I was in real danger as I heavily rely on my hearing—something I had not noticed before. But I did not really check for vehicles coming as I'm just so used to listening out for them … The task made me realise that if I were to design for those who are not as abled, things really do need to function well, not just look aesthetically pleasing. The experience needs to be easy and comfortable …

From an analysis of individual quick notes and group discussions most of the engineering students in DLI2 and DLI3 expressed deeper empathy for older people and those with age related disabilities and started to realise how difficult everyday life was for them. In terms of changes to their practice, they agreed for the need to champion and consult with differently abled users, and not design transport systems for the fittest. They could point to design features that they could improve e.g. lighting, size of fonts, acoustic signals.

**Fig. 4.7** Using the boot with reduced dexterity

DLI4 was conducted with transport professionals and academics from fields related to transport and health. They were skeptical of the approach and buy in was difficult however, once they moved on to the simulation and doing tasks they started to see the potential of the approach. Their suggestions on how to improve the journey for older people included ideas for discrete support for everyone at airports, where support staff is trained to look for people who might be having a problem, not necessarily just the elderly. They wanted somewhere quiet, and support so that they could take time out from being a carer e.g. to leave someone with a trusted member of staff while they went to the toilet.

The concept idea developed by one of the students in DLI5 revealed a more empathic understanding of the needs of elders and a more thoughtful approach. These are shown in the concept designs in Figs. 4.8 and 4.9. The first image shows a design for a walking stick handle, which is based on a handshake—representing trust, companionship, and human feeling. The second concept was for shoes that opened up, like a clam, so they were easier to put on and take off. This could be traced to the difficulties the students experienced in the low fidelity simulation when removing their own shoes and putting on someone else's. This task was designed specifically to show the problems elders have with putting on shoes, and the problems' older partners may have in helping their loved one, when they might also have a disability.

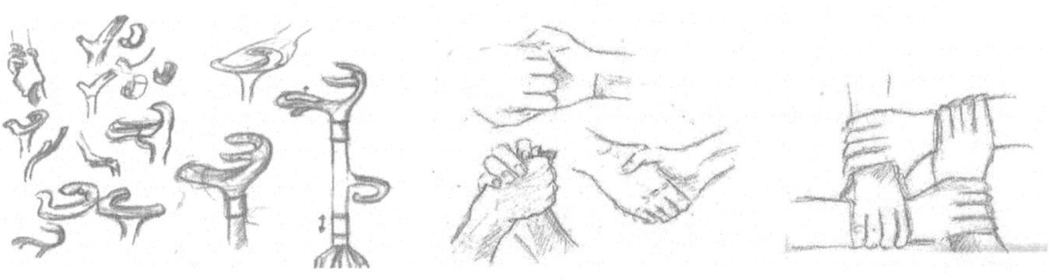

**Fig. 4.8** Movement in design direction following empathic modelling

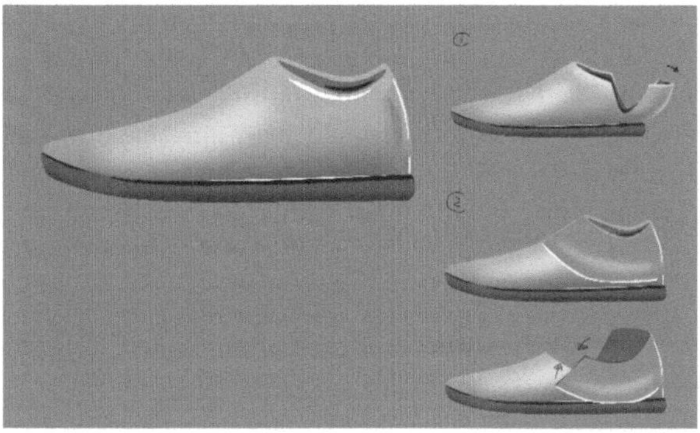

**Fig. 4.9** Initial designs for the 'opening shoe'

This task was designed specifically to show the problems elders have with putting on shoes, and the problems' older partners may have in helping their loved one, when they might also have a disability.

### *4.6.4 Iterative Design of the DLI*

#### 4.6.4.1 Adaptations to the Kit and Reflections on it

The basic low fidelity simulation kit was augmented with a Gerontology suit, tinnitus headsets and vision restricting goggles. The Gerontology suit provides a more immersive experience. However, when tried with DLI5—the overshoes had to be abandoned as they were a trip hazard, the mobility of the larger men was not restricted—even with the heaviest weights. Students could not walk without being accompanied, and security informed. Additionally, people did not react to the students in the same way as they would react to an older person.

DLI4 participants were critical of the approach. They adapted to the vision restricting glasses, by just looking through the parts they could see through. Additionally, they did not think that becoming suddenly disabled represented real life, as people's conditions gradually creep up on them.

#### 4.6.4.2 Adaptations to the Tasks

The task set was refined over time to reflect activities that people would undertake when planning or going on a journey. We included playing games (cards and word search) based on observations of what people do on journeys; the use of the mobile phone was included for trip planning tasks, and to send messages to companions. We matched both solitary and group activities, which required dexterity, hearing and vision.

We also included taking one's own shoes on and off, and then someone else's. The students seemed to gain a lot of insight from this, especially if they had limited mobility and sensation whilst putting on someone else's shoe. This replicates the situation of an elderly couple who may have to care for each other; insight that directly developed into design ideas by two students; one of whom created an easier opening shoe, and the other started to investigate the relationship between the carer and the cared for.

## 4.7 Conclusions

Developing empathy does not occur overnight nor as the result of one single intervention. Much as the gradual development of a condition and/or disability might imperceptibly affects one's own mobility, empathy occurs once seen through the eyes of experience. By presenting the participants of this study with the opportunity to experience limitations that may have been difficult to articulate, yet in a safe environment, the result is an appreciation for the effort, discomfort and changes to personal perception that many of us will experience at some point during our lifetime. Many of the students were skeptical at the outset of this study but the majority completed their experience with a better understanding of thinking outside of one's own lived experience and stepping into the shoes of another.

We have achieved this in an education setting, where students are encouraged to try new methodologies, make mistakes and learn as a collective. Once they are employed in industry however, will such behaviour be encouraged or indeed, tolerated? There is suggestion that the gradual convergence between research and design [25] is evidence of a positive change in industry for the inclusion of real user insight through methods such as co-creation, broad PPI and to some extent crowd funding. Real users have empowered themselves, designers have started to listen constructively and research is becoming visible. Once industry embraces this new opportunity, empathic design will seem like this is the only way to do things in the future; and how ever did we manage without it?

**Acknowledgements** All activities were supported by a grant from the Frank Jackson Charitable Trust. Higher-level ethical clearance was given for each instantiation of the DLI. Unless stated otherwise all activities took place at Coventry University (UK).

## References

1. Tovey, M.: Intuitive and objective processes in automotive design. Des. Stud. **13**(1), 23–41 (1992)
2. Loe, M.: Aging Our Way: Independent Elders, Interdependent Lives. Oxford University Press, New York (2011)
3. Pethick, T.: Quoted in 'Did baby boomers loft Britain out of its recession, *This is Money*'. Accessed from https://www.thisismoney.co.uk/money/news/article-2553921/Baby-boomers-helped-lift-Britain-recession-grey-pound-accounts-50-consumer-spending.html, 1-2-2019 (2014)
4. Moody, L., Mackie, E., Davies, S.: Building empathy with the user. In: Karwowski, W., Soares, M., Stanton, N.A. (eds.) Human Factors and Ergonomics in Consumer Product Design. CRC Press, Boca Raton, Florida (2011)
5. Sanders, E.B.-N., Stappers, P.J.: Co-creation and the new landscapes of design. CoDesign **4**(1), 5–18 (2008)
6. McDonagh, D.: Design students foreseeing the unforeseeable: practice-based empathic research methods. Int. J. Educ. Through Art **11**(3), 421–431 (2015)
7. Woodcock, A., Ward, G., Unwin, G., Osmond, J., Fielden, S., Ray, S.: Unlocking the potential of the younger older consumer. In: Anderson, M. (ed.) Contemporary Ergonomics and Human Factors. Proceedings of the International Conference on Ergonomics & Human Factors 2013, Cambridge, UK (2013)
8. Scott, W.D., Durling, D., Woodcock, A., McDonagh, D.: Understanding the needs of users with disabilities when working with designers. Paper presented at Design4Health 2015, held 13–16 July 2015 at Sheffield Hallam University, Sheffield, UK (2015)

9. Maslow, A.H.: Conflict, frustration, and the theory of threat. J. Abnorm. Soc. Psychol. **38**(1), 81–86 (1943)
10. Woodcock, A.: Late adopters and laggards: should we care? Int. J. Des. Manag. Prof. Pract. **7**(3), 53–61 (2012)
11. McDonagh, D.: Do it until it hurts! Empathic design research. Des. Princ. Pract. **2**(3), 103–109 (2008)
12. McDonagh, D., Denton, H.: User-centered design: developing the student designer's empathic horizons. In: Kimbell, R. (ed.) Design and Technology International Millennium Conference Sept 2000, pp. 111–115. Institute of Education, London: Design and Technology Association (2000)
13. Postma, C.E., Zwartkruis-Pelgrim, E., Daemen, E., Du, J.: Challenges of doing empathic design: experiences from industry. Int. J. Des. **6**(1), 59–70 (2012)
14. McDonagh, D.: Bringing empathy into the designing process: building in design integrity. Innov. Quart. Ind. Des. Soc. Am. 36–38 (2018)
15. McDonagh, D., Thomas, J., Khuri, L., Sears, S., Peña-Mora, F.: Empathic design research strategy: people with disabilities designing for all. In: Silva, A., Simoes, R. (eds.) Handbook of Research on Trends in Product Design and Development: Technological & Organizational Perspectives, pp. 58–79. IGI Global, Hershey, PA (2010)
16. Fulton, Suri J.: Empathic design: informed and inspired by other people's experience. In: Koskinen, I., Battarbee, K., Mattelmäki, T. (eds.) Empathic Design, User Experience in Product Design, pp. 51–57. IT Press, Helsinki (2003)
17. Mattelmäki, T., Battarbee, K.: Empathy probes. In: Binder, T., Gregory, J., Wagner, I. (eds.) Proceedings of the Participatory Design Conference, pp. 266–271. CPSR, Palo Alto, CA (2002)
18. Koskinen, I., Battarbee, K.: Introduction to user experience and empathic design. In: Koskinen, I., Battarbee, K., Mattelmäki, T. (eds.) Empathic Design, User Experience in Product Design, pp. 37–50. IT Press, Helsinki (2003)
19. Battarbee, K., Fulton Suri, J., Howard, S.G.: Empathy on the Edge: Scaling and Sustaining a Human-Centred Approach in the Evolving Practice of Design. IDEO, California (2003)
20. Schön, D.A.: Educating the Reflective Practitioner: Toward a New Design for Teaching and Learning in the Professions. Jossey-Bass, San Francisco (1987)
21. Buchenau, A., Fulton Suri, J.: Experience prototyping. In: Boyarski, D., Kellogg, W.A. (eds.) Proceedings of the Conference on Designing Interactive Systems: Processes, Practices, Methods and Techniques, pp. 424–433. ACM Press, New York, 17–19 Aug 2000
22. Kooprie, M., Visser, F.S.: A framework for empathy in design: stepping into and out of the user's life. J. Eng. Des. **20**(5), 437–448 (2009)
23. Schmidt, L., Kekel, K.: Take a look through my glasses: an experimental study on the effects of age simulation suits and their ability to enhance understanding and empathy. Gerontologist **53**, 62 (Gerontological Society of America (GSA), New Orleans) (2013)
24. Kolb, D.A.: Experiential Learning: Experience as the Source of Learning and Development, vol. 1. Prentice-Hall, Englewood Cliffs, NJ (1984)
25. Magee, P., Woodcock, A.: Thinking, design. A construct of (and for) change. Poster presented at EAD 2019, Running with Scissors, held 10–12 Apr 2019 at Dundee University (2019)

**Andree Woodcock** I have a B.Sc. in Psychology and Social Biology with an M.Sc. in Ergonomics from UCL. My Ph.D. concerned an investigation of the use of ergonomics in automotive concept design acquired from Loughborough University while I was a Daphne Jackson Research Fellow. Most of my research has been located within design departments, looking at applying user-centred design to wicked problems in the areas of health, education, regeneration and transportation. Health related grants have included design of polysensory school environments for children with ASD, the development of a patient held record system (MyCare), a user testing toolkit for SMEs (IDEAL-ALIP project) and a decision support system for SMEs moving into the assistive technology market (COMODAL). Transport related research has included looking at the in car safety and security of female car occupants, and the leadership 3 Eu projects: FP7 METPEX developing a Pan European tool to measure the quality of the passenger experience; H2020CIVITAS SUITS looking at supporting capacity in small–medium local authorities to develop sustainable transport measures and H2020 TInnGO project looking at gender inequality in transport.

**Deana McDonagh** is a Professor of Industrial Design in the School of Art + Design at the University of Illinois (Urbana-Champaign) and faculty at the Beckman Institute of Advanced Science and Technology. She is an Empathic Design Research Strategist who focuses on enhancing quality of life for all through more intuitive and meaningful products, leading to emotional sustainability. Her research concentrates on emotional user-product relationships and how empathy can bring the designer closer to users' authentic needs, ensuring both functional and emotional needs are met by products. She is a Research Fellow at Coventry University (UK), Adjunct Professor at Northwestern University (USA), visiting Professor at Arizona State University (USA) and Director of Research and Insight for Herbst Produkt design consultancy (USA). She is the Designer Entrepreneur-in-Residence (start-up incubator) at the Research Park, and Designer in Residence at the Beckman Institute for Advanced Science and Technology at the University of Illinois (Urbana-Champaign) where she provides design guidance.

**Paul Magee** is an Industrial designer trained at BIAD, UK. His design experience includes automotive styling, exhibitions, furniture and products, digital user-interfaces, infant care and medical diagnostic-technology. Paul has combined design industry experience with academic design research developing solutions that meet real user need. Collaboration is fundamental to Paul's professional approach and equally that of his personal perspective. An experienced facilitator for groups working with co-creation methods; Paul has delivered Health and Assistive Technology (AT) conference presentations on the subject. Both Paul's design work and research are developed collaboratively with users, with particular application to vulnerable user groups such as people with learning disability. His research interests include user experience, sustainable production, democratization of design, experiential insight and co-design with end users to find ways that improve lives in real terms.

**Michael Tovey** Professor of Industrial Design, Coventry University. My research interests have been in: CAD support for industrial design in the transport industry, socially aware design, and design pedagogy. The CAD work was driven by research grants in the 1990s, and the recent work on socially responsible design culminated in a book on design for transport. Through leading our Centre of Excellence in Product and Automotive Design I focused on design pedagogy, which I pursued further through leading the Design Pedagogy Special Interest Group of the Design Research Society for 9 years and producing a book on Design Pedagogy. During the 1980s I was responsible for the establishment and development of transport design, which achieved international prominence. I was Dean of the Coventry School of Art and Design for 18 years. During this time it doubled in size, achieving strong research rankings. I pioneered design research, contributing publications and holding a number of research council grants. The context for this work has been concept design in the automotive industry and developing techniques to support the design activity. From 2007 to 2010 I was Director for Design, responsible for leading and coordinating design education and design research. I am now based in the Research Institute in Transport and Future Cities, contributing to a number of research projects.

# Part II
# Usability and Uptake of Assistive Technology

# Chapter 5
# The Development of a Feature Matrix for the Design of Assistive Technology Products for Young Older People

**Andree Woodcock, Jane Osmond and Nikki Holliday**

**Abstract** This chapter describes findings from the COnsumer MODels for Assisted Living (COMODAL) project, which aimed to develop a consumer market for Assistive Technology Products (ATPs) for Young Older People (YOPs: 50–70-year olds). The project used mixed methods to explore various aspects of the ATP consumer market, together with a feature matrix for the design of ATPs, guidance on user needs and expectations from an ATP consumer market, and suggested business models. The methodologies used included literature and product reviews, market analyses, interviews, focus groups, and co-creation workshops. The triangulated findings found that currently there are significantly high rates of non-use and abandonment of ATPs, due to numerous factors including poor aesthetic appeal and ease of use, stigma, and concerns about loss of face to face care. YOPs have an appetite for the use (and private purchase) of products and services to support their independence and health, but such products must focus on the whole person and their higher-level needs, wishes and aspirations, not just a particular disability—something which the product analysis found lacking in many existing products. One of the ways the project addressed this was through the development of a product matrix to support designers and businesses to more widely consider the evolving needs of YOPs with regards ATPs. The business models provided further guidance on the implementation of such products into the consumer market. The results will be of use to businesses wishing to grow or diversify into the ATP consumer market—a key opportunity at a time when publicly funded provision of such technologies is decreasing. Ultimately, ATPs must not only assist independent living, but should also support enhanced quality of life whilst meeting higher level needs and desires.

A. Woodcock (✉) · J. Osmond
Centre for Arts, Memory and Communities, Faculty of Arts and Humanities, Coventry University, Coventry, UK
e-mail: A.Woodcock@coventry.ac.uk

J. Osmond
e-mail: J.osmond@coventry.ac.uk

N. Holliday
Faculty of Health and Life Sciences, Coventry University, Coventry, UK
e-mail: N.Holliday@coventry.ac.uk

 
A. Woodcock et al. (eds.), *Design of Assistive Technology for Ageing Populations*, Intelligent Systems Reference Library 167,
https://doi.org/10.1007/978-3-030-26292-1_5

**Keywords** Assistive technology · Ergonomics · Design · Young older adults · Design requirements

## 5.1 Introduction

An exploration of what younger older people (YOPs) (i.e. those approaching retirement and older age in the 50–70-year age range) want from consumer assistive living technology (ALT) products and the subsequent development of a design feature matrix was undertaken as part of a project carried out by Coventry University (2011–2014). Funded by the UK INNOVATE fund, the overall aim of the COMODAL project (**CO**nsumer **MOD**els for **A**ssisted **L**iving) was to develop a consumer market for Assistive Technology Products (ATPs) for YOPs. In addition, there was also a consideration of possible business models that could enhance a future ATP market. A matrix was developed to address an emergent finding of the project: that ATPs may fail to appeal to their intended users because of poor design quality and solutions.

The COMODAL project separated the market for ATPs into four segments (see Table 5.1) comprising personal, prospective, carer and non-purchasing consumers. This segmentation was used to inform a literature review, market analysis, product review, product developer survey, interviews with designers, co-creation workshops and a workshop with Age UK participants.

## 5.2 Brief Literature Review[1]

The definition of assistive technology used in COMODAL was one adopted by the Foundation for Assistive Technology (FAST) in which assistive technology (AT) is described as "… any product or service designed to enable independence for disabled and older people [14, 15]. This definition includes a wide range of products including community equipment, electronic assistive technology, telehealth and telecare, assisted living technologies (ALTs) and "assisted living products"" (ALPs).

Rowe and Kahn [17] defined successful ageing as 'growing old with good health, strength and vitality'. Psychological theories focus on 'optimization' [1] and the discovery of strengths that enhance quality of life. Information and Communication Technology (ICT) applications can provide new ways to help older people to live independently [6]. However, uptake and use of these products has not been extensive or sustained with purchase and usage influenced by factors such as design and usability, individual and societal attitudes towards assisted living technologies, and changes in health and social care provision.

Coughlin et al. [4] conducted a workshop and focus group with 30 leaders in aging advocacy and aging services from 10 north eastern states in America to better understand older adult perceptions of smart home technologies and to inform future

[1]Full literature review—see [22].

**Table 5.1** YOP segments

| 50–70-year olds | |
|---|---|
| Personal consumers | Purchase ALTs for their own use |
| Prospective consumers | Neither buy nor use ALTs |
| Carer consumers | 'Decision gatekeepers' involved in purchasing decisions for other older ALT end users |
| Non-purchasing users/consumer | Use ALTs but refer to other 'decision gatekeepers' |

research. Although the participants expressed support of ALTs they raised concerns related to usability, reliability, trust, privacy, stigma, accessibility and affordability.

Steel and Gray [19] found a significant rate of non-use of AT with rates ranging from 35 to 86.5% in some studies. They categorised the factors influencing usage as relating to:

- Client related factors—age, gender, diagnosis, education, socio-economic status, values, preferences, culture, perceived need, living arrangements, psychological readiness and a change in physical state
- AT related factors—aesthetic appeal, comfort, cost, reliability and durability, ease of use, effectiveness in meeting a need
- Assessment related factors—physical and psychological function, goals and preferences of user and caregiver/family
- Environment related factors
- Training related factors—context in which AT is provided, mode of delivery, session duration and frequency and who is involved in training
- Cost.

In terms of what might deter end users and carers from using telecare/telehealth, [5] identified the following factors:

- the potential for loss of privacy and confidentiality
- loss of social interaction where human care is replaced
- trust in the providers delivering the services, response times, maintenance and failure of systems
- stigma associated with the appearance of much of the current telecare equipment.

The literature review supported the initial hypothesis that there were barriers to both the purchase and use of ALPs. Young older people (approximately 50–70 years of age) were identified as a key market for the purchase of ALTs—either for themselves or their relatives. However, little research had been conducted on understanding the requirements of this group in terms of the design of ATPs. The following qualitative research was undertaken to understand more about ATPs and services and how they could be designed to be more attractive to the target consumer group, and subsequently the exploration and development of the consumer markets through which these products would be sold [23].

## 5.3 Qualitative Research

### 5.3.1 *Market Analysis*

Five hundred participants took part in a consumer survey and were asked to rank the importance and influence of the barriers and enablers to purchase and use of ATPs and services and the wider factors affecting purchasing decision making. The results highlighted how the consumer market has been held back by the dominance of the statutory provision of ATPs through health and social care services. The top three barriers to purchase of ATPs were cost, knowing how to choose what to buy and lack of product awareness. The top three enablers related to belief that the product would make a difference, affordability and added value, and whether it would make life safer at home. Carers were more sensitive to quality of life issues and how the person would feel about using the product. Just over 50% said they would buy a product to help in daily living, but just under 30% said they would simply struggle on [20].

To provide more in-depth market analysis to support the consumer survey, four focus groups were held by Coventry University and AgeUK. The aim of the focus groups was to explore in detail views on ALPs amongst the YOP segments of interest to the COMODAL study. The main findings for each group are summarised in Table 5.2.

Participants revealed a lack of awareness of and information about ATPs. Some were viewed as stigmatising and not meeting functional needs. Additional worries

**Table 5.2** Focus group characteristics in relation to purchasing of ATPs

| Focus group | Characteristics |
|---|---|
| Prospective consumers | Discussed benefits of ATPs for their parents, not for younger people; they were most influenced by the stigma associated with ATPs with zimmer frames (walking frames) and shopping trolleys being distinguished as particularly stigmatising. This may be overridden if the product fulfilled an essential need |
| Personal consumers | They felt that many had low awareness of ATPs, and that they were lucky to be part of social networks which enabled them to find out about them. Generally, those with good social networks had higher levels of awareness of assistive technology |
| Carer consumers | Commented most on the limitations and functionality of AT, as they have seen those they care for struggle to use various products. They were aware of the cost of ATPs, and felt that with a growing market, they would become cheaper in the future. They were similar to the Personal Consumers in their awareness of the usefulness of social networks in raising awareness |
| Non-purchasing users | Were the only group who thought ATPs should be provided by statutory services, possibly because most had been supplied by their Local Authority and had little experience or awareness of purchasing privately |

related to quality and value for money. The normalisation of assistive technology and its increasing prevalence were important enablers to purchase. Participants felt that the use of certain products by older age groups (e.g. mobility scooters) would become more acceptable. They were also keen to point out that such products could be helpful for all age groups. Mainstreaming would propagate awareness of products and their benefits as more people would be using the products (thereby improving acceptability in the public eye), and economies of scale would reduce cost. Positive images of those using assistive technology and the affirmative portrayal of disabled and older people in soap operas were important. Several enablers to the purchase process were identified with rental or "try before you buy" services and more information increasing confidence [24].

### *5.3.2 Product Review*

Desk top research was used to evaluate over 50 ATPs [25] which met the definition of assistive products provided by [10].

> Assistive technology is a generic or umbrella term that covers technologies, equipment, devices, apparatus, services, systems, processes and environmental modifications used by disabled and/or elderly people to overcome the social, infrastructural and other barriers to independence, full participation in society and carrying out activities safely and easily.

Products were selected for review by members of the research team, based on the completeness of their description provided in the URL supplied as part of the product review (see Table 5.3). Only those products which had a full description, user manual or clip showing their usage were included. The inherent weaknesses of an approach which relies on product descriptions given by the developers for sales purposes are acknowledged. However, the aim of this work was not to evaluate

**Table 5.3** Products reviewed

| Assistive product category | Products |
|---|---|
| Communication | Range of mobile phones |
| Modify the environment | Wireless mobile speakers |
| Recreation | Readers, kindle, disability scooters, future proof television |
| Housekeeping | Digital jug and scales, robotic vacuum cleaner, intelligent fridge, digital video message |
| Personal mobility | Satellite navigation, hand-held location devices |
| Personal care and protection | Emergency location system, wireless leash, intelligent water bottle, face recognition door entry, bath temperature and level alert, water pebble, community alarms, pendant, bogus call detector, pull cord, pill dispenser, bed and chair occupancy alarm, activity monitors |

Products that can be directly identified to a company have been omitted due to confidentiality the products themselves but develop a feature set against which products may be designed and evaluated.

These products fell into three distinct categories:

- Those which had crossed over from the work environment. Here companies seemed to have seen a new market and were trying to sell their product to home users without any adaptation to it. Such products may be useful e.g. secure cabinets for keys, appointment reminders, paper shredders—but in most cases the 'work aesthetic' was too strong, and such products would be more appropriate in a 'home office'. A more successful transition here would be Skype, which although emerging from a utilitarian product has crossed over because it offers affordable, useful functionality which is fairly easy to use, and has a ready formed transgenerational user base.
- Transgenerational, desirable products, such as tablets. These are produced by companies having large R and D teams, which have sophisticated business models, and can offer high levels of customizability, on—line and after sales support.
- Companies moving into the assistive technology market, but not yet meeting the needs of the user group. Such companies may not fully comply to usability standards, have not proceeded with user centred design, do not offer the levels of tailorability, instruction and customer service required from the age group and where compliance is low. Such companies might find the feature matrix to be most useful.

These products matched the categories developed by Hersh [8, 10]'s:

- Mainstream products: products designed for the 'general population' and which can be obtained from standard retail and other outlets. Unfortunately, they are frequently designed without consideration of the needs of disabled and many other groups of people.
- Design for all (European) or universal design (American) products: an approach to design which aims to make products accessible and useable by as wide as possible a range of users, regardless of factors such as disability, age, size, culture, ethnic background or class. Ideally all mainstream products would use a design-for-all approach.
- Assistive products, which are designed to remove barriers for disabled and elderly people.
- Rehabilitation products, which are designed to restore the functioning of disabled or elderly people or people experiencing ill-health.
- Medical products, which are designed to support a range of health care practices and promote healing in people experiencing illness or other forms of ill-health and who may be categorised as patients.

It was found that many ATPs focused on disability rather than the whole person, and failed to fulfil higher level needs, wishes and aspirations. This may make them less attractive and desirable and functional as addressing just one disability may mean failure to recognize other needs. There was a noticeable attempt by some companies to market existing products as assistive technologies, with little attempt to redesign them to a different usage context or user group (e.g. document shredders). Products which are suitable for medical and office environments, in most cases, do not suit the home environment.

The reviewers highlighted factors which need to be addressed in the design and delivery of ATPs and which could be included in the design feature matrix. These included:

- Locus of control in telehealth and telecare monitoring services—where agency is taken away from the user.
- Focus on lower, rather than higher order needs [16] with a lack of attention to pleasure and experiential design—few products appeared to consider the experience of the user, the attractiveness of the product, the meaning attributed to its use or the feelings which might arise during its use [12, 13].
- High value attached to service, after sales care and advice.
- Weak adherence to standards—lack of ISO or CEN authentication.
- Need for discretion, privacy and inclusivity—younger older consumers do not want products that characterize them as disabled or in need of assistance.
- Need for adaptability, flexibility and tailorability over time and context.
- Fitting into peoples' homes—many recommended products are rejected because they '*did not want their homes looking like hospital clinics*'.
- Economic feasibility—the affordability of the product to its target market, products with built-in obsolescence.
- Products should be developed based on a clear understanding of user needs.

### 5.3.3 *Product Developer Survey*

From the desktop review of ATPs, five companies agreed to take part in a survey, with most of them designing products across the age spectrum for groups with specific physical, cognitive and communication difficulties. Two companies designed for the older population, including those with disability and/or low levels of computer literacy. Two of the companies designed products based on family circumstances, which led them to see that there was a need for the product. Although only one company worked with ergonomists and universal designers, there was evidence of user involvement in the design process, either through proxy or directly. So it would appear that only some companies, at the time of the survey followed ISO 13407 (Human centred design processes for interactive systems) and ISO 9241-210 (Ergonomics and Human Systems Interaction) principles. In these documents 6 key principles are laid down relating to the need for:

1. Design to be based on an explicit understanding or users, tasks and environments.
2. Users to be involved throughout design and development.
3. Design to be driven and refined by user-centred evaluation.
4. Iterative design.

5. Design which addresses the whole user experience.
6. The design team includes multidisciplinary skills and perspectives.

Design did include user testing (by 3 companies) to ensure that the product met the needs of the market it was being designed for, and participatory design techniques. For example, Company A ran focus groups and questionnaires, Company C invited older people to look at concepts, Company D tried out new designs with existing users, and all companies responded to user (or carer) feedback.

External stakeholders included end users, marketing managers, sales people and intermediaries such as local societies, social services, internal/external designers/R&D division/some health visitors/social services, occupational therapists. This implies that where appropriate ISO (IEC 9126-1) standard was applied to ensure software quality—functionality, reliability, usability, efficiency, maintainability and portability.

The companies' main focus was on the assistive part of the technology i.e. its ability to help the user, rather than the design per se. For example, Company C, development of phones concentrated on dexterity, ease of use, flexible volume controls, high contrast screens, loud ringtones, text sizing, safety alarm and texting facilities, on/off options—and appearance was not mentioned. All products were practical and fulfilled specific needs (e.g. the ability to control the environment, to read correspondence) and were sold to private individuals (and their families), local authorities, intermediaries and health care providers.

An update of relevant ISO standard was provided in [9] who also noted the need for the design of ATPs to incorporate design principles and practice used in other consumer products, including:

- User centred design.
- Iterative multi criteria approaches considering function, form, attractiveness to all the senses, pleasure in use, usability, accessibility, performance, reliability, safety and environmental factors. These include the six as: awareness, accessibility, availability, appropriateness (usefulness), affordability and acceptability [3].
- Trade-offs between the provision of different modes of use and/or inputs and outputs and information in different formats, with a degree of redundancy, and simplicity and cost.
- Ease of upgrading, repair and maintenance, as well as robust design to reduce the likelihood of faults occurring.
- Ease and intuitiveness of use, with a minimum of documentation and training.
- Consideration of subsequent provision of information, support and repair facilities to end-users.
- A modular software architecture, to reduce the impact of any problems that occur in any one component on the rest of the design and to facilitate the later addition of further modules.
- Compliance with any relevant national and international standards or other regulation.

She [9] also noted the following differences between the design of ATPs and other (consumer) products:

- The limited size of the market for some devices (as opposed to more widespread demands (e.g. for hearing aids and wheeled mobility), which may result in non-standard approaches to design and distribution.
- The potential crossover of products to and from the mainstream.
- The difficulty of standard user interfaces for particular groups.
- The need to satisfy the needs of the end user, purchaser and funder or purchasing organization in the design of the product.

It could therefore be argued that further development of a private market for ATPs, which some of the companies interviewed as part of the research are already operating within, could see an increase in the development of ATPs which does indeed take a more standardized approach to design, and thus products which satisfy the needs *and* desires of the end use will be more prevalent, rather than products that purely meet functional needs. What consumers would want from such a more developed consumer market is discussed in more detail below.

Few conclusions can be drawn from this limited survey, especially given that the product ranges were aimed at those requiring higher levels of assistance than the YOPs. Products were developed to fulfil a particular need, with some user consultation through the design process. The effort was targeted on the functionality, perhaps without a consideration of other design qualities. However, one of the products from Company C was selected independently as an exemplary product by the designers.

This, and the earlier product review, were augmented during the evaluation of the matrix, in which its features were refined in the assessment of products catering for the assistive technology market.

### *5.3.4 Interviews with Designers*

Semi structured interviews were held with 7 product design lecturers to understand their approach to designing for this market segment. The results from the interviews were triangulated, with all designers agreeing that 50–70-year olds do not want to be singled out as needing assistance, do not see themselves as getting older, and resist being grouped with older people or those having higher levels of need. They do not want to be marginalised or to appear to be growing old. The characteristics of YOP group, from a design perspective were:

- A refusal to think of themselves as 'old' and resentment of products were targeted at the 'elderly market'. Therefore, assumptions made about YOPs needed to be constantly challenged, especially assumptions that this group were less able than when younger.
- Increased spending power and disposable income especially if they work until their late 60s, if not in their main career, then in a tertiary career.
- Increased wisdom with age which meant that this group was less likely to be bamboozled by advertising tricks; instead they were looking for value for money and 'honest products'.

- Age related changes in memory, sight and hearing impairments, the presence of previous sporting injuries, mobility restrictions, dexterity problems, reduced strength.

The lecturers felt that the YOP market would be best served through better design of mainstream products, rather than specific products targeted at older people. The group do not want to be stigmatized or singled out as in need of 'assistive' products or 'equipment', so products which are discretely tailorable are well received.

However, some generic design trends did emerge. The design lecturers felt that the aspirations of YOPs reflected their age—they want to stay healthy for as long as possible—perhaps not a noted characteristic of younger age groups; continue to work in an interesting job and remain in contact with their family. Therefore, products which aid their continuation in senior roles or allow them to take up a secondary career may be important; products which discretely adapt to changing physiological and cognitive needs are appealing (e.g. ability to change font size and contrast levels on hand held devices); technology which is simple to use and allows tailorability is as important along with the need for intergenerational/universal design.

YOPs are also more likely to value high quality and craftsmanship in products which will not become outdated or require the purchase of a new model (i.e. are timeless)—so built in obsolescence may not be appealing. They want attractive, stylish products. These accord to the traditional stereotypes of the group (where marketing hinges on timeless, classical designs). Of increasing importance are the levels of technological awareness of this group. This means that traditional pieces with dated functionality may not be appreciated. They need to be seen to move with the times. This translates into design qualities—high quality products, products which are light in weight and intuitive to use. Products which can be personalised, customized, are adaptable and can be built on were felt to be important. For example, products where the text size could be adapted, where apps could be purchased discretely and easily incorporated into a product were given as good design examples. In products such as tablets, the base unit is well recognised and does not single the user out as needing assistive technology but can be discretely tailored to meet their needs. Overall, the following design features were felt to be important for YOPs:

- Longevity in a product—this group was thought to be willing to pay more for products than a younger market, with the expectation that the product would last and also be backed up by good customer service. Related was the need for a slower recycle pattern—this group were perceived by the product design lecturers as less likely to want the 'next best thing' just because it was new.
- Appearance of products was important as well as functionality.
- Intuitive products—if it takes more than a few minutes to operate, and there is a need to read a complicated manual the product 'will go in the bin'.
- Manageability in terms of weight and grip were mentioned as were the need for logical, intuitive products.
- Good quality materials, craftsmanship, timeless design.

It is argued that YOPs, from the perspective of ATPs, can be seen as a new market segment, clearly wanting to differentiate themselves and not be included with people who are older or requiring higher levels of assistive technology. The interviews resonated with Jordan's [13] research in which he commented that '*traditionally, human factors have tended to concentrate on making products 'usable'—focussing on utilitarian, functional product benefits*'. He recognised that the feelings associated with using pleasurable products included security, confidence, pride, excitement and satisfaction, entertainment, freedom, nostalgia. In contrast, dis-pleasurable products were associated with feelings that included aggression, feeling cheated, resignation, annoyance, anxiety, contempt and frustration. The concept of usability is concerned with avoiding the negative aspects rather than producing positive emotions such as pride, excitement or surprise. Products which are highly functional and usable may not delight or excite their user or fulfil the higher-level needs which YOPs require.

The comments from designers emphasized the need for ATPs to incorporate Jordan's [12] dimensions relating to the emotional and hedonic benefits associated with product use which might be especially pertinent to older consumers. In the product review it was noted that many of the products were utilitarian and did not accommodate higher level needs. Matching these with the attributes associated with pleasure are related to helpfulness and appropriate features, good usability, aesthetics, performance and reliability convenience, size and cost.

The designers recognized that ATPs should be designed to fulfil needs related to socio-pleasure (co-operation and collaboration, community self-worth; love and friendship; helping, giving and sharing, ideo-pleasure (individual self-worth, sensual aesthetics, creativity and expression) and psycho-pleasure (cognitive arousal, progression and achievement, curiosity fulfilment) and physio-pleasure (sensual stimulation and physical arousal).

### *5.3.5 Co-creation Workshops with AgeUK*

A series of co-creation workshops were held by Coventry University and AgeUK to explore the perceptions of YOPs of the ideal consumer journey one would embark upon when purchasing ATPs [11]. Co-creation methodology was defined as an "act of collective creative, that is, creativity shared by two or more people". Sanders and Stappers' [18] guidance was chosen to create a shared vision of the ideal consumer ATP journey with the four types of YOPs and industry representatives. In total, six co-creation workshops were held across the United Kingdom, focusing on key aspects of a typical journey a consumer takes when purchasing a product, including recognising that a need exists, finding product information, making a decision to purchase, where to purchase the product and using the product in day to day life. Table 5.4 summarises the key findings regarding each stage of the consumer journey.

It is also clear that there are numerous factors which influence the decision to purchase a product, once the potential end user or their carer has become aware of that item. The findings here clearly triangulate with the findings which emerged from the development of the product matrix, highlighting the importance of the consideration of these items when designing ATPs for YOPs. Despite preconceptions regarding

**Table 5.4** Consumer journey (from [11])

| Stage of consumer journey | Key findings |
|---|---|
| Recognising that a need exists | Begins with recognition of reduction in ability/increase in disability. It may not be the older person themselves who recognises this need (e.g. may be a family member/friend/doctor). ATPs must also be recognised as a potential solution, which can be affected by current views of ATPs as stigmatizing |
| Finding product or service information | Currently, older people do not know where to seek information regarding ATPs. It should be available in physical locations often frequented (e.g. GP surgery, disability shops, supermarkets), or of knowledge to key professionals (social workers, occupational therapists, GPs, community nurses, emergency services). Social networks and social media may also be a useful source, as is television and print media advertising |
| Making a decision to purchase | Despite functionality, reliability, and potential to improve health or quality of life, aesthetic and lifestyle factors are influential. Products should be aspirational and stylish |
| Where to purchase the product or service | ATPs should be available in a variety of outlets including: mobility stores, local shops, online shops, department stores, pharmacies, catalogue shopping, supermarkets |
| Using the product or service | YOPs placed high value on post-product customer service support, product longevity, support with product upgrades. Hassle free returns policies and warranties key |

older people's use of the internet, online shopping was a popular option to purchase ATPs, in particular because it allows the reading of product reviews and offers an increased likelihood of competitive prices. Aesthetic and lifestyle considerations are a key mediator of whether or not a product is likely to be purchased [11].

## 5.4 Business Models

Further work then aimed to co-create potential new business models that would support businesses introducing or increasing sales of products into the consumer ATP market, with a view to improving currently low direct to consumer sales, as most (but not all) ATP businesses tend to sell business to business [20]. Data from the rest of the COMODAL project, supported by business focus groups and interviews with 103 ALT industry representatives fed into the development of four potential business models, which were then validated via consumer and business stakeholder workshops. It was found that both industry and potential consumers were most interested in a "Broker" business model of ATP purchase, where an expert consultant, or "broker" would support the customer in identifying their needs, and matching those needs to a particular product, or bundle of products. This model was likened positively to older people's experiences of booking a holiday through a trusted travel agent or experiencing a Personal Shopper service in a department store [20]. Other models considered in the validation groups included Diversifer (where companies previously working outside of ATP move into the area, for example a smart energy company offering remote environmental monitoring services which would enable families to use energy and appliance data to monitor and ensure the safety of loved ones [21]; Complementor (two companies co-operate to signpost each other to potential new customers, e.g. a stairlift company may signpost a customer to fall detector products if they are aware they are concerned with safety in the home) and Insurance models (where customers pay insurance premiums in anticipation of ATP expenses as they age, either sourced and paid via the insurance company, or via a 'cash-plan' reimbursement model) [20].

## 5.5 The Matrix

The matrix shown in Table 5.5 was iteratively developed, drawing on the literature review and qualitative research findings outlined above. The matrix shows the final features grouped in to 16 categories, their derivation from the research and a synthetic indication of relative importance and is provided as a quick reference tool/checklist for designers/students who maybe unfamiliar with designing for this market.

## 5.6 Conclusions

From the discussions it became clear that although this was an expanding market, not all companies followed UCD principles, ISO standard or the CEN guide. Authentication needs to be tightened to ensure standards and the development of that products are fit for purpose.

Related to this are issues around quality of support services—from initial consultation through to after sales support, so that the user knew that the product was suitable for them and was entitled to change it if they were not satisfied. Stress was placed on the training of the sales people with end users being given informed choice. This requires a change in purchasing contracts—which allow longer trial periods, money back guarantees and support through life time of the product.

There was still a lack of appreciation of the context of use, and how the user would feel when using the ATPs. Would they feel empowered? Have freedom to use it? Or would they feel controlled? Were they regarded as passive? Who derives benefit from services such as telemonitoring? This led to a discussion of ethical issues around the

**Table 5.5** ATP Design Matrix showing relative importance of design features

| Category | Features | Importance for YOP/ALT | Derivation |
|---|---|---|---|
| Confidence building | Users should know what operation to do next: either through prompts, conformance to stereotypes or intuitive interfaces | 👍 | LR |
| | Provision of timely and appropriate feedback | 👍 | LR |
| Dependability | The product should be tolerant of user error | | LR |
| | The product should be able to correct itself | 👍 | LR |
| | Fault should be able to be detected | | LR |
| | The user should be able to individualize the product | 👍👍 | I |
| | The product should have high levels of security | 👍 | PR |
| Trustworthiness | Available when needed | | LR, PR, I |
| | Reliable | | LR |
| | Safe | | LR |
| | Maintainable by user | 👍👍 | LR, PR |
| | Offer confidentiality | 👍 | LR, I |
| | Have high integrity | 👍 | LR |
| Acceptability | Usability: high levels of usability | | LR |
| | Learnability: easy to learn | | LR |
| | Cost: affordable to the consumer with no hidden extras | 👍👍 | LR, I, E, CS |
| | System compatibility | 👍 | LR |

(continued)

**Table 5.5** (continued)

| Category | Features | Importance for YOP/ALT | Derivation |
|---|---|---|---|
| | Efficiency: easy to achieve tasks | 👍 | LR, PR, I |
| | Responsiveness: machine should respond quickly and appropriately to user input | 👍 | LR |
| | Aesthetics: should appeal to the intended age group | 👍👍 | LR, I, PR |
| | Lifestyle compatibility | 👍👍 | LR, I, PR, E, CS |
| Fit for purpose | Beneficial: user should derive tangible and intangible benefits from the use of the product | 👍👍 | LR, I, PR, E, CS |
| | Reliable | 👍👍 | LR |
| | Safe | | LR |
| | Useful | | LR |
| | Accessible | 👍 | LR |
| | Accreditation and compliance with ISO standards | 👍 | LR, PR, E |
| | Dependable and provide having extended warranties/money back guarantees | 👍👍 | PR, CS |
| Does not highlight disability | Lack of stigmatising aesthetic | 👍👍👍 | I |
| | Devices that contribute to positive self image | 👍👍👍 | I, E, CF |
| | Mainstreaming of prod-ucts/transgenerational design | 👍 | I, E, LR |
| Befitting the home environment or usage context | Unobtrusive | 👍👍👍 | PR |

(continued)

**Table 5.5** (continued)

| Category | Features | Importance for YOP/ALT | Derivation |
|---|---|---|---|
| | Appropriately styled | 👍👍👍 | I |
| Empowering | Providing independence | 👍👍👍 | I, E |
| | Giving peace of mind | 👍👍👍 | I, E |
| | Enjoyable | 👍👍👍 | I, E |
| | Freedom to use the product, or not use it | 👍👍 | I, E |
| User interface related | Reduced amount of output needing to be processed | 👍 | LR |
| | Short messages | 👍 | LR |
| | Low functionality systems | | LR |
| | Few options presented | | LR |
| | Tailorable/flexible—add more possible | 👍👍👍 | LR |
| | Structured progression (to new features) | 👍 | PR, E |
| | Self-assessment before moving to higher functionality | 👍 | PR, E |
| | Audio enhancement/flexibility | | LR, PR, I |
| | Hearing aid compatibility | 👍 | LR, PR, I |
| | Minimal twisting motions | 👍 | LR, PR, I |
| | One button/one menu | 👍 | LR, PR, I |
| | No hidden controls | 👍👍👍 | LR, PR, I |
| | On-screen menus with visuals | 👍 | LR, PR, I |
| | Flexible volume controls | 👍 | LR, PR, I |

(continued)

**Table 5.5** (continued)

| Category | Features | Importance for YOP/ALT | Derivation |
|---|---|---|---|
| | Resizable text | 👍 | LR, PR, I |
| | Vibration | | LR, PR, I |
| | Voice recognition | | LR, PR, I |
| | Large screen controls | 👍 | LR, PR, I |
| | Zoom controls | | LR, PR, I |
| | On/off buttons | 👍 | LR, PR, I |
| | Safety alarm | | LR, PR, I |
| | Loud ring tones | 👍 | LR, PR, I |
| | Big buttons | 👍 | LR, PR, I |
| Design factors | Build quality should be high | 👍 | I |
| | Good centre of balance (physically stable) | 👍 | I |
| | Degree of honesty | | I |
| | Longevity: no built-in obsolescence | 👍👍👍 | I |
| | Value for money for the consumers | 👍👍👍 | I, LR, PR, CF, |
| | Good quality materials | 👍 | I |
| | Timeless design | 👍 | I |
| | Show high levels of craftsmanship | 👍 | I |
| | Quality | | I |
| | Thoughtful design | | I |
| | Stylish design | | I |
| | Subtle | | I |
| | Straightforward and predictable design | 👍 | I, LR |
| | State of the art | 👍👍 | I |
| | High finish quality | | I |

(continued)

**Table 5.5** (continued)

| Category | Features | Importance for YOP/ALT | Derivation |
|---|---|---|---|
| | Aesthetically pleasing | | I, LR |
| | Desirability | | I, LR |
| | Aspirational | | I, LR |
| | Transgenerational | | I, PR, LR |
| ATP related | Tone of the product should be appropriate | 👍👍 | I |
| | Design should not identify the user as being old | 👍👍👍 | I, E, PR, CF |
| | Should not be called an assistive technology product | 👍👍👍 | E, PR, CF |
| Usability | Logical operations, functions and design | | LR |
| | Product should be intuitive | | LR |
| | Easy to use | | LR, I |
| | Should not require high levels of dexterity to operate | 👍 | LR, I |
| | Functional | | LR |
| | Good customer service | 👍👍👍 | LR, PR, CF, E |
| | Easy to set up | 👍👍👍 | LR |
| | Clarity of instructions should be high | 👍 | LR |
| | Suitable for those unfamiliar with technology | 👍 | LR, CF, PR |
| | Should comply with all safety regulations and be safe to use in all instances | | LR, PR, E |
| | State of the system should be clear and explicit | 👍 | LR |
| | Good human interface/Human (user)-centered | | LR |
| | Adoption of ucd principles | 👍 | LR, E, PR |

(continued)

**Table 5.5** (continued)

| Category | Features | Importance for YOP/ALT | Derivation |
|---|---|---|---|
| | Adaptable to user needs, and changing user needs | 👍👍👍 | LR, I, E |
| | Products and instructions designed to fit the cognitive map of the users | 👍👍 | LR, I |
| Flexibility | Bolt-on facilities; system should be expandable | 👍👍 | PR |
| | Customisable | 👍👍👍 | PR, I |
| IT related | Seamless operation in many environments | | LR, PR |
| | No built in obsolescence | 👍 | I |
| Pragmatic issues | Social access: allowing people access to societies, friends and neighbours | 👍👍👍 | I, E, CF, |
| | Affordable | 👍👍👍 | I, E, CF |
| | Privacy | 👍👍👍 | I, E, CF |
| | Non invasive | 👍👍 | E, |
| | User maintainability | 👍👍👍 | E, PR |
| Quality of support services | Trusted | 👍👍👍 | E |
| | High quality after service support | 👍👍👍 | E |
| | Training of support and sales staff | 👍👍👍 | E |
| | Provision of trial periods and money back guarantees | 👍👍👍 | E |

Key: *LR* literature review, *I* interviews, *PR* product Review, *E* evaluation, *CS* consumer survey and focus group development and use of ALTs (e.g. [2] and the locus of control. For example, in telemonitoring, movement sensors will detect that someone has fallen out of bed and alert friends/emergency services of a potential injury, but it will not deal with the surrounding context—e.g. why the fall has happened—without other systems being put in place. Telemonitoring also affects levels of privacy and can be seen as an unwanted intrusion or restriction by those being observed. Additionally, care needs to be taken that the new technologies will not increase social isolation. In the past neighbours, family and the community might look in on older neighbours. Telemonitoring, installed without a wider understanding of the social context, may reduce the need for visits, by providing remote monitoring. Applying innovations in ICT (mobile computing, social networking etc) to this market will be extremely important, especially given the emphasis placed on remaining in contact with family members (as mentioned in the interviews with industrial design staff.

The predominant medical model of disability and the emphasis on functional ergonomics was seen to have restricted the development of products which are aspirational, can empower and provide dignity. Very few procures were considered as having considered the user experience. For [7] this includes 'the entire set of effects that is elicited by the interaction between a user and a product, including the degree to which all our senses are gratified (aesthetic experience), the meanings we attach to the product (experience of meaning), and the feelings and emotions that are elicited (emotional experience).' Blythe et al. [2] commented that occupational therapists found that the products they recommended for patients were rejected because people did not want their homes looking like hospital clinics. With more people retaining independence into older age but requiring more care and AT there is a real need for developers to start designing products which suit the environment that people actually live in, not for hospitals or the ideal home so frequently depicted in idealisations of future scenarios. Thus, the consumer market and consumer business models may be a more appropriate context in which to design ALT products.

Interviews with designers and AgeUK representatives highlighted the need for products which cater for a range of disabilities—which are not constant over time. Some products reviewed did include end user configurability, but very few take a holistic view, catering for the needs of the whole person, rather than just one disability. For example, dexterity reduces as well as vision—in many applications fonts can be customised, but no provision is made for loss of sensitivity/dexterity of fingers in keyboard and mouse design.

The matrix can be used to start discussions and thoughts about the design of ATPs with junior designers and those new to the design of ATPs. The business models co-created can then support the introduction of these products into the consumer ALT market. Although many of the factors are ones that are relevant for any product design scenario, the development of ATPs for YOPs and their relatives requires a more detailed consideration of the actual usage context and the evolving needs of the end users. ATPs are designed not only to assist independent living, they should also support enhanced quality of life. Merely following the medical model, or transferring designs from hospital, public health and work settings into the home environment, not only fails to recognise the needs of the whole person but also the potential of design.

### *5.6.1 The Consumer Journey for ALT and Developed Business Models*

Products which do incorporate Jordan's [12] dimensions of emotional and hedonic benefits are more likely to be situated in the consumer market, rather than the statutory market which currently dominates ATP provision. Indeed, some of the companies interviewed for the development of the product matrix are already selling to private consumers, rather than bulk purchases to local authorities and health organisations. Historically, ATPs and services have often been provided by statutory agencies, however this is increasingly not the case, leaving increased numbers of people with preventative and acute care needs unmet by the state. For example, in the United Kingdom, there has been annual decreases in the number of people receiving assistive equipment and technologies from their local authority and the National Health service since 2010 [15]. Following feedback from the COMODAL user surveys and focus groups, there is indeed an appetite amongst older people to purchase products and services to support their independence and prevent further ill-health [24].

**Acknowledgements** This chapter describes findings from the Consumer MODels for Assisted Living project, led by Coventry University in partnership with AgeUK and Grandparents Plus, as part of the Assisted Living Innovation Platform and funded by the Technology Strategy Board (now Innovate UK).

## References

1. Baltes, P.B., Baltes, M.M. (eds.): Successful aging: perspectives from the behavioural sciences. University Press, Cambridge, MA (1990)
2. Blythe, M.A., Monk, A.F., Doughty, K.: Socially dependable design: the challenge of ageing populations for HCI. Interact. Comput. **17**, 672–689 (2005)
3. CEN 2003 [internet] CEN workshop agreement CWA 14661. Guidelines to standardisers of ICT products and services in the CEN ICT domain (cited 2010 Aug 12). Available from ftp://cenftp1.cenorm.be/PUBLIC/CWAs/e-Europe/DFA/cwa14661-00-2003-Feb.pdf
4. Coughlin, J.F., D'Ambrosio, L.A., Reimer, B., Pratt, M.R. (eds.): Engineering in Medicine and Biology Annual Conference. 'Older Adult's Perceptions of Smart Home Technologies: Implications for Research, Policy and Market Innovation in Healthcare', August 2007, Lyon, France. Online available at http://www.ncbi.nlm.nih.gov/pubmed/18002331 (2007). IEEE
5. Damodaran, L., Olphert, W.: User responses to assisted living technologies (ALTs)—a review of the literature. J. Integr. Care **18**(2), 25–32 (2010)

6. Commission, European: Active ageing and independent living services: the role of information and communication technology. Office for Official Publications of the European Communities, Luxembourg (2008)
7. Hekkert, P.: Design aesthetics: principles of pleasure in design. Psychol. Sci. **48**(2), 157–172 (2006)
8. Hersh, M.A., Johnson, M.A.: On modelling assistive technology systems part 1: modelling framework. Technol. Disability **20**(3), 193–215 (2008)
9. Hersh, M.A.: The design and evaluation of assistive technology products and devices part 1: design. International Encyclopedia of Rehabilitation, Center for International Rehabilitation Research Information and Exchange (CIRRIE) (2010)
10. Hersh, M.A.: The design and evaluation of assistive technology products and devices part 1: design. Available [online] from http://cirrie.buffalo.edu/encyclopedia/en/article/309/ (2011)
11. Holliday, N., Ward, G., Fielden, S.: Understanding younger older consumers' needs in a changing healthcare market—supporting and developing the consumer market for electronic assisted living technologies. Int. J. Consum. Stud. **39**, 305–315 (2015)
12. Jordan, P.W.: Displeasure and how to avoid it. In: Robertson, S. (ed.) Contemporary ergonomics. Taylor and Francis, London (1996)
13. Jordan, P.: Human factors for pleasure in product use. Appl. Ergon. **29**(1), 25–33 (1998)
14. King's Fund: Consultation meeting on assistive technology. FAST, London (2001)
15. Kings's Fund: Key challenges facing the adult social care sector in England. Kings Fund, London (2018)
16. Maslow, A.H.: Motivation and personality. Harper and Row, New York (1954)
17. Rowe, J.H., Kahn, R.L.: Successful aging. Dell, New York (1998)
18. Sanders, E., Stappers, P.: Co-creation and the new landscapes of design. CoDesign **4**(1), 5–18 (2008)
19. Steel, D.M., Gray, M.A.: Baby boomers' use and perception of recommended assistive technology: a systematic review. Disabil. Rehabil. Assist. Technol. **4**(3), 129–136 (2009)
20. Ward, G., Fielden, S., Muir, H., Holliday, N., Urwin, G.: Developing the assistive technology consumer market for people aged 50–70. Ageing Soc. **35**(5), 1050–1067 (2017)
21. Ward, G., Holliday, N., Awang, D., Harson, D.: Creative approaches to service design: using co-creation to develop a consumer-focused assistive technology service. Technol. Disability **27**, 5–15 (2015)
22. Ward, G., Holliday, N., Prothero, L., Woodcock, A.: COMODAL literature review. Available on request from authors (2011)
23. Ward, G., Holliday, N., Woodcock, A., Prothero, L.: Consumer MODels for assisted living (COMODAL)—work package 1: understanding users and the barriers and enablers to ALT. Available on request from authors (2011)
24. Ward, G., Ray, S.: Unlocking the potential of the younger older consumer: consumer preferences and the assisted living market. Available [online] at www.comodal.co.uk (2014)
25. Woodcock, A., Ward, G., Ray, S., Unwin, G., Osmond, J., Fielden, S.: Younger older consumers of assistive technology products. In: Contemporary Ergonomics (2013)

**Andree Woodcock** I have a B.Sc. in Psychology and Social Biology with an M.Sc. in Ergonomics from UCL. My Ph.D. concerned an investigation of the use of ergonomics in automotive concept design acquired from Loughborough University while I was a Daphne Jackson Research Fellow. Most of my research has been located within design departments, looking at applying user-centred design to wicked problems in the areas of health, education, regeneration and transportation. Health related grants have included design of polysensory school environments for children with ASD, the development of a patient held record system (MyCare), a user testing toolkit for SMEs (IDEAL-ALIP project) and a decision support system for SMEs moving into the assistive technology market (COMODAL). Transport related research has included looking at the in car safety and security of female car occupants, and the leadership 3 Eu projects: FP7 METPEX developing a Pan European tool to measure the quality of the passenger experience; H2020CIVITAS SUITS looking at supporting capacity in small–medium local authorities to develop sustainable transport measures and H2020 TInnGO project looking at gender inequality in transport.

**Dr. Jane Osmond** is a Research Fellow for the Research Institute: Transport and Future Cities at Coventry University. Jane's research focuses on equality, particularly equality of access to transport and also higher education. Jane is involved in research on two EU projects which are considering the relationship between gender/diversity and the use of transport.

Focusing on equality of access to higher education, Jane has been researching the impact of harassment and hate incidents (including sexual assault) on students. She has developed two online modules: one for staff in relation to taking disclosures from students, and one for students that covers bystander training. She has also been instrumental in the development of online reporting system, to allow students (and staff) to report incidents.

Jane's voluntary work includes being a volunteer director for C.O.V. (www.creativeoptimisticvisions.co.uk) a social enterprise that uses Protective Behaviours to help less advantaged people, including those who are excluded from school and women in prison.

**Nikki Holliday** After receiving my B.Sc. (Hons) Psychology at the University of Leeds, and completing a Postgraduate Certificate in Social Science Research Methods at the Open University, I began working as a researcher at the Health Design and Technology Institute (HDTI) whilst completing an M.Sc. in Assistive Technology. I am currently a Design Manager at the Centre for Intelligent Healthcare at Coventry University. I specialise in managing and informing technology product and service design through user-centred techniques such as co-creation, co-production, usability studies, focus groups and interviews, and utilising rapid prototyping to support data collection and idea exploration with users. I have extensive research experience with service users, carers and a wide range of health and social care staff. I have experience of working on nationally funded projects such as the UK Technology Strategy Board (now Innovate UK), the NHS, and the Ministry of Defence. I am interested in projects which utilise the possibilities of existing and future technologies to address the "wicked" problems facing us today, such as the ageing population and increasing pressure on health and social care systems. I am passionate about the interplay of research and design, which can be used to create technologies that are not only functional, but also desirable to those that may benefit from them.

# Chapter 6
# Assistive Technology (AT) Usability and Adoption: Future Drivers

**Avril D. McCarthy and Louise Moody**

**Abstract** The aim of this chapter is to demonstrate the importance of usability in the development, adoption and effectiveness of assistive technologies (AT). Usability engineering as well as related approaches, including human factors, inclusive design and acceptability are explained. The importance of considering device usability is highlighted through examples of adverse events that have resulted from issues associated with medical device design. The chapter goes on to introduce key medical device regulatory frameworks and standards. These have been produced to guide developers of medical devices (including applicable assistive technology) to improve user acceptance, device usability and safety. Usability engineering methods are listed with examples given of how these have been applied in the development of assistive technology devices within the Devices for Dignity portfolio of projects.

**Keywords** Usability · Assistive technology · Ageing · Safety · Adoption · Market failure · Medical device · Regulations and standards · User acceptance

## 6.1 Introduction

The UK Department of Health released its '*NHS 10 Year Plan*' in January 2019 [5]. The plan recognizes the ageing demographic and the need for technology to assist people to live at home for longer. The World Health Organization (WHO) defines

A. D. McCarthy (✉)
Clinical Engineering, Sheffield Teaching Hospitals NHS Foundation Trust, Glossop Road, Sheffield S10 2JF, UK
e-mail: avril.mccarthy@nhs.net

L. Moody
Centre for Arts, Memory and Communities, Faculty of Arts and Humanities, Coventry University, Coventry, UK
e-mail: l.moody@coventry.ac.uk

A. D. McCarthy · L. Moody
NIHR Devices for Dignity MedTech Co-Operative, Sheffield Teaching Hospitals NHS Foundation Trust, Glossop Road, Sheffield S10 2JF, UK



A. Woodcock et al. (eds.), *Design of Assistive Technology for Ageing Populations*, Intelligent Systems Reference Library 167,
https://doi.org/10.1007/978-3-030-26292-1_6

assistive technology (AT) as '*any external product whose primary purpose is to maintain or improve an individual's functioning and independence and thereby promote his or her well-being*'. The WHO International Classification of Functioning, Disability and Health (ICF) acknowledges the importance of assessing functionality across the lifespan [38] and measures health and disability at both individual and population levels [42, 18]. With an ageing global population and a rise in non-communicable diseases, the WHO predicts that more than 2 billion people will need at least one assistive product by 2030, with many older adults needing two or more [41]. Ageing does not necessarily equate to, or only apply to, senior citizens. People of all ages experience the ageing process and each will have their own personal experience of the functional impairments that it can bring. Children with disabilities can be affected by maturation transitions in a similar way to the functional changes that affect an older person as they age. Similarly, the effects of obesity and lack of exercise mean that for an increasing number of adults, functional ageing is experienced earlier. Obesity leads to increased rates of diabetes and kidney disease, with associated impairments potentially including impaired vision, peripheral sensing, dexterity, poor mobility and even limb loss. As a result assistive technologies, such as mobility aids, are in greater demand at an earlier age.

A '*World Report on Disability*' [40] empirically demonstrates the unmet global need for assistive technology (and disproportionately so for older people). Many who could benefit, have little, or no access to AT. An international report on improving access to [39] identified the following five challenges:

1. Research and Development: more R&D is needed on basic, low-cost technology products to complement research in high-end technologies such as robotics.
2. Standards and regulation: there is a need for countries to adopt suitable regulatory mechanisms to ensure that assistive products on the market meet the relevant standards and are safe and effective.
3. Manufacturing: quality, quantity and sustainability of AT manufacturing. The lack of context-appropriate product design was noted, where parts were not replaceable locally e.g. hearing aid batteries and wheelchair tyres, where a lack of available spares can lead to abandonment.
4. Selection, pricing and reimbursement: Donations of low-quality, or used AT products causes issues where repair, maintenance or replacement is not possible locally.
5. Procurement and supply: Large scale or national procurement to gain economies of scale discounting is often lacking, as are waivers on import duties to reduce costs.

Where AT is available, users abandon a startling high proportion of assistive products. Estimates run as high as 75%, with hearing aids as common example of a device with high rejection rates [30, 31]. It is argued that this can be addressed through better design and consideration of human factors issues.

### 6.1.1 Improving the Acceptability of AT

Abandonment of available AT is often related to the user acceptance of the product. By acceptability, we mean '*the user satisfaction and willingness to use the product*' [27]. The technology acceptance model [6] was developed for predicting user acceptance of technology and considers a range of factors including usability. Usability is the extent to which something is able or fit to be used [28]. A research project dating from 2005 [14], investigated the acceptability of AT to older people through in-depth interviews of 67 AT users over the age of 70 years. The study found a complex model of acceptability, in which a 'felt need' for assistance combines with 'product quality', whereby the AT is perceived as working properly, reliably and safely. Examples were given of stair-lifts that had broken down, shower chairs that felt slippery, overly sensitive or poorly located, smoke alarms that triggered in response to overdone toast and wobbly grab rails, as reasons for users declining to use AT. Regrettably, over ten years later, there is a still a high proportion of AT that is not accepted and is abandoned by users.

Within the medical device sector, design has typically considered the primary device user (often a healthcare professional) or the primary disability or condition (if intended for a lay user), and not necessarily the range of users or functional limitations and environments of use that impact effective device usability. For example, a design team for a powered wheelchair may consider stability and postural support but may omit to consider related issues such as continence or cognitive load required to use it. Understanding the wider human factors issues and the relationship between product use 'demand' (especially key combinations of capability loading) required for effective use and device exclusion are important [37]. The cognitive and emotional response to a product are also important. For example a product may be usable, but may make the user feel stigmatized leading to rejection. It is crucial that the development methods adopted take into account the needs and wants of the end user.

The National Institution for Health Research (NIHR) in the UK has recently published a Themed Review report on the use of AT for older people, particularly those with complex conditions [13]. A key question it addresses to commissioners of AT is:

- '*Have our technology solutions involved service users in their testing and design*?'

While to older people and carers, it asks:

- *Which tasks and activities do I find difficult and which are most important to me*?
- *Which technologies might help*?

Furthermore a White Paper from the British Standards Institution [BSI] gives guidance on engaging stakeholders in the home medical device market and reinforces that it is critical for device safety and effectiveness, particularly for complex or unfamiliar technology [2].

Inclusive design (sometimes referred to as universal design) has aimed to consider population diversity in the design of consumer products to permit effective use by people across a diverse and broad a range as possible of functional abilities and limitations. Good design, that takes account of diverse needs can result in products that work better for everyone. A commercial benefit can therefore be to extend the addressable (available) market size. An example is of US company 'Oxo' that created its 'Good Grips' kitchen utensils in response to users with arthritis struggling to use the conventional design of a potato peelers Its success in improving utensil usability through better ergonomic handles and materials selection resulted in huge commercial success. For designers wishing to aspire to Oxo's mantra of '*Making everyday life easier*', colleagues at the University of Cambridge in the UK, have created an '*Inclusive Design Toolkit*' [36] including 'capability loss' simulation tools to help designers experience common functional limitations that older people experience such as poorer vision and hearing, limited reach and dexterity and impaired mobility.

### *6.1.2 Human Factors and Usability Engineering*

There are a number of related approaches and processes that encourage an increased focus on user needs, requirements, capability and limitations including human factors, ergonomics, usability, usability engineering. The Medicines and Healthcare products Regulatory Agency in the UK (MHRA) who offer guidance on the definition and safe use of assistive technology refer to '*human factors*' as how a person will interact with the systems surrounding them, including the technology they use. They define Human factors as:

> the scientific discipline concerned with the understanding of interactions among humans and other elements of a system, and the profession that applies theory, principles, data and methods to design in order to optimize human well-being and overall system performance. (International Ergonomics Association)

Human factors draw on knowledge from a broad range of subjects including psychology, engineering, and physiology in order to guide the design process and ensure that solutions meet the needs, requirements, capabilities and limitations of the user. As a discipline, it is well-developed having been applied in high-hazard industries such as defence, nuclear, petrochemical and aviation for some time. The impact on reducing error, minimizing risk and capitalizing on technological development has been effectively demonstrated and recognized by international regulators such as the Federal Aviation Administration (FAA) in the USA and its equivalent, the European Union Aviation Safety Agency in the EU.

The importance of human factors in healthcare is increasingly recognized, especially within safety-critical clinical settings that require effective teamwork, such as in operating theatres [26]. There has been an international drive to embed the approach throughout healthcare systems with human fallibility recognized as a key factor in patient safety errors since 2000 in the USA [4]. More recently, applying human factors approaches in the design of healthcare workplaces and practices work is being advocated through NHS policy [26]. Emerging standards and guidance in this area are an important development.

The application of human factors aims to ensure products and systems can be effectively and safely used. The multidisciplinary approach takes into account a range of features of the intended user, such as their age, size, physical and cognitive capabilities and training. Human Factors also takes a systems approach considering the wider context of use and where and how the product will be used, and how it will integrate within the environment or wider system. The design of a piece of assistive technology for example may need to function within a hospital environment, within a mode of transport as well as the home environment. It will also need to take account different scenarios of use, for example the level of training of the user, competing distractions, use in different settings etc.

There are many and complex human factors and usability issues that can lead to a lack of acceptance and abandonment of AT that should be considered through the design process. The potential effect of users choosing not to, or being unable to use a product correctly is potentially very serious. The MHRA define the process of achieving usable products that address user needs and fit with their practices as '*usability engineering*' [16]. Usability engineering has been increasingly recognized as a key driver and approach in the design of safe and effective medical devices and will be discussed further in respect to the standards and methods of application. In the next section, we demonstrate the impact poor design and neglect of human factors and usability can have through adverse events associated with medical devices.

## 6.2 Adverse Events Due to Human Factors and Usability Issues

Within the UK, the MHRA is the body responsible for monitoring and investigating adverse incidents involving medical devices, including assistive technologies. In 2012, assistive technologies were the third most commonly implicated device type, with approximately 1400 adverse events reported (behind surgical and patient monitoring and vascular fluid management devices). Examples are provided from cardiovascular assistive devices, to therapeutic oxygen, to more typical assistive technologies such as wheelchairs. The examples seek to highlight the importance of a considering a wider of context of use during the development process.

### *6.2.1 Self-management of Diabetes Alert*

One such example from December 2014, affected small numbers of patients self-managing their diabetes using an ambulatory insulin pump and led to one incident of severe hypoglycemia due to incorrect selection and delivery of bolus amount

[15]. The root cause was the design of the user interface. There were two different effects of a down arrow button press, contingent on which menu the button was accessed. If accessed via the 'Express Bolus' button, the down arrow would scroll to zero units and stop. Whereas within the main menu, one button press of the down arrow, enabled the user to change the bolus dose inadvertently from zero units to the maximum programmed—by default a bolus setting of ten units. The risk was exacerbated in situations where the default had not been programmed to the individual's insulin needs in advance by a healthcare professional. The alert notice flagged the importance of updating the maximal bolus value, and of making the users aware of the potential for use error.

### 6.2.2 Hoists and Wheelchairs

A search of the MHRA database using the codes BHSE 08/00 and 08/03 (fall, fell, drop, slip) relating to patient hoists, revealed that 78 incidents had been reported to the MHRA in the period January 2011 to December 2014. The incidents resulted in three deaths and nine severe injuries [24]. The review led to a *Stage One: Warning Risk of death and serious harm by falling from hoists* being sent to all providers of NHS funded care [24] and publication of a Health and Safety Executive safety notice directed to users of self-managed ceiling hoists [8]. The analysis identified five root causes for the incidents, the first four of which were primarily human factors/usability engineering issues, namely:

1. Failure to follow correct manual handling procedures including instructions on number of carers required to perform manoeuvre.
2. Inadequate assessment of service user ability/disability and selection of equipment.
3. Sling straps incorrectly fitted to hoist and/or the person being lifted.
4. Wrong size or type of sling used.
5. Unclear responsibility for equipment maintenance.

Searching the FDA MAUDE database under product class '*Wheelchair, powered*' and Event Type '*Death*' showed that there were 14 deaths associated with powered wheelchair use between 01/Jan/2018 and 31/Dec/2018, of which 7 were directly related to use or user error resulting in falls, collisions or burns. These were related to either use of the wheelchair in an inappropriate environment (e.g. at height or confined areas), lack of use of safety belts, or unqualified modifications leading to an electrical fire. A further search of MAUDE using the terms '*Use of Device Problem*', and Product Class '*Wheelchair, powered*' and '*Wheelchair, mechanical*' revealed 5 further incidents that resulted in injury (3 for powered, and 2 for mechanical wheelchairs. All incidents could be attributed to user error, although in one fall from a powered wheelchair, the root cause appeared to be poor accommodation of the user's high muscle tone that meant the lap belt was unsuitable in use, implying inappropriate prescription of the wheelchair.

### 6.2.3 Patient Safety Alerts

The NHS manages the *National Reporting and Learning System (NRLS)*, a central database of critical incident reports in order to improve patient safety. It writes and publishes *Patient Safety Alerts,* some of which are produced jointly with the MHRA. Examples include inadvertent misuse of non-invasive ventilation equipment leading to three patient deaths. The warning highlighted patients bringing their own device from home into a hospital environment as a particular concern if staff had not been trained in its use [25]. This reflects the importance of considering the range of environments in which a device will be used as part of the risk management process, both from the manufacturer's stance in developing instructions for use and device labeling and for the institutions in which devices are introduced and used. Other medical device alerts published by the MHRA arise as a result of field safety notices issued by manufacturers where a trend of device incidents, or near misses emerges from post market surveillance. An example is of a spreader bar on a patient hoist that could potentially detach, dependent on the range of motion required during usage, increasing the risk of a fall during patient transfer [17]. Such alerts are intended to warn users of the devices already on the market and to highlight risk mitigation solutions. In more severe instances, the mitigation may be to apply a modification in the field (e.g. part replacement or software update), or devices may be recalled by the manufacturer.

### 6.2.4 Events Highlighted by MAUDE

Within the U.S.A., the U.S. Food and Drug Administration (FDA) manages the '*Manufacturer and User Facility Device Experience*' *(*MAUDE) database. A search on between 01 January 2018 to 31 December 2018 using the following: search terms 'Product Problems: '*Use of Device Problem*' and Event Type '*Death*', revealed 53 deaths in the USA associated specifically with medical device use problems rather than any of the many other reasons.

To differentiate two similar terms: '*use error*' and '*user error*'. Both terms relate to mistakes that occur during the usage of technologies. The former implies that the mistake could be attributed to poor design or human error while the latter *implies* that the mistake is the fault of the user. As might be expected, the majority of use error related deaths were associated with high risk devices such as implantable cardiovascular (n = 30) or orthopedic devices (n = 5). One related to a mixing error of dialysate for a hemodialysis machine, and another to accidental misuse of an implanted medicine delivery pump. The pump failure resulted in a patient with tetraplegia not receiving anti-spasticity treatment, which although subsequently addressed, resulted in the patient's death shortly after.

Of the cardiovascular devices, the most commonly implicated were left ventricular assist devices (LVAD) where the root cause of failure was cited by manufacturers as user error. At first glance, this type of device would not be considered as a piece of assistive technology. However, its safe usage delivers significant quality of life and independence to users with heart failure, and requires daily user involvement and interaction (as opposed to a 'fit and forget' device). The users, generally older and often with multiple-morbidities, including confusion, failed to connect their devices to a power source, with subsequent device failure resulting in death.

In 2011, a clinical publication by a professor of biomedical engineering acknowledged that the most important factor in LVAD adoption by patients was overwhelmingly the quality of life improvement offered. The author also advocated the importance of human factors in the design of reliable LVAD systems, particularly with regard to ergonomics of peripherals, reduced power requirements, robust feedback-control, and reduced invasiveness [1]. By 2017, academic bioengineers working in a heart and vascular institute reviewed malfunctions (device not working or not working correctly) in 200 rotary LVADs implanted at a single center and highlighted that >60% of LVAD system malfunctions were because of batteries, controllers, and peripheral cables [11]. The authors recommended enhanced usability testing beyond simple reliability testing, noting that bench testing could under-estimate the required robustness required for at-home use by lay-users. The importance of patient and lay-caregiver education in proper device component care was also emphasized.

The database also highlights issues with AT which require the patient or a lay carer to manage them at home, in particular compliance with instructions for use and adoption of safe behaviors. Examples include three deaths related to using oxygen from a cylinder or via an oxygen concentrator at home; two involved patients smoking; the inevitable explosions, burns and other injuries ultimately led to their deaths. Another death related to a fall from an active patient lift system (intended as a standing a raising aid) in a nursing home. Although the manufacturer had made it clear that the use of the equipment for people with poor postural management was not appropriate, the partially paralyzed patient had refused to accept the use of any alternative lifting equipment such as passive hoist equipment. These cases highlight the need to design AT that is acceptable to users to reduce the risk of inappropriate use.

Even if a device is used as intended, limited consideration of usability engineering can lead to unintended consequences as shown in the following two examples involving powered wheelchairs. The first, an anonymized US example of a powered wheelchair incident reported in October 2018, highlights this:

**Report logged on the FDA MAUDE**

- Device Problem Use of Device Problem
- Event Date 09/01/2018
- Event Type Injury.

**Event Description**

*'The consumer was coming out of the bathroom and bumped the touch screen, changing the drive to outdoor mode. She didn't notice the change of drive modes, so the chair drove faster than expected into the wall. The consumer had to go to the emergency room (er), receiving an X-ray and stitches to her ankle. She was admitted overnight for evaluation after the incident.'*

**Manufacturer Narrative**

- *'There was no malfunction of the device. The system design allows the end-user to change drive modes with the touch screen while driving. The end user unintentionally changed drive modes while indoors, causing the chair to go faster than expected in tight surroundings. The touch screen can be disabled in the setting menu to prevent use while in a selected drive.*
- *Per the end users request, the dealer is changing the joystick to a non-touch screen option.'*

While it is true to say that there was no malfunction, it is predictable that a user could accidently bump a touch screen while the chair was in use. Rather than relying on a dealer changing the chair to a non-touch screen option, it is reasonable to expect that the designers would have engaged with a range of potential users to uncover potential use risks before committing to the final design.

A second incident involved a powered wheelchair from a different manufacturer.

**Report logged on the FDA MAUDE**

- Device Problems Use of Device Problem; Improper or Incorrect Procedure or Method
- Event Date 04/10/2018
- Event Type Injury.

**Event Description**

*'On 19-April-2018, the manufacturer was notified by the importer that one of its dealers had been notified of an incident that took place on (b)(6) 2018.*
*The dealer reported that the end user's left foot slid off from the back-left corner of the footplate. Since he has no sensation in the lower extremity of his body, he continued to drive until he heard a snap, that he saw, it was on his ankle.*
*As a result of this incident, there were 2 broken bones in left ankle.*
*Power center mount front rigging with heel straps were used in the system as the end user had a history of legs issues on wheel chair.*
*In addition to that, the dealer installed calf straps to the calf pad for the safety of the end user to avoid any adverse event. The dealer stated that the calf*

*straps were not in use when this incident happened and the system did not malfunction.'*

**Manufacturer Narrative**

- *'This system was ordered by the dealer ((b)(4)), according to the end user specification. Front rigging of the system was changed from 70° swing away hangers, which he was using in his old chair, to power center mount with heel straps as there were leg issues for the end user in the past. Dealer also added calf straps to the calf pads to avoid incidents like this.*
- *Unfortunately, the end user wasn't using the calf straps and 'it seems that his feet were not properly supported by the heel straps at the time of event'.*
- *The dealer has also reported that there was no malfunction of the chair.*
- *It is noted that the motion concepts user manual provided with this system, includes the following warning:*
  - ***'before operating your wheelchair, always ensure footplates/platforms are in their full down position and your feet are properly supported by the heel straps'***
- *However, in this case, this precaution was not followed by the end user.'*

In the second scenario, the manufacturer is quick to deflect the causation of the device incident to user error, primarily the failure to use the calf straps. While it is not possible to be sure of the detail of this case, one senses that the root cause was not quite so simplistic. Clearly, the wheelchair user suffered from loss of sensation from the lower limbs and had experienced issues with legs not locating on the foot rests previously so was at higher risk of a reoccurrence. Despite the user not requesting calf straps, the dealer acknowledged the risk by including them to mitigate the risk of the legs coming off the powered foot mount. However, it is not apparent whether this change to the user specification was achieved in consultation with the user, or whether the user was capable of using the calf straps. Whatever the actual cause of the incident—accidental or failure to engage with the design—this type of incident highlights the importance of assessing that an assistive technology is suitable and is set-up appropriately for an individual to use. Ideally, they should have had received information appropriate to their needs of how to operate the AT safely and effectively before the AT is made available for their use.

The impact of increasing numbers of people living with multiple and often complex conditions, coupled with device failures resulting in death or injury associated with use errors, has led to a regulatory drive for proactive consideration of usability during design—rather than being in reaction to device failures [34, 3]. More effective summative testing with typical device users in the development stages may have enabled earlier identification and avoidance of the examples discussed above.

## 6.3 Regulation and Standards

A WHO report [39] cites standards and regulation as a challenge for assistive technology, noting the lack of adequate regulation and oversight globally and highlighting the need for countries to adopt regulatory mechanisms to ensure that assistive products are safe, effective and legal to be on the market. The WHO further identified a need for quality and safety standards that are appropriate for diverse settings, especially rural environments and not only high-income settings. Of the top 50 global AT priorities highlighted by the WHO in 2018 [41], some 38% clearly also qualify as medical devices (e.g. wheelchairs, prescribed visual and hearing aids, prosthetics and orthoses), with a further 12% falling into a 'borderline' medical device category (e.g. some types of communication aids). In countries that regulate medical devices, adherence to regulatory processes and application of appropriate standards is critical to market adoption. For those ~50, mostly low-income countries that do not have regulatory frameworks for medical device approval [32], users who access AT are at higher risk from unsafe products and systems. Irrespective of what is mandated, users are more likely to have a good experience of an AT product—one that is safe and meets their needs—if AT developers employ the guiding principles of the usability standards.

### *6.3.1 Medical Devices and Key Regulatory Changes*

As noted above, many types of AT will also meet the definition of a medical device. Medical devices are highly regulated products. Within the EU the definition has recently been extended by the Medical Device Regulation (MDR) (Article 2:1 MDR) [7] (in effect from 26th May 2020). The emboldened text below indicates changes from the previous EU Medical Device Directive (MDD):

*'medical device' means any instrument, apparatus, appliance, software,* ***implant, reagent****, material or other article intended by the manufacturer to be used, alone or in combination, for human beings for one or more of the following* ***specific medical purposes****:*

- *diagnosis, prevention, monitoring, prediction, prognosis, treatment or alleviation of disease,*
- *diagnosis, monitoring, treatment, alleviation of, or compensation for, an injury or* ***disability****,*
- *investigation, replacement or modification of the anatomy or of a physiological* ***or pathological*** *process* ***or state****,*
- ***providing information by means of* in vitro *examination of specimens derived from the human body, including organ, blood and tissue donations****,*

*and which does not achieve its principal intended action by pharmacological, immunological or metabolic means, in or on the human body, but which may be assisted in its function by such means…*

As part of the update of the MDD [7], the European Union has responded to the increasing prevalence of healthcare Apps and other digital health technology by explicitly implementing a new classification rule (11) for software that must be considered when applying the definition to general medical devices (Annex VIII of the MDR: Section 6.3. Rule 11).

There are core international standards that are expected to be applied *en route* to gaining regulatory approval as a medical device. One of the fundamental standards is ISO 14971:2007 (EN 14971:2012 in the EU) that encompasses the application of risk management to medical devices. It requires developers to identify hazards, estimate and evaluate the associated risks, to control the residual risks, and to monitor the effectiveness of the controls. The overarching principle (Clause 6.5.2 of ISO 14971) is to minimize harm by employing the following safety principles as prioritized in the order below:

(a) *inherent SAFETY by design;*
(b) *protective measures in the MEDICAL DEVICE itself or in the manufacturing PROCESS; and*
(c) *information for SAFETY.*

Developers of medical electrical equipment have a plethora of parts to consider of the IEC/EN 60601 series standard intended to ensure safe usage. Collateral standard Part 1–6 of IEC/EN 60601 specifies a process for a manufacturer to analyze, specify, design, verify and validate usability, as it relates to basic safety and essential performance of medical electrical equipment, but only for normal use. For medical device software, IEC/EN 62304 (software lifecycle management) was introduced in 2006 to complement ISO 14971 in managing software-specific risk. Its release acknowledges that software is often an integral part of medical device technology and also that software can qualify as a medical device in its own right.

### *6.3.2 Application of Usability Engineering to Medical Devices (ISO/IEC 62366)*

In addition to IEC 60601:1-6 for medical electrical equipment, the international standard **IEC62366M—Application of usability engineering to medical devices** was introduced in 2007 and updated in 2015 [3]. The standard along with supplementary guidance [9] seeks to specifically address human factors issues and usability for the development of medical devices:

- IEC 62366-1:2015 Part 1: Application of usability engineering to medical devices and,
- IEC/TR 62366-2:2 Part 2: Guidance on the application of usability engineering to medical devices.

The key differences between the 2007 and 2015 usability standards are discussed further elsewhere [35]. The update was initially prompted by safety incidents involving anesthesia equipment where the root causes were use errors by qualified healthcare professionals, and latterly incidents with infusion pumps. In a five-year period the FDA received over 56,000 reports of adverse events associated with infusion pumps leading to numerous injuries and deaths [33]. The update reflects the importance of designing not only to meet safety requirements but also to design in usability across the range of potential users. It also presents a format for conducting, and then documenting the use engineering process in a way acceptable to the international medical device regulators, including the US Food and Drug Administration (FDA) in the United States and the European Commission's Medical Device Regulation (MDR) within Europe. A BSI White Paper gives guidance on the updated regulatory requirements for medical device usability within the EU [29].

The supplementary Part 2 guidance to the 62366 standard [9] provides practical advice and examples on how to apply the standard. It is not intended to be prescriptive, but to be advisory for developers and notified bodies, recognizing the importance of human factors in managing patient safety. Within the standard, usability is defined by the IEC 62366 standard as being: '*the desirable end-product of applying USABILITY ENGINEERING from the beginning and throughout the MEDICAL DEVICE design PROCESS*'. (Definition 3.17; BS EN 62366-1:2015). The standard makes the important point and reinforces that '*USABILITY does not normally arise just from the well-intentioned application of common sense in design*'. Taking information from the guidance an overview of the usability engineering process and stages is provided in Fig. 6.1.

The guidance details 21 different usability engineering methodologies which are mapped to the clauses of the standard with examples of which method to use and when. In addition to the detailed formal guidance document, the MHRA has created a succinct overview guidance document [16] for the interpretation of the standard. The MHRA's document includes a helpful table with a subset of 14 of the 21 key usability methodologies cited in the Part 2 guidance, highlighting where they are best suited.

### 6.3.3 Applying the Standard: Examples from Devices for Dignity

Devices for Dignity (D4D) is a national MedTech and In vitro diagnostic Co-operative (MIC) funded by the National Institute for Health Research (NIHR). D4D acts as a catalyst within the UK's NHS for the development of new medical devices and healthcare technologies that promote dignity and independence for people with one or more chronic health conditions. Devices for Dignity specializes in the development of AT and has sought to apply a design approach and a range of methods for ensuring device usability. D4D also has a crossing-cutting MedTech Theme that advises on quality, risk management and regulatory requirements to support medical device projects within its portfolio. Whilst more detailed explanation of human

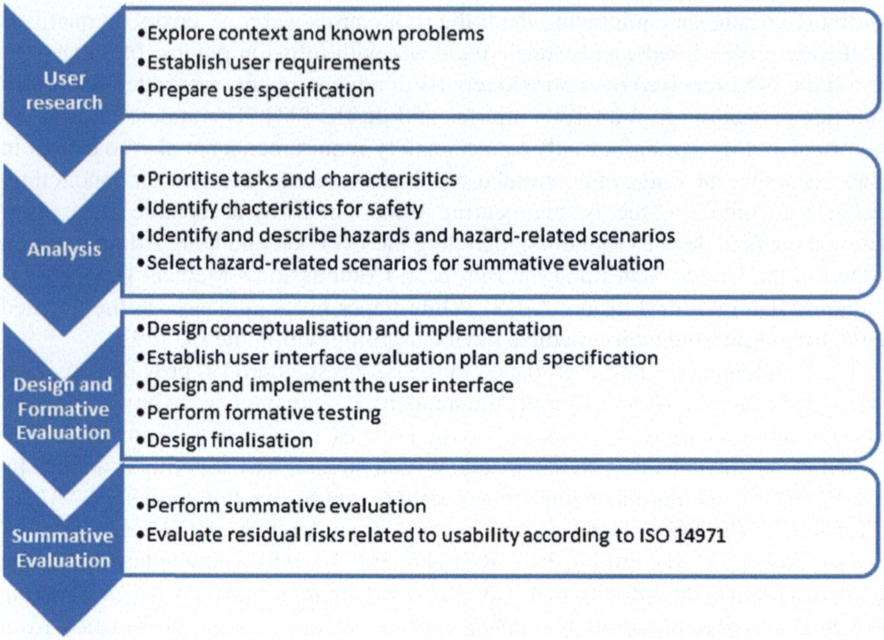


**Fig. 6.1** Illustrative usability engineering process and stages. Adapted from content in IEC TR 62366-2:2016, p 26 [9]

factors methods and their application is available [33] it is useful to see how the methods have been employed in the development of medical technology and AT. Table 6.1 lists the methods detailed in both ISO/IEC 62366 parts 1 and 2. The methods are mapped to the stages of usability engineering and illustrative examples are provided within D4D's project portfolio. Two D4D project examples in Table 6.1 use interviews and questionnaires. For some types of assistive technology, the user may find it difficult or time-consuming functionally, or challenging emotionally, to share their views via techniques that rely on observation or group participation. In an example of the former, 18 users of a range of commercial voice output communication aids (VOCAs) were interviewed one-to-one while 43 other VOCA users chose to respond to a questionnaire to elicit their perceptions of VOCA devices. By matching the method to the communication constraints and preferences of the user, important design deficits were uncovered. The process informed three areas for future design attention: specific aspects of VOCA device design; the consideration of the wider picture around the person; and the personal context in which someone uses their device [10]. As an example of the latter, incontinence is an factor in people becoming socially isolated and incontinence rates increase in the ageing population. Another project sought to ask users of leg-worn urine drainage bags about their experiences to inform future design and again one-to-one interviewing, either in person or by telephone, was offered. Many of the experiences shared were highly personal,

**Table 6.1** Suggested use of usability engineering and design methods (in blue) with illustrative examples of application in D4D projects

| Method / Technique | Usability Engineering and Design stages: User research | Analysis | Design conceptualisation | Design implementation | Formative evaluation | Design finalisation | Summative Evaluation | Post-production analysis | Illustrative examples of use from D4D project examples |
|---|---|---|---|---|---|---|---|---|---|
| Advisory panel reviews | ■ | ■ | ■ | ■ | ■ | ■ | ■ | ■ | All of D4D's projects make use of advisory panels reviews |
| Brainstorm use scenarios | | ■ | ■ | | ■ | | | | Helpful when devices may be used in different environments such as hospital or home-based dialysis with a range of carers (including self-care) to assess likely impacts on technology usage |
| Cognitive walkthrough | ■ | | ■ | | ■ | | | ■ | Employed within the evaluation of a healthcare App for recording lower urinary tract symptoms to support diagnosis |
| Contextual enquiry | ■ | ■ | ■ | | | | | ■ | Employed within the development of the HeadUp cervical orthosis with a combination of semi-structured interviews and video /diary of people with MND managing with their current orthoses in their daily lives |
| Day in the life analysis | ■ | ■ | ■ | | | | | | Used in co-design workshops (e.g. in unmet needs workshops). Participants have mapped out a 'day in the life' to identify and prioirtise needs. |
| Expert reviews | | | ■ | ■ | ■ | ■ | ■ | | All projects involve clinical and expert patient input to guide the design and evaluation of device development projects |
| Failure Modes and Effects Analysis (FMEA) | ■ | ■ | ■ | ■ | ■ | ■ | ■ | ■ | Typically this is used by manufacturers in developing new products (employing risk analysis matrices as per ISO 14971) and to investigate incidents involving an existing product. Used in the design of a complex electrical stimulation device for rehabilitation |
| Fault Tree Analysis(FTA) | ■ | ■ | ■ | ■ | ■ | ■ | ■ | ■ | Used in the design of the DAFNE Type 1 diabetes educational website to ensure that expected and correct flags were presented to users |
| Focus groups | ■ | ■ | ■ | ■ | ■ | ■ | | | Prosthetics sockets and limbs for children. The D4D-managed project 'Starworks' highlighted that children are rarely involved in stating or prioritising their own needs. D4D undertook 4 focus groups across the UK with children using limb prosthetics and their parents to address the failures: https://devicesfordignity.org.uk/starworks_cp/ |
| Functional Analysis | ■ | ■ | ■ | ■ | ■ | | | ■ | Testing to specified standards e.g. retention force of a urinary catheter balloon (BS EN 1616:1997) |
| Heuristic analysis | ■ | | ■ | | ■ | ■ | | ■ | To identify usability issues with a new urinary drainage catheter design |
| Interviews | ■ | ■ | ■ | ■ | ■ | ■ | ■ | ■ | To identify the needs of users of leg worn urinary drainage bags: [19] Moody & McCarthy, 2015 |
| Observation | ■ | ■ | ■ | | ■ | | ■ | ■ | Observation of daily routines has been used to identify needs leading to a storyboard of challenges to be addressed by designers. Clinicians also inform D4D of new unmet needs based on observations from clinical practice. |
| Participatory design | ■ | | ■ | | ■ | | | | Used to co-create new products such as a cervical (neck) orthosis for people living with Motor Neurone Disease (MND) http://eprints.whiterose.ac.uk/94616/16/WRRO_94616.pdf |
| Perception, Cognition, Action (PCA) analysis | ■ | ■ | ■ | | ■ | | ■ | ■ | Used when trying to ascertain causes of error - often in complex device such as insulin pumps or in assessing the likely factors associated complex issues e.g. medication adherence |
| Personas | ■ | ■ | ■ | | ■ | | | | Urinal device for women with functional incontinence. Personas were used to aid designers in considering the range of issues that women might face and also to de-personalise the issue for focus group participants. |
| Questionnaires / surveys | ■ | | ■ | | ■ | | ■ | ■ | A Urinary catheter survey to prioritise catheter-related issues to determine which projects to select.<br>A dignity survey used to identify unmet needs across the D4D clinical themes. |
| Scenarios | ■ | ■ | ■ | | ■ | | | | Used to guide the development of a discrete device for measuring urine output at home |
| Simulation | ■ | ■ | ■ | ■ | ■ | | ■ | | Use of an Advanced Catheterization Training simulator to test a new catheter design |
| Standards reviews | | | ■ | ■ | ■ | ■ | ■ | | Used throughout D4D device development projects |
| Task Analysis | ■ | ■ | ■ | ■ | ■ | ■ | ■ | ■ | Detailed task analysis of a new compact hemodialysis device (SC+) intended for home use (https://onlinelibrary.wiley.com/doi/full/10.1111/hdi.12757). In other projects task analyse has been used to describe posture, movements, senses employed, consideration of whether a tool was being used as intended. |
| Think-aloud | ■ | | | | ■ | ■ | | ■ | Early prototype of bladder diary App initially with clinicians in the design phase |
| Time and motion studies | ■ | ■ | ■ | ■ | ■ | | | | This is not a method that D4D has employed to date. |
| User testing / usability testing | ■ | | | | ■ | | ■ | | Undertaken with clinical staff to identify potential usability issues with a new catheter design using an anatomical catheterisation training model |
| Working with existing sources | ■ | ■ | | | | | | | All projects use existing sources to understand needs and practices |
| Workload assessment | ■ | ■ | ■ | ■ | ■ | | | | This is not a method that D4D has employed to date. |

including tales of embarrassing device failures (relating to poor usability) or of the impact of the technology on personal relationships [19]. The examples demonstrate that reaching those users more likely to experience usability issues, takes effort and careful consideration of the most appropriate methodology with which to engage. It is tempting to recruit the most easily accessible, or use established and confident user groups, or even to bypass this step, but future market failure, and user frustration due to poor usability, is the risk.

## 6.4 Future Design and Economic Drivers

When looking to the future, the boundaries between consumer products, assistive technologies and medical devices intended for lay use will become increasingly blurred as governments and healthcare providers promote care closer to home and self-management of long-term health conditions and disabilities. Users of devices in this space are increasingly the person with the condition or disability, or an informal carer, rather than a trained healthcare professional. This produces a dichotomy between consumer and medical products, which is influenced by those responsible for product purchasing decisions, the size of their budgets and the desirability of the products. The sale price of a consumer product will be pitched at the intended market and typically has the advantage of economies of scale to reduce production costs and often will be designed to 'look good' to attract buyers. In contrast, a medical device is likely to be designed for a considerably smaller market and is designed primarily to be functional, safe and to meet applicable standards. Looking good is not typically a priority, especially if it adds to the purchase price. Regrettably, like desirability, usability has often been seen as a cost-adding 'nice-to-have' rather than an essential requirement. Returning to the example of the passive hoist, while the person could have been moved safely using it, it removed the individual's control and was undoubtedly perceived as being stigmatizing. There are many examples of medical assistive technologies languishing in people's homes unused because they are not accepted by their users or they fail to adapt to evolving usability needs. Healthcare providers and prescribers need to consider the impact of AT device non-use in their cost modelling.

An example demonstrated within a UK Project that was supported by the National Institute for Health Research (NIHR) and delivered by a NIHR established organization, Devices for Dignity, is of a novel modular cervical orthosis (neck collar) [12]. Collars are commonly prescribed for people with long-term neurological conditions such as stroke or motor neurone disease but many prescriptions result in non-use or only short-term use as traditional devices fail to meet user needs or respond to changing levels of muscular strength and support needs. A cost-effectiveness study of a new design of orthosis employing a cost-utility analysis examined four different use scenarios. Even in 2011, it was shown that when using a funding threshold of £20,000 per quality adjusted life year, a collar that produced a large effect and large impact on rates of dis-use, could be cost-effective up to ~£1000, some 10–15 times more than typical collar prices. However, many prescribers will not consider cost-effectiveness and overall value, unless National Guidance is provided, and will revert to the minimal base purchase price.

Within the United Kingdom, the National Institute of Clinical and Healthcare Excellence (NICE) produces medical technologies guidance to help people in the NHS make efficient, cost-effective and consistent decisions about adopting new medical technologies. It supports innovation, transformation and improves healthcare delivery. It also produces social care guidelines, the primary role of which is to provide recommendations on "what works" in terms of both the effectiveness and cost-effectiveness of social care interventions and services [20], including for older people with multiple conditions requiring social care needs [21]. Assistive technology (including telecare, telehealth and equipment) is included amongst a wide range of relevant topics. Examples of guidance or earlier stage '*medtech innovation briefings*' (where evidence is less certain) for AT include guidance for an exoskeleton for rehabilitation of walking [23] and electronic AT for adults over 25 years with cerebral palsy (CP) [22]. For CP, NICE supported that the critical outcomes for this question were participation, function, independence and health-related quality of life. Due to lack of evidence, the committee was unable to recommend any specific electronic assistive technologies because these would be individualized to the person's needs. It recommended that adults with cerebral palsy were referred to existing services with expertise in electronic assistive technology where appropriate. The need for appropriate training was emphasized citing that such training could also help in learning to identify any technology malfunction if it occurred.

## 6.5 Conclusions

Assistive technology takes many forms, and increasingly includes new types of technology evolving in response to increased demand. As well as increased demand for AT, there is a need for increased use by 'lay users', to address multiple health conditions in a home environment. The technologies may be developed by innovative companies some of which may be new to the field and to the regulatory frameworks and standards expected. This chapter has highlighted some of the important standards that are intended to help increase the usability of medical devices and AT to ensure safe usage. The chapter has not attempted to cover the myriad of specific standards that could apply to the diverse range of assistive products available. Instead, it draws attention to those key standards and guidance that should help developers that create AT products that also qualify as medical devices. It is argued that this is important for the user acceptance of the technology, ease of use and safety as well as market success.

Case study examples have been outlined of critical incidents that have occurred in relation to AT design. They also highlight that much more needs to be done to ensure that assistive technology developers and providers are meeting the often complex and challenging requirements of, and for, those most in need. Usability engineering will become ever more important as self-management and shared management of long-term conditions increases due to pressures on healthcare systems worldwide. Designers and manufacturers need to consider the needs of ever more complex user needs in the home environment where people are living longer and with increasing numbers of multi-morbidities.

The safety and value of medical devices and assistive technology is reliant on them being accepted and used, and working as intended. As highlighted by standard IEC 62366:2015, this requires that human factors are taken into account during the design and evaluation of the products. Usability standards and processes have been outlined that reinforce the importance of involving all anticipated users of the technology throughout its development, and the importance of a thorough process of iterative design and testing. The MHRA has issued guidance for manufacturers and developers and notified bodies responsible for assuring the quality of devices [16]. It is anticipated that the guidance is of use to a much wider community for example physicians, NHS, NICE, and other stakeholders. The standards and guidance put forward a range of human factors and usability methods available for use that are outlined in the standards and regulations. They vary in the ease of employment and the stage of development to which they are best suited. We have sought to provide examples of how they have been applied in AT development projects within the Devices for Dignity portfolio as a means of guiding readers in their potential application when seeking to apply the available standards and guidance.

**Acknowledgements** Drs McCarthy and Moody are Theme Leads for the National Institute for Health Research (NIHR) Devices for Dignity MedTech and In vitro Diagnostic Co-operative (MIC). The work of the Devices for Dignity MIC is funded by the National Institute for Health Research. The views expressed are those of the authors and not necessarily those of the NHS, the National Institute for Health Research or the Department of Health and Social Care.

## References

1. Antaki, J.F.: Insanity of left ventricular assist therapy doing the same thing and expecting different results. Circ. Heart Fail. 4(6), 680–681 (2011). https://doi.org/10.1161/CIRCHEARTFAILURE.111.964650
2. Bayer, K., Mitchell L., Gardner S., Pine, R.: Engaging stakeholders in the home medical device market: Delivering personalized and integrated care. British Standards Institution (BSI) Presentation. https://www.bsigroup.com/LocalFiles/EN-AU/ISO%2013485%20Medical%20Devices/Whitepapers/WP5___BSI_md_home_medical_devices_whitepaper_UK_EN.pdf (2014). Accessed 31/12/2018
3. British Standards Institution. Medical devices—part 1: application of usability engineering to medical devices. BS EN 62366-1:2015. https://shop.bsigroup.com/ProductDetail/?pid=000000000030331076 (2015). Accessed 02/01/2019
4. Corrigan, J.M., Donaldson, M.S., Kohn, L.T. (eds.): To Err is Human: Building a Safer Health System. Institute of Medicine (US) Committee on Quality of Health Care in America. National Academies Press (US), Washington (DC). http://www.nap.edu/catalog/9728.html (2000). Accessed 31/12/2018. ISBN-10: 0-309-06837-1

5. Department of Health: The NHS Long Term Plan, January 2019. https://www.longtermplan.nhs.uk/wp-content/uploads/2019/01/nhs-long-term-plan.pdf (2019). Accessed 11/01/2019
6. Davis, F.D.: Perceived usefulness, perceived ease of use, and user acceptance of information technology. MIS Q. **13**(3), 319–340 (1989)
7. European Commission: Regulation (EU) 2017/745 of the European Parliament and of the Council. https://eur-lex.europa.eu/legal-content/EN/TXT/HTML/?uri=CELEX:32017R0745&from=EN (2017, April 5). Accessed 10/01/2019
8. Health & Safety Executive: Risks to Service users known to self-hoist and transfer using ceiling track hoists. Safety Notice: FOD WSW 2-2010. http://www.hse.gov.uk/safetybulletins/ceilingtrackhoists.htm (2010, February 3). Accessed 03/01/2019
9. International Organization for Standardization: Medical devices—Part 2: Guidance on the application of usability engineering to medical devices. IEC/TR 62366-2:2016. https://www.iso.org/standard/69126.html (2016). Accessed 02/01/2019
10. Judge, S., Townend, G.: Users' Perceptions of Communication Aid Design. https://www.researchgate.net/profile/Simon_Judge/publication/264314439_Users'_Perceptions_of_Communication_Aid_Design_-_D4D_Project_Report/links/53d8ae4b0cf2e38c633188da/Users-Perceptions-of-Communication-Aid-Design-D4D-Project-Report.pdf (2010). Accessed 04/04/2019
11. Kormos, R.L., McCall, M., Althouse, A., Lagazzi, L., Schaub, R., Kormos, M., Zaldonis, J.A., Sciortino, C., Lochard, K., Kuntz, N., Dunn, E., Teureberg, J.J.: Left ventricular assist device malfunctions: it is more than just the pump. Circulation 136, 1714–1725 (2017). https://doi.org/10.1161/circulationaha.117.027360
12. Latimer, N., Dixon, S., McDermott, C.J., McCarthy, A.D., Tindale, W., Heron, N., Shaw, P.J.: Modelling the cost effectiveness of a potential new neck collar for patients with motor neurone disease. HEDS Discussion Paper 11/10, 2011, ScHARR. Available at http://www.shef.ac.uk/polopoly_fs/1.74441%21/file/HEDS-DP_11-10.pdf (2011). Accessed 18/01/2019
13. Lamont, T., Evans, T., Ford, A., Hanss, K.: Help at Home—Use of assistive technology for older people. Published by NIHR Dissemination Centre. http://www.dc.nihr.ac.uk/themed-reviews/Help-at-home-WEB.pdf/https://doi.org/10.3310/themedreview-03345 (2018, December). Accessed 03/01/2019
14. McCreadie, C., Tinker, A.: The acceptability of assistive technology to older people. Aging Soc. 25, 91–110. https://doi.org/10.1017/s0144686x0400248x (2005)
15. Medicines & Healthcare products Regulatory Agency (MHRA): Risk due to incorrect selection and delivery of bolus amount. (MDA/2014/019). https://www.gov.uk/drug-device-alerts/medical-device-alert-paradigm-ambulatory-insulin-infusion-pumps-risk-of-hypoglycaemia#problem (2014). Accessed 29/12/2018
16. Medicines & Healthcare products Regulatory Agency (MHRA): Human Factors and Usability Engineering—Guidance for Medical Devices Including Drug-device Combination Products. Version 1.0. https://assets.publishing.service.gov.uk/government/uploads/system/uploads/attachment_data/file/645862/HumanFactors_Medical-Devices_v1.0.pdf (2017, September). Accessed 31/12/2018
17. Medicines & Healthcare products Regulatory Agency (MHRA): Medical Device Alert MDA/2019/004 Issued: 30 January 2019 at 11:00 Arjo Minstrel passive floor lift (portable hoist)—risk of spreader bar detachment from lifts WITHOUT a scale. https://assets.publishing.service.gov.uk/media/5c5081dced915d7d4877c01f/MDA-2019-004.pdf (2019). Accessed 01/03/2019
18. Meyer, C., Grenness, C., Scarinci, N., Hickson, L.: What is the international classification of functioning, disability and health and why is it relevant to audiology? Semin. Hear. 37(3), 163–186 (2016, August). https://doi.org/10.1055/s-0036-1584412
19. Moody, L., McCarthy, A.D.: Experiences of leg bag users and emerging design priorities. J. Wound Ostomy Cont. Nurs. 42(2), 176–182 (2015). PMID 25734462
20. National Institute for Health and Care Excellence: The remit of NICE Social Care Guidelines. https://www.nice.org.uk/Media/Default/About/what-we-do/NICE-guidance/NICE-guidelines/social-care-guidelines/Social%20Care%20Guidelines%20Remit.pdf (2013)

21. NICE—National Institute for Health and Care Excellence: Older people with social care needs and multiple long-term conditions. NICE Guidance NG 22. Available at https://www.nice.org.uk/guidance/ng22/resources/older-people-with-social-care-needs-and-multiple-longterm-conditions-pdf-1837328537797 (2015, November 4). Accessed 10/01/2019
22. NICE—National Institute for Health and Care Excellence: Draft for Consultation: Cerebral palsy in adults [D3] Interventions that improved functions and participation: electronic assistive technology. July 2018 (GID-NG10031 expected publication date 19/01/2019)
23. NICE—National Institute for Health and Care Excellence: Ekso exoskeleton for rehabilitation in people with neurological weakness or paralysis. Medtech innovation briefing. Published www.nice.org.uk/guidance/mib93 (2017, January 18). Accessed 19/01/2018
24. NHS England: Patient Safety Alert: Stage One: Warning Risk of death and serious harm by falling from hoists. Reference NHS/PSA/W/2015/010. https://improvement.nhs.uk/documents/714/psa-falling-from-hoists-1015_UNQZYZO.pdf (2015a). Accessed 02/01/2019
25. NHS England: Patient Safety Alert: Risk of severe harm and death from unintentional interruption of non-invasive ventilation. Reference Number NHS/PSA/W/2015/003. https://www.cas.mhra.gov.uk/ViewandAcknowledgment/ViewAttachment.aspx?Attachment_id=102109 (2015b). Accessed 10/01/2019
26. NHS National Quality Board: Human Factors in Healthcare: A Concordat from the National Quality Board. https://www.england.nhs.uk/wp-content/uploads/2013/11/nqb-hum-fact-concord.pdf (2016). Accessed 10/05/2019
27. Nielsen, J.: Usability Engineering. Academic Press, 1–358 (1993). ISBN-13: 978-0125184052
28. Norman, D.A.: The Design of Everyday Things. Basic Books, 1–327 (2002). ISBN 978-0-465-05065-9
29. North, B.: The growing role of human factors and usability engineering for medical devices. What's required in the new regulatory landscape. White Paper Series, Published online by British Standards Ltd. https://www.bsigroup.com/en-GB/medical-devices/resources/whitepapers/ (2016). Accessed 03/01/2018
30. Petrie, H., Carmien, S., Lewis, A.: Assistive technology abandonment: research realities and potentials. In: Miesenberger, K., Kouroupetroglou, G. (eds.) Computers Helping People with Special Needs. ICCHP 2018. Lecture Notes in Computer Science, vol. 10897. Springer, Cham (2018). https://doi.org/10.1007/978-3-319-94274-2_77
31. Ravneberg, B.: Usability and abandonment of assistive technology. J. Assistive Technol. 6(4), 259–269 (2012) https://doi.org/10.1108/17549451211285753
32. Regdesk: List of Countries with NO Formal Regulatory Approval Process. https://www.regdesk.co/50-countries-with-no-medical-device-regulations/ (2017). Accessed 09/01/2019
33. Stanton, N.A., Salmon, P.M., Rafferty, L.A., Walker, G.H., Baber, C., Jenkins, D.P.: Human factors methods: a practical guide for engineering and design. CRC Press (2017)
34. Stein, R.: Usability—Continuation and Update from 2016 Webinar. British Standards Institution (BSI) Presentation. Richard.Stein@BSIgroup.com: https://www.bsigroup.com/globalassets/meddev/localfiles/en-gb/webinars/bsi-md-usability-webinar-october-17-rs.pdf (2017, October 12). Accessed 31/12/2018
35. UL: Comparison of IEC 62366-1:2015 and IEC 62366:2007 + AMD1:2014—The Major Differences. https://library.ul.com/wp-content/uploads/sites/40/2018/01/BNG-UL17-Comparison-of-IEC-WP-011118.pdf (2018). Accessed 31/12/2018
36. University of Cambridge: Inclusive Design Toolkit. http://www.inclusivedesigntoolkit.com (2017). Accessed 10/01/2019
37. Waller, S.D., Bradley, M.D., Langdon, P.M., Clarkson, P.J.: Visualising the number of people who cannot perform tasks related to product interactions. Universal Access in the Information Society 12, 263–278. https://www.repository.cam.ac.uk/bitstream/handle/1810/252706/Waller_et_al-2013-Universal_Access_In_The_Information_Age-AAM.pdf?sequence=5&isAllowed=y (2013). Accessed 10/01/2019
38. World Health Organization (WHO): International Classification of Functioning, Disability and Health (ICF). https://www.who.int/classifications/icf/en/ (2001). Accessed 02/01/2019

39. World Health Organization (WHO): Improving access to assistive technology: Report by the Director-General. EB142/21. http://apps.who.int/gb/ebwha/pdf_files/EB142/B142_21-en.pdf (2017). Accessed 02/01/2019
40. World Health Organization (WHO): World report on disability. Geneva. http://www.who.int/disabilities/world_report/2011/en/ (2011). Accessed 4 Jan 2019
41. World Health Organization (WHO): Assistive technology. https://www.who.int/news-room/fact-sheets/detail/assistive-technology (2018). Accessed 02/01/2019
42. World Health Organization: How to use the ICF: A practical manual for using the International Classification of Functioning, Disability and Health (ICF). Exposure draft for comment. WHO, Geneva. https://www.who.int/classifications/drafticfpracticalmanual2.pdf?ua=1 (2013). Accessed 02/01/2019

**Avril D. McCarthy** is based at Sheffield Teaching Hospitals NHS Foundation Trust in the UK where she leads the innovation team within one of the largest NHS Clinical Engineering departments. A lead clinical scientist with a broad background across a number of clinical areas, she specializes in the development of medical technologies to support people living with long term conditions, particularly those that impact an individual's independence and dignity. She is the MedTech Theme lead for the NIHR Devices for Dignity MedTech Co-operative, having been one of its founders in 2008, and is an active member of the Medical Technologies Advisory Committee of the National Institute for Health and Care Excellence (NICE).

**Louise Moody** is Professor of health design and human factors in the Centre for Arts, Memory and Communities at Coventry University. She is also Human Factors Theme Lead in the NIHR Devices for Dignity MedTech Co-operative. She has a BSc in psychology and sports science, and a PhD in electronic and electrical engineering from the University of Birmingham. After working as Human Factors Researcher and Consultant, she joined Coventry University in 2006, initially as Senior Lecturer in intelligent product design in the Department of Industrial Design. Since then, she has held a number of management roles within the School of Art and Design as well as leading a research theme related to arts and well-being. Her research is focused on the design and development of products, interventions and services to benefit health and well-being. She employs a range of research methods as well as art-based approaches to ensure that new solutions are functional as well as being desirable and acceptable to end-users and stakeholders. Her focus is on ensuring the involvement and representation of diverse user needs and requirements from the early stages of user need assessment, through to the design, development and evaluation of solutions. Her current projects are particularly focused on assistive and self-management technologies.

# Part III
# Dignity and Ageing

# Chapter 7
# Experiencing Aging: Analogue Versus Virtual

**Deana McDonagh and Debra Reardanz**

**Abstract** The aim of this chapter is to discuss and compare the impact of an aging suit compared to the virtual experience of aging with the goal of expanding a person's understanding and felt experience of aging. The authors are both immersed within the area of aging and disability, with one focusing on physical simulations, while the other has adopted virtual simulations to support the development of empathic understanding for others. This chapter shares the experience of each author as they immerse themselves into the simulation currently favoured by the other, so that they can compare both the physical and the virtual from the perspective of the felt experience. Felt experience refers to the personal understanding gained when experiencing the experience of another person's experience. As our demographics are shifting towards a population of elders, and disability is no longer perceived to be a barrier to quality of life, it becomes the responsibility of developers of services and products to ensure that they understand real people, to ensure that real solutions are offered. On a cognitive level the average person can begin to appreciate how aging and disability could impact another person, however, the authors would profess that until you have dived deeper into that person's reality (physically or virtually) your understanding is limited. It is argued that experiencing the experience of another person triggers reflection, repositioning of worldviews and a deeper sensitivity than would have been possible before the physical and/or virtual experience. While this approach does not replace the process of aging for individuals, it does provide a valuable tool in the researcher's blended research tool kit.

**Keywords** Experience · Empathy · Aging · End of life · Gerontology suit · Virtual reality

---

D. McDonagh (✉)
School of Art + Design, University of Illinois, Urbana Champaign, IL, USA
e-mail: mcdonagh@illinois.edu

D. Reardanz
Clark-Lindsey Retirement Village, Urbana, IL, USA

 
A. Woodcock et al. (eds.), *Design of Assistive Technology for Ageing Populations*, Intelligent Systems Reference Library 167,
https://doi.org/10.1007/978-3-030-26292-1_7

## 7.1 Introduction to Experiential Learning

> The only way to experience an experience is to experience it. [1]

This chapter focuses on the relative value of a gerontology suit and virtual reality system to increase empathic understanding of elders within our community in order to better serve them and ensure more appropriate design decisions for this age group. The authors are professionally located within design research new product development and managing a retirement community. The gerontology suit provides a physically tangible experience of aging, while the virtual reality system provides by definition provides a virtual experience.

Together the authors shared their different approaches to developing empathy, in order to expose the other to a different way of gaining deeper understanding into aging, disability and death. Simply put, one of the authors transcended her analogue approach to experience a virtual reality, while the other author traded her digital tools for the analogue felt experience in order to compare and contrast the merits of each approach. Their professional interest in aging is as a design educator (who trains design students to develop empathic understanding of the future consumers), and as a CEO of a retirement community who needs to equip her staff (e.g. nurses, carers) with an appreciation of the final stages of life.

> We are constantly learning the difference between empathy and sympathy. Patients don't want us to feel sorry for them, they want us to understand what they are going through. [2]

Empathic design refers to a design thinking approach that focuses on the functional and emotional needs of users when products, services and environments are designed. It respects the importance of functional needs as well as the often-underserved emotional needs (e.g. cultural, social, and aspirational). The developed world is moving to an empathy economy, where consumers are seeking positive experiences rather than merely utilitarian means of completing tasks. If a product, services and environments do not resonate with the consumer on these more emotional levels they can become underused, misused and/or literally abandoned.

As Höök states "… you cannot design as though the body is a mere machine, an object that can be perfected by the technologies we strap onto it or surround it with" [3]. This is why empathic modelling (e.g. adjusting one's ability to hear, see and move) is one way in which we can go outside our own comfort zones to experience another person's experience in order to ensure design outcomes are more appropriate, intuitive and resonate with real people. Providing insight into tangible challenges that consumers' experience has tended to focus on only physical challenges. With the development of virtual reality designers now have an opportunity to become immersed within an experience at a level that has not previously been possible and expand their design research toolkit. Empathy has been referred to as "imaginative reconstruction of another person's experience" [4]. Empathic modelling is not always positively perceived. de Vignemont and Singer challenged empathic modelling as a "vicarious adoption of someone else's experience" [5]. What is important is that gaining empathy is not understood to replace a person's real experience, but rather provide some insight so that positive action (by others) can be taken.

The authors come from two very different backgrounds. Deana is an industrial designer, a researcher, and an educator and Debra is the CEO of a retirement community serving a population of older adults with the average age of more than 87 years. Within their roles, Deana needs to fully understand consumers while Deb needs to fully understand her residents. Both Deana and Debra are lifelong learners who understand the value of growing empathy for others—both in terms of doing their jobs better and becoming better humans. One fateful day, Debra attended a University of Illinois Lunch and Learn hosted by Deana. She was presenting to a full room of faculty, students, and young entrepreneurs on the topic of empathic design. An introduction was made, and in a matter of minutes, Deana had invited Debra to experience the GERT (gerontology) suit [6]—a system she had been using to simulate a range of physical challenges ranging from arthritis, mobility issues through to diverse vision impairments. Debra countered with an offer to visit her at her retirement community. The purpose of Deana's visit would be to participate in a virtual reality experience that allows the participant to embody an older man from the time he first receives his terminal diagnosis through his death, surrounded by family.

In addition to their interest in learning, Debra and Deana share the common traits of curiosity and a desire to try new things. They enthusiastically accepted one another's invitation and within a few weeks, both had deepened their connection to and understanding of the challenges and fears that many older adults face. The following is a personal telling of our experiences and breaks with traditional academic writing norms in that we have chosen to write in first-person due to the personal nature of the experience.

## 7.2 Impact of the Environment

The products that we surround around ourselves with, to the environments in which we spend our time, along with the way in which we complete tasks, all contribute to our daily experiences. Industrial Design is a relatively new profession (compared to the sciences and engineering) that is involved in product design, user experience, user computer interaction and more recently the role of design in supporting a more empowered and independent individual.

Historically industrial design considered styling (e.g. form and colour) but more recently, the relationship the individual (user) has with a product, environment and service, has a significant impact on their productivity and their personal sense of independence. What is extremely clear is that we cannot create "yesterday's solutions" [7] for the emerging population any longer because such products will not solve tomorrow's problems. Reducing stigma and responding to emerging needs requires significant insight and empathic understanding. We literally need to rethink aging [8] and reframe our attitudes how we support elders.

As the demographic is changing within our communities, so is the socio-economic landscape. Elders are already the dominant customers/consumers [7]. Services models, business models, products and environments (residential and retire) will need to be rethought. With the increase in elders living longer, with their powerful financial sway "... societies will function differently" [7].

## 7.3 Retirement Community

I (Debra) have worked at Clark-Lindsey Village [9] for over 20 years and have had the pleasure of knowing hundreds of older adults. Clark-Lindsey is a life plan retirement community located near to the University of Illinois Urbana-Champaign campus. It provides a full continuum of housing, services, and amenities, for approximately 275 adults age 62 years or older. There are multiple levels of living available including independent living (with and without support services), assisted living memory care, long-term care skilled nursing and short-term post-acute skilled nursing. The average age of the independent living population is 87 years. Due to the advanced age of the resident population, there are many who live with one or more disabilities that affect activities of daily living. Both residents and staff comprise a community in which support is routinely offered. Without a clear understanding of the effects that a disease may have on a person, the caregiver is not always effective in addressing the real needs of the resident. Additionally, misunderstandings about the resident's circumstances can sometimes lead to impatience and affects one's ability to maintain compassion. 'Behaviours' is a word that is used to describe when a resident is not doing what is expected of them. Many times, the resident's actions make perfect sense when one is able to understand how the resident is experiencing the environment. A resident who refuses to eat may have a visual impairment that makes eating difficult. Hurried staff members may misinterpret the refusal as stubbornness.

Residents choose to live at life plan communities for a variety of reasons but one of the most frequent reasons cited is a fear about losing independence as a result of future disabilities associated with aging. Living in a community in which they are well known as an individual and where services can be customized to support their unique life goals is an important factor in their sense of overall well being.

Such a service-enriched environment depends on the skills and talents of many employees. At Clark-Lindsey, the workforce is comprised of over 300 employees requiring skills in nursing, environmental services, dining, administration, social services, clerical, and transportation. Interaction with residents is encouraged throughout all departments, and one major factor in the success of the relationship is for the employee to have an understanding of a resident's experience as an 80+ year old. Since the majority of Clark-Lindsey's employees are under the age of 35, this understanding is not easy to accomplish. Finding a systematic way to grow employees' empathy for common experiences of the older adult is a major priority for the organization.

### 7.3.1 Part A: Virtual Reality

I have known for a long time that a key indicator of resident and patient overall satisfaction is residents' satisfaction with their relationships and interactions with staff. An important component to successful staff/resident relationships is compassion. For some time, we had been looking for an alternative to the two education methods typical in the field of aging services: (1) labour-intensive and task-oriented 'skills day' or (2) passive online learning, requiring lots of reading or listening and little engagement. At the urging of a colleague, I drove to Chicago to be introduced to virtual reality technology as a potential solution. Using the virtual reality to deliver education seemed to be a way for staff to share meaningful experiences that demonstrate what it might be like to live with different health conditions—so that all of us may deliver more compassionate, empathic care. The Embodied Labs virtual reality (VR) [10] experience was introduced to the Clark-Lindsey employees in March 2018. In the nine months since its adoption, over 230 employees have experienced at least one VR module, in addition to other community members including clinical students and other community members. This includes employees from all departments, both clinical and non-clinical.

To date, three sessions have been offered:

1. *A Journey Through Alzheimers Disease* where the learner embodies Beatriz, a middle-aged Latina woman, as she progresses through early, middle, and late stage Alzheimer's Disease.
2. *We are Alfred* where the learner embodies Alfred, a 74-year old African–American man with macular degeneration and high frequency hearing loss, as he spends time with family, visits the doctor, and receives a diagnosis.
3. *End of Life Conversations* where the learner embodies Clay Crowder, a 66-year old veteran with stage IV, incurable lung cancer and his experience from receiving a terminal diagnosis through transitioning to hospice care to an active dying visualization of the body, including the effects of pain medication and end-of-life symptoms in his final days (refer to Table 7.1).

The standard process is to host the VR session in a small group setting of no more than five people, including a trained facilitator. The sessions are held in a quiet room with few distractions and privacy from passers-by. All participants are given a standard set of instructions that includes an invitation to stop the VR experience at any time they choose. The facilitator's role is to assist the participant in the set-up and use of the VR system, as well as to provide information pertaining to the session. Following each module, the facilitator starts a discussion pertinent to the session and encourages the participants to share their thoughts about the experience, including how they are feeling and what they noticed while embodying the older adult. Ideally, a pre and post assessment is also administered.

The VR sessions vary in length and are comprised of two or more modules to allow for discussion in between modules. In total, each session is less than 1 h long, including time for discussion. The in-person Train-the-Trainer education was provided by the software company ahead of the VR launch and took approximately four hours for the participants to complete.

**Table 7.1** Three stages of the end of life virtual reality simulation

| | | |
|---|---|---|
| **Stage 1**:<br>Receiving a terminal diagnosis | Oncologist delivers the terminal diagnosis while you have close family with you |  |
| **Stage 2**:<br>Transitioning to home hospice care | Hospital visit followed by the hospice experience in the home environment |  |
| **Stage 3**:<br>Last days are visualised | Visual representation of what active dying and experiences of the end of life from the participant viewpoint |  |

## *7.3.2 Part A: Reactions to Virtual Reality*

Despite my unfamiliarity with the technology, the VR experience had an immediate impact on my understanding of several aspects of aging that I did not have before. My improved understanding of macular degeneration had me changing how I approached someone living with that vision impairment, making sure to not stand in the centre of his or her line of sight. My experience as Beatrice living with dementia gave me newfound understanding of the importance of a quiet environment and a patient caregiver. My most intense connection was to Clay and his end of life experience, bringing me to tears as I remembered my own mother's passing following a long battle with cancer.

Data collected from the over 200 participants indicate similar impact. Post assessments received from participants show an overwhelmingly positive response to the VR training experience. When asked whether or not the experience improved his/her understanding of how a person experiences living with a particular disease, the vast majority indicated "to a great extent". Similarly, the vast majority of participants indicated that the VR experience had positively impacted his/her confidence level for interacting with someone living with the disease.

From a dining services employee: "I think it (the VR Experience) brings a *greater level of empathy*. Because of this, I think I will be more aware and patient with people struggling with conditions such as this."

From a nursing employee: "Knowing they can't see and hear as I do helps me to understand better why they do what they do. It will *give me more patience* because their behaviours make more sense to me now."

From a business manager: "I was my mother's caregiver when she was dying from breast cancer and I wish I knew then what I know now. This VR experience would have prepared me to be a *better listener* and *more effective advocate* for my mother. I think I would have been more patient too. I know that I was receiving this training as it pertains to my job but has also had *an important impact on my personal life* too."

The facilitators had a unique vantage point, as they were able to watch the participants' reactions throughout the range of small classes. They heard the participants' immediate response during the experience and in the group discussions that followed.

The facilitators noted that there was a wide range of reactions from participants. The most common "ah ha!" moment for participants was when they were able to see how a resident's behaviour made sense in light of how they themselves were feeling during the VR experience. For many participants, this was their first experience with VR technology and so there was some nervousness and a lot of excitement about trying out the equipment. That seemed to improve engagement with the training over traditional online format. Participants who have also done more traditional aging simulations—where you suit up—indicated that the VR experience felt more immersive. Specifically, the VR experience seemed to inhibit the participant's cognitive ability such that it made the experience more realistic. While many participants related their VR learnings back to their specific job, it was not uncommon for their discussion to include personal stories of being a caregiver of a family member or friend. It seemed like the connection was strongest when the disease or condition had affected the participant personally—beyond their job.

I asked the facilitators what they thought was most important in the participant having a meaningful experience:

> The small group setting is so important to the experience. Some sessions might be ok to do autonomously—like the one on macular degeneration and hearing loss. But the dementia and end of life sessions require the support of the facilitator and fellow participants in order to fully understand the experience. People's reactions to those modules seemed to be more intense and included strong anger or deep sadness that needed the support of the group to work out.
>
> During each session, the participant would be asked to verbally respond to a situation. I'd guess that only 50% of the people spoke out loud. Some would talk throughout the session, even when they weren't prompted, and others stayed silent throughout. It was important as a facilitator to allow the participant to experience the session in whatever way they were most comfortable. They needed to feel safe.
>
> I was occasionally surprise by people's reaction as being so different from my own. In one instance, a participant shook with anger over how the VR daughter was treating the older adult. Others were overwhelmed with emotion, including tears. Every time I watched another group of participants, I learned something new because often a participant noticed something in the experience that I had completely missed. It was sometimes difficult not to insert my own opinions of the experience, but I knew it was important that the participant be able to 'own/their own experience.

The facilitators noted that not all sessions went well. Participants' unfamiliarity with the technology sometimes caused frustration when the technology did not work as expected. Some participants felt uncomfortable at times due to motion sickness brought on by the VR. The oculus devices are designed to fit over a person's eyeglasses, but some participants indicated discomfort when wearing eyeglasses. And finally, like any use of technology, there were occasional technological glitches that interfered with the participant's ability to become fully immersed in the experience.

Deana, signed up for the session featuring 'Clay' and his *end of life* experience. She brought with her a depth of experience concerning other aging simulation models, but not specifically with virtual reality and not specifically with 'end of life' as the primary focus. She entered the makeshift classroom—a shared office just off the busy skilled nursing facility hallway—with excitement and trepidation. She was introduced to Clay's experience in three stages (refer to Fig. 7.1).

Deana's reflections on her virtual reality experience:

> I am more accustomed (and comfortable) with taking others outside their comfort zones in order for them to develop deeper understanding of the wider population. My career has focused on eliciting the marginalized, neglected or overlooked 'voice' to ensure that products, services and environments are more appropriately designed for all. Developing courses and workshops around design for ability rather than disability has significantly improved a range of design outcomes in the areas of consumer products, biomedical and educational environments.
>
> The invitation to experience the virtual reality 'end of life' was both exciting and daunting. I did not anticipate how the session would truly impact me. Initially, the equipment put me off. The headpiece is relatively heavy and needs to be supported around and over your head. As I had asked for someone to observe me, I was therefore aware of being observed. This adds pressure to 'hold it together' as you embark upon a socially and emotionally difficult journey. To begin with I was somewhat self-conscious, but this diminished as the simulation continued through the first stage.
>
> What I had not anticipated was how I would respond to looking at 'my hands' during the latter stages of the simulation (refer to Fig. 7.2). Looking at my hands with the blood flowing away and turning them blue/purple, which indicated the body slowly shutting down, was extremely emotional for me. The experience had hit home for me. It has managed to permeate my consciousness. I had transformed from a research participant to a person with very little time left alive. As the simulation finished and I removed the equipment, I was aware of all the

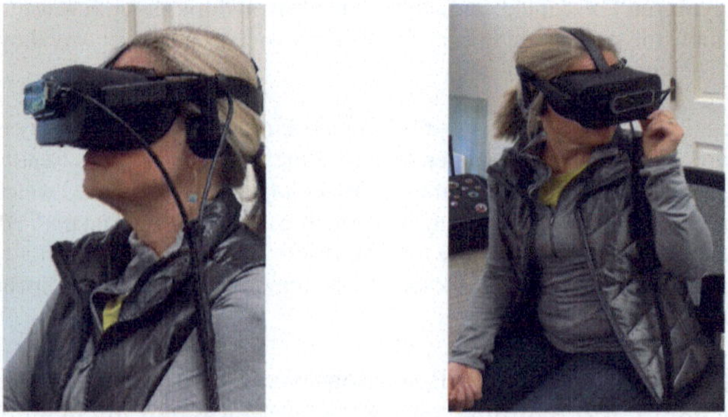

Setting up the VR head gear    Going through the simulation

**Fig. 7.1** Going through the end of life simulation

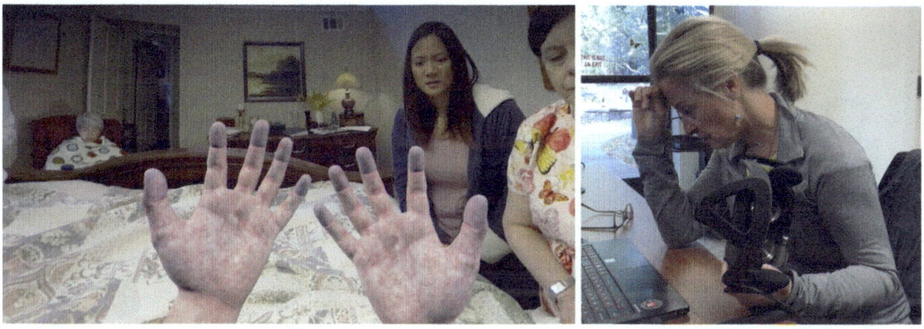

**Fig. 7.2** Emotional response at the end of the simulation (left) Blood draining from hands, (right) emotional impact 'grieving' felt immediately upon completion of the simulation

> observers being silent waiting for my reaction. It took a few moments for me to pull myself together. The grief was overwhelming and tangible. This had transcended an exercise and become a tangible felt experience. It was clear that this approach managed to impact me differently than previous empathic modelling approaches. The physical impact of the GERT suit (e.g. body weights, gloves and neck support) is anticipated and is felt as one is putting all the contributory parts of the system onto your body. With the VR approach one becomes immersed through the process and the emotional response develops slowly, this is possibly by the impact was somewhat of a surprise at the end of the process.

### *7.3.3 Part B: The Aging Suit*

The aging suit is a system that can be worn as individual elements or as a whole to simulate a range of physical challenges ranging from arthritis, mobility issues through to diverse vision impairments. The goal of the system is to provide able-bodied individuals with the experience of physical challenge(s) to enable a deeper understanding of the felt experience of age-related disabilities and developing disabilities.

There are several aging suits available (e.g. AGNES, Senior Suit) [11, 12]. For the purposes of this study, the GERT (gerontology) suit was employed [10]. The GERT suit simulates impaired vision (goggles/glasses) with a range of eye conditions (e.g. glaucoma, cataracts), limited mobility (e.g. wrist and ankle weights, cervical collar, knee and elbow wraps); impaired hearing headphones (e.g. tinnitus), weighted (torso) vest (e.g. simulate surplus body weight), reduced sensitivity in finger (e.g. arthritis, diabetes), reduced confident in walking (e.g. padded shoes to reduce ability to feel ground) and for the more adventurous the back pain simulator (e.g. sharp metal cones strapped to one's back) and the Chronic Obstructive Pulmonary Disease simulator. The elements of the suit can be worn individually or combined to choreograph a specific experience for the participant.

Figure 7.3 illustrates a range of participants that have all experienced the suit and taken away similar and different insights. For the subject who is 6′ 4″ (1 m 87 cm) who has always identified as having significant physical strength, the suit provided

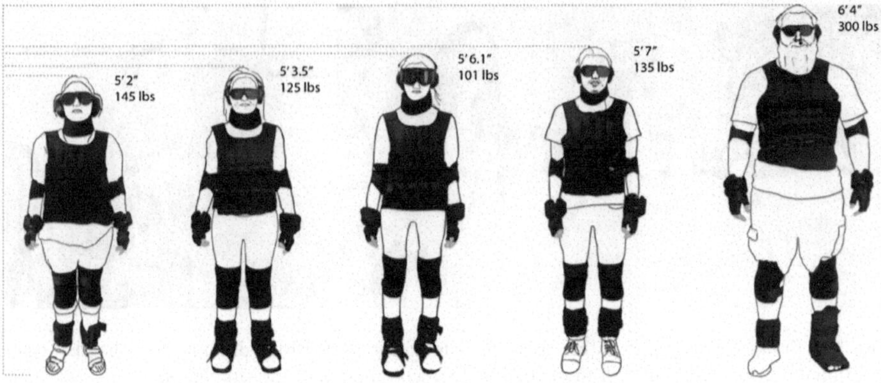


**Fig. 7.3** Range of users that respond differently to the various challenges

the unfamiliar experience of vulnerability through the vision impairment glasses and the tinnitus headphones. The participant who weighs 101 lb (48.81 kg) was weighed down by the 40 lb (18.14 kg) body weight resting on her torso. This significant additional weight had the most impact on her as it provided the unfamiliar experience of carrying additional body weight that slowed her down and resulted in her exerting additional effort while carrying out activities of daily living.

It is important to ensure participants are comfortable when wearing the suit, and we strongly recommend that there is an able-bodied person alongside them while they wearing any elements of the suit. Both authors experienced the GERT suit within the context of their personal residences (homes). The rationale for trying the suit on within one owns personal environment is to immediately experience how the material landscape that you have personally choreographed becomes unusable, unsafe and too challenging once a disability or reduce ability develops.

Reflection is key to an empathic research approach that relies upon the individuals connecting with their own feelings. Time needs to be dedicated to expressing ones thoughts prior to, during and after the experience. The most impactful part of the empathic modeling process is the conversation that follows. While learning comes through the general awareness of others' experiences, reflection transforms into a general impression that you carry forward into supporting your decision-making. If you do not reflect upon the experience, you are less able to name or draw upon the learning later in a meaningful manner.

Deana's reflections on her GERT experience:

> Having practiced Industrial Design and studied aging for over 30 years, I had anticipated that my home environment would be a straightforward space to navigate with the suit on, without any significant challenges. It is relatively time consuming to place all the elements of the suit on and I did require assistance, especially with the 40 lb (18.14 kg) vest that covers one's torso. As a person weighing 125 lb (57 kg) this represented a significant increase in weight I had to manage. It impacted my ability to sit, stand and move around.
>
> What became immediately clear what that every movement had to be carefully considered, as spontaneous movement was not now possible. Also having someone in the environment

become less significant because I could not hear or really see them, though I was mindful that they were capturing my movements photographically.

The first challenge I experienced in the kitchen environment was the microwave and then the cooker control knobs. From the photographs (refer to Fig. 7.4) it is clear that I had difficulty seeing (due to the glaucoma goggles) and I needed to steady myself with each action. What is somewhat alarming is that my hand is located next to heating ring on the top of the hob (stove top). This could potentially lead to severe burns especially as I was wearing surgical gloves that reduced my finger sensitivity. The furniture that populated my home, which was chosen for its aesthetic appearance, proved increasingly difficult to sit in and ultimately get out of.

Navigating the lower level of my home was gone tentatively, not only because of limited mobility, vision and hearing, but my cat was wondering around and I was afraid to trip up over him (refer to Fig. 7.5). Using my fingers tips and hands I navigated the circumference of the space. Having recently purchasing a new television, I was keen to see how I could operate it. The television control was sleek and had black buttons on a black surface therefore I was unable to operate it at all. Very disappointing. When I simulated sitting on my toilet, it became clear that I could not see or feel the toilet paper, and more alarming was the fact I literally could not turned around to wipe myself. My independence had evaporated. This exercise was beginning to get to me. When I tried to make a cup of tea which represents one of my activities of everyday living that gives me joy, I realized my white cups that are placed on a clear glass shelf within a white cupboard, within a white kitchen surrounded by clear drinking glasses provided the final straw. I was unable to locate the glass teapot, or cup or even figure out how to fill my teakettle.

Microwave controls | Cooker control knobs | Getting out of low furniture

**Fig. 7.4** Experiencing physical challenges within home environment (Part 1)

Navigating room and handling black on black television remote control device | Loss of sensitivity in the finger tips renders engaging with toilet paper a challenge | Use of glass shelves and white crockery becomes a challenge

**Fig. 7.5** Physical challenges within home environment (Part 2)

Over all the experience was overwhelming emotionally and physically. I found myself sitting for several minutes upon completing the task in order to pull myself together. Simple activities of everyday life that I do not give a second thought to and often complete at the same time as doing over tasks had become to difficult. I was overwhelmed and to a degree in shock. Having studied disability and aging for so long, the shift from *felt sense* to *felt experience* was extremely impactful.

Deb's reflections on her GERT experience:

> This was not my first aging simulation, so some of the suit's features were not a surprise. In particular, the restricted eyesight and hearing were familiar, as we had used similar devices at a simulation at Clark-Lindsey a few years ago. However, the vest (especially the spikey contraption that generated a great deal of discomfort on my back) was new to me (refer to Fig. 7.6). What I noticed immediately was that it took so much more energy to do even the simplest of tasks. Getting out of the chair, or moving from one room to the next—things I don't even think twice about in my 46-year-old body—had me asking myself, "Is what I'm getting up for "worth it"? It's going to take quite a bit of effort, and I wonder if I could just do without." Similarly, the lack of hearing made everything seem more difficult and required a lot of concentration (and energy) to catch the few words that I could. It wasn't too long into the experience that I stopped asking Deana to repeat herself and assumed that she would correct me if I misunderstood. It occurred to me that this lack of understanding due to my hearing deficit could easily be mistaken for some sort cognitive deficit, and I wondered how many times I had jumped to that wrong conclusion with the older adults I have encountered.
>
> I have a relatively small house, which came in handy as I used my hands to feel my way from one place to the next. Part of this was due to compromised eyesight, but I think there was a general lack of confidence in my balance that had me reaching for the walls, countertops, and back of chairs. I wasn't actually leaning on them for support or pushing off them to move forward, but they were part of my navigation plan throughout the house. This brought back memories of when residents were upset that the new handrails in Clark-Lindsey's long hallways were updated with a version that only ran along one side of the hall (due to cost considerations) and created intentional gaps at each art piece (for aesthetic reasons). Several residents with vision impairments were very vocal about their disappointment in that design decision and now I had a better understand as to why they were upset.

Making the bed — Placing kettle on stovetop — Low level storage cabinet

**Fig. 7.6** Experiencing physical challenges within home environment (Part 1)

> My time in the kitchen seemed particularly precarious. When I reached down to grab the tea pot from a lower cabinet, I worried that I wouldn't be able to get back up and made a mental note not to stoop that low again. Whenever there was not enough colour contrast between the item I was grabbing and the counter or shelf, I struggled to find the ideal way to grasp the item and relied on my memory to know where to find a particular mug. When I went to put the teakettle on the stovetop, I could not decipher any of the oven knobs by sight and depended solely on my memory. In addition, the buttons on the microwave were indecipherable as was the timer I set to remind me that the tea was on the stove (while I could eventually hear the high pitched whistle of the kettle, I was not aware at all of the loud boiling or the steam that was being generated by the kettle). I gained new empathy for how many new challenges are created when a person with sight impairment moves to a new kitchen. Had I been put in front of an unfamiliar stove or microwave, I would not have been able to accomplish this simple task. And if my mug or silverware were on another shelf or in another drawer, it would have been a big challenge to find what I was looking for. I had new appreciation for the gigantic undertaking a move to a new apartment or home would be if my sight were impaired, particularly in the kitchen (Fig. 7.7).

> When I ventured to the bedroom to make my bed, I notice a couple of things. First, my current slobbish habits of leaving shoes or dirty laundry at the side of my bed would be a recipe for a fall. Also, the act of making the bed itself, which included folding a medium-sized throw blanket, took a lot of work. I was well aware of the vest's spikes digging into my back and my lack of finger dexterity and tactile definition made it difficult to get the task done. When I was finally able to complete the task, I recognised it as sloppy work and I took no satisfaction in the final product. Also, due to the sight impairment, I no longer noticed the dust that I knew was there on the surface of my dressers. And I wondered if this might be a blessing (ignorance is bliss?) or one more lowering of standards that might have to occur. If I was unaware of the dust, I suppose it wouldn't really make a difference. But if Deana had let me know that she saw the dust, I'm sure I would have been embarrassed and it would have been one more reason to hesitate at inviting others to my house.

The investment in VR Education equipment is relatively modest. The equipment costs, per station is approximately \$1800 (£1400) with the potential for approximately \$400 (£310) in optional accessories. Essential VR equipment includes: a VR headset, touch controllers, oculus proximity sensors, VR-ready laptop, and hand tracking device. Optional accessories include: carry-on sized travel case and hygienic VR cover. For our purposes, a room was designated as the VR lab and due to size

Experiencing tinnitus while watching television — Microwave controls — Choosing a tea bag

**Fig. 7.7** Experiencing physical challenges within home environment (Part 2)

constraints, is limited to four VR stations. The investment in staff training is the more substantial component of the VR budget, and ongoing. Each employee spends approximately one paid hour in the training. Every group of four trainees requires 1.5 h of facilitator labour (including extra time to set-up the equipment). Total labour cost per session is estimated to be $5000 (£5880) (for approximately 275 people). When combined with on-going software costs, the estimated cost per employee per session is $50 (£38).

## 7.4 Discussion

The GERT suit and the virtual reality model each have their own advantages and disadvantages. The following table highlights variants between the two approaches that need to be considered when investing both time and resources into pursuing either (refer to Table 7.2).

What would really be a significant contribution to understanding the aging consumer would be a discrete gerontology suit that integrated virtual reality. Ideally, when experiencing the last stage of end of life, having one's extremities cooled down (e.g. cold hands and feet) would bring in a physical element to that experience. In addition, if the VR system could be customised to reflect/mirror the participant's age, gender, ethnicity and so on, the level of immersion may be enhanced. Equally, with the aging suit, if the physical constraints could integrate a cognitive impairment (e.g. dementia) the overall experience may prove more compelling. The designer's research toolkit is evolving to include gerontology suits and virtual reality equipment. They are moving away from relying on secondary data (e.g. marketing intelligence reports, ergonomic data) to generating their own primary data (e.g. practice-based design research) to support a new depth and level of intimacy of user-centered design [13, 14].

What is clear from this exchange of research approaches (one going from analogue to digital, and the other going from digital to analogue) is that there is much to be learned even when you consider yourself an expert in your field of enquiry. The potential implication is significant. Nurses, nursing assistants, assisted living staff, family care partners, and more would be able to drastically change their knowledge, beliefs, and attitudes about aging adults [15].

Deb shared the following with the residents via this recounting of her personal experience:

> I know that every aging experience is unique and that not every disability that was part of this simulation is inevitable with age. Admittedly, I get impatient with people who seem more interested in discussing their current medical conditions than learning or experiencing something new. You can't work in this field for 21 years without recognizing some less appealing realities of aging—the possibility of loss of physical or cognitive abilities, or even the loss of friends and family. I have tended to think that the antidote to those losses is to try and find the "new" that might feed our human need for joy and growth and connectedness. That may still be true for some people in some situations; however, after experiencing the GERT suit and the Embody Labs virtual reality, I have a better appreciation for the amount

**Table 7.2** Comparison of virtual reality versus aging suit

| | Virtual reality | Aging (analogue) simulation |
|---|---|---|
| Applicability to home setting | This experience of VR was in a home setting but not the authors' own home setting | GERT simulation can occur in any place, including the authors' own home |
| Level of realism | Emotional strain | Physical, mental and emotional strain on the person |
| Experience range | VR can only simulate cognitive impairment | GERT can only simulate physical impairment |
| Costs | The VR system used in this research costs approximately $2500 (£1940) with minimal training for the facilitator | The GERT suit is available for purchase at an approximate cost of $4000 (£3100). However, low cost versions can be created from relatively easily available materials (e.g. masking tape, wooden slats, safety goggles) |
| Time required | VR sessions and discussion are approximately 1 h in length. A facilitator can facilitate up to 10 people | Each author's experience lasted for approximately 90 min to ensure the novelty factor wears away |
| Customisable | Not customizable but does have different modules | Mix and blend impairments |
| Storage | Dedicated room | Minimum storage (large suitcase) |
| Portability | Moderately portable but requires a power source | Very portable (though heavy) |
| Generalisability | Simulation is constant for each participant | Empathic modelling impacts each participant differently |

of energy that is devoted to simply getting out of bed and "doing" regular life. I imagine that as I get older and adapt to additional losses of physical and cognitive ability, a greater proportion of energy will be spent on the essential tasks of daily living.

Active compassion is a job skill that is absolutely essential to elder care. Both the GERT suit and the virtual reality experience have allowed my staff and me, to grow our understanding and our empathy of the challenges faced by the people we serve.

## 7.5 Conclusion

Ultimately a blended approach to seeking further understanding of the people we hope to understand cannot be achieved by a fleeting moment in time. One needs to consider continually developing the 'empathy muscle' that will remind, refocus and reflect on other peoples' experiences. Both authors related their experience of disability, aging and end of life with people within their personal social networks. The two empathic modeling tools (virtual and analogue) do not substitute real experience but they are still valuable tools nonetheless. It is clear that both approaches require significant dedicated time, not only to complete the activities but also to reflect upon possible learning. The individual needs to feel a personal shift in their perspective/viewpoint.

What would be ideal is for a combination of both virtual reality with the physical constraints offered by the GERT suit. A discreet system so that you could be worn without being noticeable to others would get closer to a more realistic experience of living with a disabilities, especially as the current VR and GERT are so noticeable when being worn. What is clear from these experiences is that we have a long way to go before we can fully appreciate active aging and active dying. Through this process two professionals have introduced a new way of looking, learning and knowing. Learning needs to be a lifelong commitment. The authors began this experiment as respected strangers and we have now become friends with a shared passion for empathic understanding of others.

## References

1. Pullen, G.: Design Meets Disability. MIT, Cambridge (2009)
2. Radford University: Exercises in Empathy Highlight School of Nursing's Immersion Week Activities. https://www.radford.edu/content/radfordcore/home/news/releases/2014/september/school-of-nursing-geren-immersion.html (2018). Accessed 21 Jan 2019
3. Höök, K.: Designing with the Body: Somaesthetic Interaction Design. MIT, Cambridge (2018)
4. Nussbaum, M.: Upheaval of Thought: The Intelligence of Emotions. Cambridge University Press, Cambridge (2001)
5. de Vignemont, R., Singer, R.: The empathic brain: how, when and why? Trends Cogn. Sci. **10**(10), 435–441 (2006)
6. Clark-Lindsey. www.clark-lindsey.com
7. Coughlin, J.F.: The Longevity Economy: Unlocking the Fastest-Growing Most Misunderstood Market. Hachette Book Group, New York (2017)
8. Loe, M.: Aging Our Way: Independent Elders, Independent Lives. Oxford University Press, New York (2011)
9. Virtual Reality Software: Embodied Labs. www.embodiedlabs.com
10. GERT suit. www.age-simulation-suit.com
11. AGNES suit. www.agelab.mit.edu/agnes-age-gain-now-empathy-system
12. Senior Suit. www.senior-suit.com
13. McDonagh, D.: Design students foreseeing the unforeseeable: practice-based empathic research methods. Int. J. Educ. Art. **11**(3), 421–431 (2015)
14. Woodcock, A., McDonagh, D., Osmond, J.: Developing empathy for older users in undergraduate design students. Des. Technol. Educ. Int. J. **23**(2), 24–39 (2018)
15. Washington, E., Shaw, C.: The Effects of a VR Intervention on Career Interest, Empathy, Communication Skills and Learning with 2nd Year Medical Students. https://embodiedlabs.com/impact/ (2018)

**Deana McDonagh** is a Professor of Industrial Design in the School of Art + Design at the University of Illinois (Urbana-Champaign) and faculty at the Beckman Institute of Advanced Science and Technology. She is an Empathic Design Research Strategist who focuses on enhancing quality of life for all through more intuitive and meaningful products, leading to emotional sustainability. Her research concentrates on emotional user-product relationships and how empathy can bring the designer closer to users' authentic needs, ensuring both functional and emotional needs are met by products. She is a Research Fellow at Coventry University (UK), Adjunct Professor at Northwestern University (USA), visiting Professor at Arizona State University (USA) and Director of Research and Insight for Herbst Produkt design consultancy (USA). She is the Designer Entrepreneur-in-Residence (start-up incubator) at the Research Park, and Designer in Residence at the Beckman Institute for Advanced Science and Technology at the University of Illinois (Urbana-Champaign) where she provides design guidance.

**Debra Reardanz** has served as the President and CEO of Clark-Lindsey since 2009. Prior to that, she served as the Clark-Lindsey's Vice President/Chief Financial Officer. Clark-Lindsey is a Life Plan Community that includes 275 residents and 260 employees. In 2017, Clark-Lindsey opened Illinois' first privately owned Green House Homes and advanced regulatory readiness for a more resident centric, small home care model. She currently serves on the Green House Advisory Council. Clark-Lindsey received the 2016 LeadingAge Excellence Workplace Award recognizing its culture of high resident and employee engagement and the 2016 Community Impact Award from LeadingAge Illinois. In 2013, Clark-Lindsey partnered with Masterpiece Living to further advance its vision of becoming a centre for successful aging where lifelong learning and continued growth is supported by the entire community. As a result, Clark-Lindsey was recognized as only one of eight Successful Aging Centers in 2015. Deb is a graduate of the 2012 LeadingAge Leadership Academy program, an association which represents aging services providers throughout the nation and served as coach and facilitator for the LeadingAge Illinois leadership program. She currently serves as Past Board Chair of LeadingAge Illinois. Deb has an MBA from Eastern Illinois University, and a Bachelor degree in Accountancy from the University of Illinois. She is a Licensed Certified Public Accountant and a Licensed Nursing Home Administrator.

# Chapter 8
# Dignity in Ageing: Living Well for Longer

**Katherine Jeays-Ward, Vitaveska Lanfranchi, Avril D. McCarthy, Liz Pryde, Lise Sproson and Wendy B. Tindale**

**Abstract** This chapter presents context for the need for better designed technologies within and outside of healthcare settings to support the increasing needs of ageing populations, and provides a model and exemplars through which technology development can have a greater impact.

**Keywords** Dignity · Ageing · Technology · Medical devices · Health · Independence · Methodology · Innovation · Design · User involvement · Co-design

## 8.1 Introduction

This chapter addresses the importance of dignity and independence as considerations in the successful development of new technologies to support an ageing population. It uses a national organisation based in the UK, the National Institute for Health Research (NIHR) 'Devices for Dignity—D4D' Medtech and In vitro Diagnostic Co-operative (MIC), as an exemplar and outlines its methodology with project examples to highlight its benefits. Established to support collaborative acceleration of medical technologies to market for patient benefits, it operates within a wider network of 11 NIHR MICs, each with a different focus. There are many other national and international organisations that successfully support and accelerate medtech development,

K. Jeays-Ward · V. Lanfranchi · A. D. McCarthy · L. Pryde (✉) · L. Sproson · W. B. Tindale
NIHR Devices for Dignity MedTech Co-Operative, Sheffield Teaching Hospitals NHS Foundation Trust, Glossop Road, Sheffield S10 2JF, UK
e-mail: liz.pryde@nihr.ac.uk

A. D. McCarthy
Clinical Engineering, Sheffield Teaching Hospitals NHS Foundation Trust, Glossop Road, Sheffield S10 2JF, UK

V. Lanfranchi · W. B. Tindale
University of Sheffield, Western Bank, Sheffield S10 2TN, UK

 
A. Woodcock et al. (eds.), *Design of Assistive Technology for Ageing Populations*, Intelligent Systems Reference Library 167,
https://doi.org/10.1007/978-3-030-26292-1_8

but few, if any, with this focus. D4D typically operates across the early to mid-technology readiness levels (TRL) with other UK organisations, such as the NHS Vanguards and Test Beds focussing on large scale evaluation and adoption of technologies at later TRLs.

## 8.2 The Need for Technology Development to Support the Health Needs of Ageing Populations

Three quarters of 75 year olds have more than one long-term condition, rising to 82% of 85 year olds, according to one study in the UK [1], and 81.5% of those aged 85 or older in the US [2]. The older people are, the more likely they are to be living with multiple long-term conditions and disabilities. Thus multi-morbidity becomes the norm rather than the exception as the ageing process continues [3]. For many of these individuals, the combinations of their long-term conditions can have a significant impact upon the way they are able to live their lives and their level of independence, both of which can affect quality of life [4]. If available assistive technology does not adequately support their independence and ability to live with dignity as they choose, there is a technology gap to address which is important in itself. When coupled with economic drivers, stretched health services, and our increasing ageing population, the case for appropriate technology development to fill the gaps and support all care and home environments of healthcare becomes compelling. Examples where such gaps have been, or are being addressed, will be covered later in the chapter.

Typically when ageing is discussed there is an assumption that this applies to older people, which is often defined as those over 65 years old [5]. Within D4D, we consider people's needs across their life course, accepting that ageing is a continuous process from birth onwards. It is worth noting that, whilst the majority of younger people are essentially healthy, for the minority who have specific healthcare needs, for instance impaired mobility as a result of cerebral palsy, changes such as growth spurts can present new challenges to reaching maturity in a dignified way and transitioning to independence and adulthood for individuals, and their families and carers. In addition, children with disabilities such as lower limb loss may experience accelerated ageing effects as a result of abnormal loading through prosthetic limbs. As with older people, these obstacles have the potential to be sensibly approached and addressed through the development of health technologies that assist delivery of services, support clinical practices, decision making, and treatment, and that help individuals manage their conditions and remain independent.

## 8.3 A Sense of Scale—How Many People Are Affected by Long-Term Conditions and Disabilities?

At the start of 2019 there were around 67 million people living in the UK [6]. In England, healthy life expectancy is currently 63.6 years for women and 63.1 years for me [7]. Life expectancy has been rising during the last century and projections by the Office for National Statistics predict an increase in the number of people aged over 60 in the UK from 14.9 to 21.9 million people between 2014 and 2039. This trend reflects global projections; according to the Population Reference Bureau, which has been producing World Population Data Sheets since 1962, by 2050 16% of the world's population will be aged 65 or older; in 2018 the figure was 9%. For the world's more-developed countries the projected proportion of their populations over 65 is 27%, which will put a huge strain on health (and other) services [8]. At the end of 2018, 2.9 million people in England had long-term conditions affecting their lives and that of their families [9].

Some common multi-morbidities include kidney disease and neurological conditions. Kidney disease is associated with a tremendous economic burden. High-income countries typically spend more than 2–3% of their annual health-care budget on the treatment of end-stage kidney disease, even though those receiving such treatment represent under 0.03% of the total population [10]. Around 3 million people in the UK have chronic kidney disease, and its impact is growing as people live longer with increasing risk factors such as obesity and diabetes.

Stroke is the largest cause of adult disability in England; there are 1.2 million UK stroke survivors, of whom almost two thirds leave hospital with a disability. The cost to society of stroke is estimated at £2.6 billion annually [11]. There are over 600 neurological conditions, which accounts for 14% of social care budgets, and 3.5% of NHS spending. Neurological conditions result in substantial loss of quality of life and are associated with a significant carer burden [12].

There are over 3 million people in the UK suffering from urinary incontinence for a wide range of causes [13]. Additionally, incontinence can be associated with isolation, increased risk of falls and as a symptom following stroke (Fig. 8.1).

Frailty also increases incrementally with age, and is a better indicator of an individual's likelihood of reducing independence and consequent increasing care needs than chronological age alone. The contracted requirement for GPs in England to identify frailty in patients over 65 reflects this; earlier identification of people at increased risk of injury, disease, and poor health outcomes allows for preventative measures and support to be put in place. This offers further opportunity for technology to be developed and put in place to support people's independence and prolong good health.

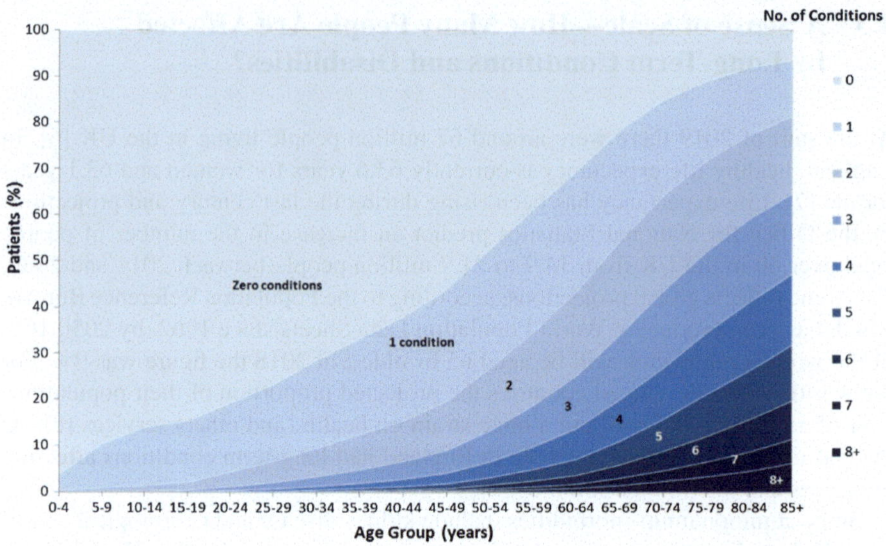


**Fig. 8.1** Number of chronic disorders by age group [1]. Image provided by Professor Bruce Guthrie, University of Edinburgh

## 8.4 Challenges Arising from Long-Term Conditions and Multi-morbidities

Despite the wide-ranging prevalence of multi-morbidities, much of health and social care provision has been developed from the basis of research into single conditions. Disease-specific care pathways and care planning algorithms, although evidence based, can lead to patients becoming overburdened with multiple hospital appointments. Similarly technology prescription or development does not always take into account the complexities introduced by multi-morbidities. As many randomised controlled trials exclude patients with multi-morbidity, little strongly-evidenced clinical guidance has been generated to help clinicians prioritise diagnostic, treatment and management options appropriately across multiple or even conflicting guidelines [14, 15].

Awareness of this failure to provide truly holistic and patient-centred services has added to the drivers for patient education and empowerment so that they can become more active in self-managing their health, service use and general wellbeing. Again, appropriately developed technologies that take greater account of patient and health service needs can offer opportunities to support patients both within and away from clinical settings.

## 8.5 Health Impacts Beyond the Healthcare Setting

In clinical settings in the UK the past two decades has seen a progressive shift of focus from largely clinically-led care, to care in which patients are active, decision-making participants, whose choices and dignity are generally respected and supported. There are and will always be ways in which clinical services can be more effectively and efficiently delivered for the benefit of all, including through the appropriate development and deployment of technology.

For most people with chronic health conditions however the greatest impact of their conditions is on their day to day lives. Away from traditional healthcare settings the impact on what may seem simple activities can be significant at both individual and societal levels.

According to The Work Foundation, a provider of analysis, evaluation and policy advice employment rates for people with long-term conditions are persistently low compared to other groups within the workforce [16]. Work is not the only aspect of people's lives that can be affected. There is a great deal of well-articulated anecdotal evidence available about the ways in which living with one or more long-term conditions affects people's lives if one can find ways to access it. In 2015 the NIHR Devices for Dignity Healthcare Technology Co-operative (now a MIC) sought to draw specific aspects of this evidence into a technology development context through a national survey and patient-led event, both supported by multiple charities. The purpose was to elicit greater detail of lived experiences, including perceptions of dignity, across a range of long-term conditions, and to consider how this may influence the development of medical and assistive technologies for the benefit of patients, carers, health services and technology development teams. The findings reflected many experiences that were shared across different morbidities, and focused mainly around the practical impact that long-term conditions can have on people's ability to: undertake activities of daily living, be mobile, run a home and a family, communicate, sleep, control pain, and maintain mental health [17].

There are over 2 million older people in the UK providing unpaid care, more than half of whom have a health condition or disability themselves [18]. Caring has both positive and negative effects for both the care giver and recipient; as well as enabling greater independence and deepening relationships it can be physically and emotionally demanding, tiring, time consuming, and can itself lead to stress, injury, and poorer health.

Some long-term conditions can be isolating, especially those that affect mobility, communication, and/or continence. Loneliness is increasingly being recognised and acknowledged as a global public health problem associated with poor physical and mental health, decreased quality of life, and life expectancy; it increases risk of premature death by up to a quarter [19], and is a risk factor for coronary heart disease and stroke [20].

## 8.6 Quantifying the Value and Impact of Loss of Dignity

Quantifying the impact on individuals of perceived loss of dignity in a form that can be considered within economic decision-making is currently not possible. Dixon et al. [21] at the University of Sheffield in collaboration with Devices for Dignity sought to answer the question '*What is dignity*?' as a start towards achieving a solution. The driver behind the question was the lack of a standardised questionnaire that could be used to evaluate dignity as a health outcome measure.

A comprehensive review uncovered a vast range of definitions of dignity and conceptualisations of it in the published literature and referred to within healthcare systems, yet without consensus. A health outcome measure that has been widely adopted across the globe for economic evaluation within the pharmaceutical and medical device industries is the EQ-5D; its descriptive system comprises five dimensions: mobility, self-care, usual activities, pain/discomfort and anxiety/depression [22]. It does not assess perceptions of dignity and thus it can be argued that it omits an important factor in the multi-faceted nature of what defines 'quality of life'. Without a validated and accepted means to 'value' in economic terms dignity or its loss, its importance to technology developers, and thereby to technology users, is often neglected. Therefore, the authors set out to form a conceptual mapping of dignity. They proposed an update of Sen's [23], '*capabilities approach*' moral framework (Fig. 8.2), where dignity (K) was integrated as both an independent and potentially modifying factor on well-being (G, I), as shown in Fig. 8.2.

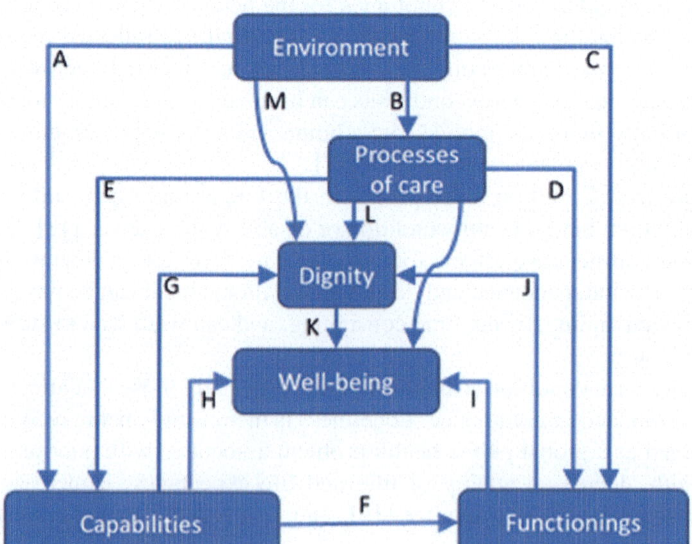


**Fig. 8.2** Updated Sen capabilities framework incorporating dignity. Adapted from Dixon et al. [21]—used with permission

Two approaches for integrating dignity into health economic evaluations were proposed:

1. Considering the EQ-5D and the dignity score separately but with an '*exchange rate*' between quality adjusted life years (QALYs) and more specifically, the EQ-5D scores that lie beneath them and the dignity score. The authors suggested that this could be produced by a separate valuation exercise. Alternatively:
2. Adding a dignity domain to the current EQ-5D. This would still require a valuation exercise to create a new tariff for a new EQ-6D, but would have the implementation advantage of integrating dignity within a single questionnaire.

The proposal would be to add a direct dignity question that would be compatible with the format of the EQ-5D. The possible wording of the additional question is as below:

**Dignity**

I feel that I live with dignity

I feel that I live with some dignity

I feel that I live with very little dignity

Further work to define dignity was undertaken at the D4D National Patient Voices patient-led event [17]. Despite the excellent groundwork undertaken, the potential for a 'dignity QALY' is still work to be completed. The concept of living with dignity is one that most would agree is important. However, achieving consensus in its definition is not easy as it can be interpreted in many different ways, thus adding considerable complexity to the task.

## 8.7 Opportunities for Technology to Support Care Across the Spectrum of Healthcare Provision, and Beyond

Given the increasing proportion of our global populations who are seeking support for their health needs and to support their independence, the use of technology to support health and health-related needs is inevitably on the increase. Organisations involved in developing health guidelines such as the National Institute for Health and Care Excellence (NICE) in the UK are increasingly looking for treatments and technologies that offer improved 'value'—the balance of health outcome versus cost versus existing treatment options. Personalised medicines provide better outcomes than traditional 'one size fits all' treatments—these are therapies and tools tailored to individual people, having been selected based on that individual's needs, with personalisation sometimes even extending as far as being based on an individual's genetic profile. Technologies can play a pivotal role in tailoring medicines, and in supporting self-management aspects of care.

Medical device regulations now place a requirement upon developers to ensure the 'usability' of technologies; an excellent way to ensure usability of new technologies is by ensuring that the individual experiences of potential users are factored into the development process from the outset. When developing either guidelines, new services, or new technologies for managing long-term conditions it is important to acknowledge that contextual health inequalities exist [24], and to address them, or at the very least to avoid exacerbating them; again, the consideration of a range of patient and carer experiences can be greatly beneficial.

Every patient experiences health systems in different ways. Care needs vary with time and severity, and as such individual experiences may incorporate primary care (GPs and community nurses), secondary care (hospitals), tertiary care (highly specialised services), and care homes across the public and private sectors. Each level of service has historically functioned under separate budgets, making it less likely that investment in technology development and provision will be made according to the needs of individual patients as they travel through their whole healthcare pathway. There is general recognition that such silos need to be overcome with various approaches in train to create more partnership working across traditional boundaries.

Whilst variations in care systems further complicates the development and provision of technologies that support people's dignity and independence, it also offers a huge range of circumstances and opportunities for technology to be developed within the appropriate contexts to effectively support people. An area of particular growth is digital technologies; the proportion of older people who are technologically able is increasing. Many therapies, especially self-management and monitoring tools, can be supported by digital technologies, and they can enable otherwise isolated people to stay connected both with health professionals, and with their families and friends.

## 8.8 The NIHR and Its Infrastructure

The NIHR (National Institute for Health Research) is the research arm of the NHS, and was formed to create a health research system in which the NHS supported outstanding individuals, working in world-class facilities, conducting leading-edge research focused on the needs of patients and the public [25]. Since its inception in 2006 the NIHR has identified the need for large-scale changes in the nature of research and service delivery within the NHS and consequently funded a series of initiatives to address this. NIHR Devices for Healthcare Technology Co-operative (D4D) was funded as one of two pilot initiatives in 2008. After the pilots were independently reviewed and deemed successful [26], a further round of HTCs were funded 2013–2017, and were themselves succeeded by 11 MedTech and In vitro Diagnostic Co-operatives (MICs), which are funded 2018–2022. Within its NIHR MIC variant, D4D continues to build on its existing infrastructure, which has a wealth of expertise in the innovation, development, commercialisation and appraisal of medical and healthcare technologies.

Medical devices are often developed in response to technological advance and perceived market demand without taking due account of key requirements including clinical pathway fit, and acceptability by the patient or clinical user. According to the Department of Health in 2011, innovation can be described as 'an idea, service or product, new to the NHS or applied in a way that is new to the NHS, which significantly improves the quality of health and care wherever it is applied' [27]. D4D has developed a partnership-working methodology that brings together patients, carers, clinicians, researchers, and industry in order to create innovative technologies to target and address real healthcare needs across the spectrum of care providers and environments.

D4D started with a mission of enhancing dignity and independence for people with prolonged ill-health. The D4D model addresses technology development through eight themes, three of which have a clinical focus (Diabetes; Renal; Long Term Neurological Conditions) with a further five as cross-cutting and enabling themes (Human Factors; Assistive Technology & Connected Health; MedTech; Rehabilitation Technologies; Integration & Impact). Typically, health conditions are managed in isolation, which is not always effective for patients or in determining the suitability of technology to help and empower them. D4D recognised the important impact of overlapping complications and illnesses on quality of life and ability to live and function independently. The model looks at how people are restricted and the challenges that arise from combinations of conditions, which impact independence and quality of life. These include poor mobility, limb function, impaired speech, depression and pain. Part of the model is to prevent, as well as reduce or treat such limitations, by tackling these issues earlier through using intelligent technology, informed by those that need it—patients and the public. D4D addresses the health technology needs of children and adults with prolonged illness or disability, with particular focus on people ageing well. To do this, teams are established, including clinicians, medical technology designers, developers and businesses to engage with patients, carers and the public on problems that matter to them most, and which will deliver most impact, for both patients and the NHS.

The D4D patient focus is 'Living my life well for longer'—recognising the challenges of overlapping complications of illnesses on quality of life and the ability to live, work and function independently. D4D has 3 dimensions for how it addresses unmet need:

1. WHAT: focus on health conditions that cause significant burden to individuals and the health system
2. WHY: assess holistically how these conditions prevent people from living as full a life as they choose
3. WHO: consider the context of use and who the technology users are

This way D4D can fully understand the challenges—only by understanding all these aspects can the development of a technology that has a significant impact on the problem it addresses commence.

## 8.9 D4D Development Model

D4D's innovation model (Fig. 8.3) follows a series of stages from '*Identification of Unmet Need*' through to '*Dissemination and Adoption*'. We recognise that the innovation pathway is non-linear and we can often be approached by companies, and indeed academics and clinicians, at mid points in the pathway. Our innovation model is depicted in the following diagram demonstrating the journey from need to adoption.

- **Stage 1**: Identification of unmet needs. D4D seeks and receives unmet needs from many sources; clinical themes and networks, charities, public website, patient surveys, industry and proactive themed calls.
- **Stage 2**: Validation of unmet needs. We have developed a mechanism to validate the identified unmet needs quickly through our established networks of experts—clinical, industrial, patients, carers and charities.
- **Stage 3**: Project/Product design. The ability to provide clear insight into 'the problem to solve' is a core strength of D4D. Once a need is validated, D4D will bring together its theme leads and invited experts (including clinical specialists, GPs, patient leads, designers, engineers, health economists and NHS business managers) to assess potential projects from all D4D angles.
- **Stage 4**: Product development. These projects are beyond the funding capacity of D4D, and will be financed through grant funding bids to major funders. Funded projects will have active industry partners and outcomes will typically be an advanced prototype or CE Marked product (as required within Europe).
- **Stage 5**: Evaluation. The prototypes or CE Marked products as appropriate are then ready for suitable clinical studies to demonstrate performance and effectiveness as required.
- **Stage 6**: Dissemination and adoption. Projects will actively pursue collaborations with other parts of the innovation landscape, such as (e.g. the network of 15 regional Academic Health Science Networks (AHSNs) to take these products towards successful commercialisation and adoption.

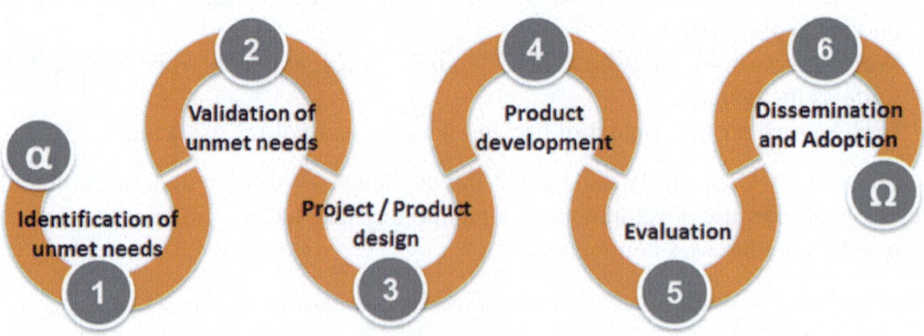


**Fig. 8.3** The D4D development model

Throughout the innovation pathway our model has always been advanced in our approach to Patient and Public Involvement and Engagement (PPIE). Effective partnership with patients and the public has been embedded in our work programme since inception and brings a breadth of insight, experience, knowledge and skills.

## 8.10 D4D Case Studies Illustrating the D4D Development Models

### *8.10.1 (Stages 1 and 2)—Identification of Unmet Needs and Validation of Unmet Needs*

Stage 1 is the identification of unmet needs and Stage 2 is the validation of unmet needs, these two stages are very closely linked and to illustrate this we have used the 'REACH' unmet need as a case study example.

Whilst attending a motor neurone disease (MND) charity event a group of patients set the challenge to develop a device to augment upper limb function. The statement from them was:

> We may take it for granted that we can blow our nose or sip a cup of coffee, but simple activities such as these can be difficult for people with motor neurone disease (MND). The challenge for individuals with MND is adapting to loss of function. It is the same for a range of other people, including those that have suffered from stroke or lack forearm strength in general.

While MND and stroke can affect people at younger ages, the disorders occur more typically in older people, typically those over 50 years [11]. Between now and 2035 the Stroke Association estimate that the number of stroke survivors aged 45 and over is expected to rise by 123%. The impact of such neurological conditions can be exacerbated further with age. This is as a result of the general decline in muscle mass and thus strength that people experience with ageing unless they undertake strength training to counter the tendency.

The Sheffield-based team from the charity approached D4D to take the challenge forward, recognising the unmet need had relevance to a wide variety of neurological conditions. D4D then established a collaboration with the Sheffield Institute for Translational Neuroscience (SITraN) to understand the specific need before determining how best to take the project forward. This detailed uncovering and understanding of what would be required to address the need and how success would be measured is critical to formulating a design specification that can be interpreted meaningfully by design engineers. In the quality management system standard ISO 13485:2016 used by medical device manufacturers these represent the key '*design inputs*' to initiate the design and development phase and are those that determine the resulting '*design outputs*'. This aspect of the process is commonly referred to as the '*requirements capture*' phase.

During these stages the D4D team worked with clinical and engineering colleagues in SITraN and Sheffield Teaching Hospitals NHS Foundation Trust to host a number of workshops with people who experienced the need together with those that cared for them at home, and healthcare professionals. The aim was to further understand the underpinning detail of the unmet need and validate it before continue to the next stage: project/product development. The due diligence would include checking that no solution existed already that would meet the specific need for the intended users and within their chosen environment of use. If a potential solution is found, or horizon scanning indicates a product close to market, further iteration is completed to determine the novel features that would be required to avoid a developing a 'Me Too' device.

It became apparent that the need was for people with neurological conditions that result in varying degrees of upper limb mobility restriction and primarily for use at home. People who were affected highlighted many problems associated with everyday living such as:

- Getting dressed and undressed
- Eating and drinking
- Attending to personal hygiene including washing and drying off
- Scratching an itch
- Reading
- Using a computer
- Communication

What they wanted to see out of the project was a wearable device that was functional, easy to use, including donning and doffing, and was lightweight and discrete. Furthermore, they wanted the resulting device to not require electrical or battery power i.e. they wished it to function passively. Part of the validation process identified that, although there were assistive devices available on the market that met the functional requirement, they failed to meet the additional requirements of being wearable, lightweight and discrete.

The agreed ultimate intended outcome arising from these stages was to develop a device to assist people with upper limb weakness, caused by neurological disorders such as Motor Neurone Disease (MND) and stroke. People living with the effects of MND or stroke can experience limited usefulness of their upper limbs, which severely reduces their quality of life. A key issue is that as users become weaker because of the disorders, they become less able to counter the effects of gravity and lifting the arms is a particular challenge. Upper limb weakness also impacts a carer, who has to assist with basic tasks, such as taking a drink from a cup, or scratching the patients face when they have an itch. The project aimed to develop a device to augment the remaining strength of the patient to give them greater independence at performing simple daily living tasks or more advanced tasks such as typing. This would improve the quality of life for the patient and decrease their reliance on the carer.

While mimicking the kinematic chain of movements of the upper limb is highly complex to achieve in full, especially while meeting the other user requirements, it became clear through the workshops that even limited functional assistance of certain joints and movements would be beneficial to users. People do not typically focus on how they achieve a motor task as it becomes part of autonomous motor memory, so determining on which to focus was not easy. By undertaking careful questioning within the workshops to break down the tasks into discrete movements and guiding participants through the process, of 12 potential upper limb movements, 3 were prioritised by the users. These were:

1. Elbow flexion and extension
2. Hand power grip
3. Hand precision grip

Completing these stages would result in translating the user requirements into robust design inputs answering questions such as:

- Functional: specifying the joints and joint ranges to be assisted and level of assistance e.g. provide 20% reduction in perceived effort.
- Easy to use: Defining how this would be measured e.g. in terms of time or being intuitive of use, particularly if they is also residual cognitive impairment or fatigue as a result of the condition suffered
- Donning and doffing: Whether this needs to be independent, carer-assisted and the required time to achieve.
- Lightweight: specifying a target weight
- Discrete: Describing how this would be defined e.g. fit under clothing or appear to look as clothing.

On concluding the validation stage there was a specification to start the design and development stage of the D4D model. The proposed device would aim to meet the unmet need by reducing the effect of gravity on the upper limbs, allowing the patient to use their residual strength to regain or retain a limited use of their arms for longer.

This principle replicates the effects of hydrotherapy, which gives people with upper limb weakness movement of their arms again. In people living with weakness following stroke, rather than MND, we envisage that the REACH device not only as an assistive device but also facilitating rehabilitation.

### *8.10.2 (Stage 3)—Project/Product Design*

Stage 3 is Project/Product Design where we seek to form a team with the requisite skills and commitment needed to collaborate effectively to progress the project. This team will be made up of multiple stakeholders, namely, but not exclusively; people living with the condition, carers, designers, academics, researchers, clinicians, manufacturers, commercial partners and those with regulatory, business development and other specialist expertise. This is a real strength of D4D given its national approach and extended networks of expertise from its many strategic partners. Key to getting the right team together involves identifying the right technology development partners, involving users and purchasers and collaborating early with industry [28]. Once the project team is identified they will work together to develop the project further, including the identification of, and applying to, various funding bodies to provide funding to develop the project with an ultimate aim of producing a product to deliver patient benefit.

Many people develop neck muscle weakness as a result of various neurological conditions common in the ageing population (e.g. stroke, MND), leading to pain and restricted movement, as well as problems with swallowing, breathing and communication. Such symptoms impact greatly on the ability to live independently and can lead sufferers to perceive that they have also lost dignity.

Ideally a neck collar (cervical orthosis) would help alleviate these problems. However, neck collars that were available prior to the start of the project were of limited use for people with MND, and were often rejected by patients as being over-restrictive of movement and uncomfortable or comfortable but providing inadequate support. A key need was for a personalisable and adaptable collar that could be tuned to provide support at the level and position needed. This project built on work the team had already done with users in identifying their problems with current neck supports and what their requirements were for an ideal neck support. A few early conceptual designs that had been developed with patients were transformed iteratively into prototypes, which included computer modelling of the designs (shape and material), and user focus groups on the designs, with hands on interaction with the materials.

Having had the need highlighted to us directly from patients and carers with the view that existing cervical orthoses were inadequate in terms of function and support, D4D formally identified the need and went on to validate and confirm the need through our national networks of patients, carers and clinicians.

To undertake the product design and subsequent development, a collaborative team was established between D4D, the SITraN team at The University of Sheffield, Lab4Living at Sheffield Hallam University, Barnsley Hospital, Sheffield Teaching Hospitals and the Motor Neurone Disease Associate (MNDA) charity. The collaborative was then expanded to bring additional patient representation to the team which then went on to secure funding through the UK's NIHR Invention for Innovation (i4i) Programme.

While undertaking this stage, D4D will always create a collaboration that focuses on putting the patient first; our vision is to prevent, reduce and treat limitations that people living with long term conditions face. We include patients and carers at every stage of technology development to ensure those technologies are informed by those who need them and those who use them. For this project, people living with the neurological condition/s and those directly affected were integrated into the research team to be actively involved in multiple aspects of the proposed project; designing the research, management of the research, developing participant information resources, undertaking and analysing the research, contributing to the reporting of the study and dissemination of research findings.

The aim of the project was to bring together patients, carers, designers, physiotherapists, engineers, orthotists and neurologists to develop a new orthotic collar. D4D took a lead role with regard to project management and clinical trial management to ensure the project delivered against its objectives, as well as the regulatory arrangements in achieving a CE Mark for the device. We were also active in managing the IP aspects of the device and in exploring the most suitable business/manufacturing partners to ensure the product was licensed with a suitable return on investment to the NHS while offering the product to patients and an affordable price.

The project design stages that we undertook for this project were:

- WP1—Iterative, co-design process
- WP2—Manufacturing and CE marking
- WP3—Extended user evaluation
- WP4—Ascertain commercial feasibility
- WP5—Forward development

Whilst the cervical orthosis developed focussed on MND as its case condition, the team were also aware that it would benefit individuals with weakness of the neck muscles due to other diseases, such as stroke, myopathy, dystonia and multiple sclerosis. In addition to the 5000 people in the UK living with MND, there are also an estimated 100,000 people living with multiple sclerosis and 250,000 people living with the effects of stroke.

### 8.10.3 (Stage 4)—Product Development

Lower urinary tract symptoms (LUTS)—including urinary incontinence—can affect people of any age but increase in prevalence and severity as people age. Approximately 30% of men aged 50 and older have moderate to severe LUTS. This is a very large group potentially requiring treatment. Age is an important risk factor for LUTS and the prevalence of LUTS increases as men get older [29]. The impact of incontinence on self-esteem and dignity can be considerable and may even prevent sufferers visiting their GPs for help. People suffering LUTS often feel embarrassed talking to someone about their symptoms and in many cases they consider these symptoms as a 'normal' consequence for age. For this reason, the consultation with a GP is usually done a long time after symptoms appear. An effective initial assessment is helpful to provide an accurate diagnosis, discard or identify potential serious pathologies of the urogenital tract and decide the most appropriate treatment. If the patient is diagnosed and treated, a regular assessment is recommended to monitor the progress and to confirm the effectiveness of the treatment. NICE Guidelines in the UK, recommend that diagnosis includes a urinary frequency volume chart for men [29] and includes a 3-day bladder diary in the initial assessment of women with urinary incontinence or overactive bladder [30].

NG123 further states '*There is likely to be an increasing need for surgery for urinary incontinence and pelvic organ prolapse because of the ageing population.*'

Such charts or diaries are notoriously unreliable when provided in paper form and can influence adversely the diagnostic process. To aid the collection of LUTS event data, D4D was approached to advise a company in the development of an electronic bladder diary to aid data capture and subsequent data processing. Initially the advice focussed on the usability of the design of the system, with colleagues from the Engineering Design Centre (EDC) at the University of Cambridge. The valuable feedback from the EDC prompted design modifications to improve the usability of the user interface. D4D was subsequently asked to undertake independent testing of the amended system and its diagnostic algorithm to ensure that GPs would receive appropriate and reliable indicative diagnoses from the processed data. The testing provided evidence in support of CE Marking of the system for the EU market. Product development is rarely a linear process and such collaborative working supplying complementary expertise is particularly valuable for small and medium enterprises (SMEs)—which make up the majority of medtech companies in the EU and which have limited resources.

### *8.10.4 (Stage 5)—Evaluation*

There are occasions where clinical teams and/or industry partners approach D4D with a fully formed product, requesting help for clinical evaluation of its effectiveness within the NHS and/or guidance and support on:

- Help with trial design and application for research funding and ethical approval
- Regulatory processes such as compilation of technical evidence for CE assessment and marking
- Identification of commercial partners for manufacture of the product at scale
- Preparation for adoption and uptake within UK health and social care systems

One project illustrating the strength of D4D in this stage of evaluation of devices is our work with Ampcare LLC in evaluating their '*Ampcare Effective Swallowing Protocol*' (ESP) in the treatment of dysphagia.

Dysphagia (swallowing difficulty) is commonly experienced in older people, since it is a symptom common to many conditions (for example stroke, head/neck cancer, dementia, progressive neurological diseases as well as developmental and congenital conditions).

Post stroke, up to 50% of patients experience dysphagia [31], and the associated aspiration (penetration of food and/or fluid into the airway) which can lead to increased mortality, longer hospital stay, pneumonia, reduced quality of life and significant NHS costs. Being unable to eat and drink safely impacts severely on quality of life; disrupting or preventing patients and families engagement with everyday home and social activities, cultural and faith practices [32, 33].

Since an initial paper in 2001 on the use of electrical stimulation for swallow rehabilitation [34], there have been conflicting results regarding treatment efficacy [35]. A number of different devices and differing protocols for exercises and the shape, and placement of electrodes had resulted in confusion. Calls were therefore made for further research in order to provide clinical guidance [36, 37].

Ampcare LLC developed new electrodes, to more evenly distribute electrical stimulation, a novel neck brace in order to provide resistance for exercises and a carefully described treatment protocol. These elements were combined to form the Ampcare Effective Swallow protocol (Ampcare ESP™).

The specially designed electrodes are placed under the chin in order to deliver 5 s pulses of transcutaneous electrical stimulation to targeted (supralaryngeal) muscles. During stimulation, patients are instructed to carry out evidence-based swallow rehabilitation exercises. The duration of the pauses between stimulation episodes can be altered in order to provide progressively more opportunities for swallow practice events per treatment session.

The Ampcare ESP™ equipment and programme had secured FDA clearance in the USA and the device was being used successfully to treat patients in the United States. However, this treatment approach was not available to patients in the UK as further research had been called for regarding its safety and efficacy. NICE (2014) had at that time therefore advised that electrical stimulation should only be used for dysphagia intervention in a research context [37].

Clinical experts at Sheffield Teaching Hospitals NHS Foundation Trust (STH), the D4D team and its wider networks, and Ampcare LLC worked collaboratively in order to enable evaluation of this innovative treatment programme with stroke patients in the UK.

Independently of the company, clinical and research design experts designed a parallel randomised controlled trial. D4D and its STH collaborators contributed towards securing research funding for this, and were instrumental in securing CE Marking of key components of the equipment, and acquiring the appropriate NHS ethical approval to carry out a trial at 3 hospital and community sites.

Each of these aspects would have been challenging for Ampcare LLC to have achieved alone as a small non-UK based company.

Participants were randomised either to:

1. use of the Ampcare ESP™ protocol
2. usual dysphagia care.

Study results showed that a greater proportion of the Ampcare ESP™ cohort improved compared to those receiving usual care and also that the ESP treated group made greater measurable progress and were more satisfied with their treatment outcomes [35].

Ampcare ESP™ has now been adopted for NHS use with stroke patients in Sheffield Teaching Hospital NHS FT and in other UK NHS Trust sites, with ongoing audit of healthcare outcomes and impact on service redesign which is being fed back to NICE in order to help inform further guidance for clinicians.

### 8.10.5 (Stage 6)—Dissemination and Adoption

Like LUTS, chronic kidney disease (CKD) also increases in prevalence and severity as people age. Its occurrence is commonly associated with obesity, diabetes and cardiovascular disease. For those with the most severe kidney function loss, renal replacement therapy (RRT) or a transplant may be the only options for survival. Two main types of RRT are available: peritoneal dialysis (PD) and haemodialysis (HD). For the latter, the majority receive 3 sessions of HD in hospital or a satellite centre per week. Although sustaining of life, the regime is highly intrusive on people's lives due to the requirement to travel for treatment and to undergo typical 4-h treatment sessions. It is often described by recipients as 'living to dialyse' rather than 'dialysing to live'. Dialysis at home may be an option for some but the complexity and large size of hospital HD machines restricts its uptake. Within the UK, the target for home dialysis is 15% but this is generally closer to 5–6% of those receiving HD. To try to address this deficiency and to increase the choice for those living with CKD, Devices for Dignity collaborated with British fluidics company (Quanta) which believed that its technical expertise could be used to disrupt traditional HD machine design. The aspiration was to create a machine that would be much easier to use to promote self and shared care, and to be considerably more compact so that it could fit into a wider range of homes, also eventually offering portability to allow for travel. From early in the project's evolution D4D provided valuable clinical and technical expertise during the company's transition from the industrial to healthcare sector. The success of the company in attracting significant and substantial investment to support the machine's journey to market provides an excellent example of how early de-risking and support from an organisation like D4D can accelerate technology development and future adoption. The device has been trialled in four major centres in the UK to evaluate efficacy, safety and performance, with 1346 successful treatments across 60 participants with mean age 60 years (range 22–85 years) [38]. Importantly, the company has also generated independent usability evidence (from 32 users, mean age 44 years, range 19–76 years) in centres in Canada and 1 in the UK to substantiate its claim to be suitable for self-care use [39].

## 8.11 Conclusions

This chapter has highlighted the importance of collaborative medical technology development with the need for representative user involvement throughout, using the D4D MIC as an exemplar organisation. The six stages of the D4D development model were described with relevant project examples to demonstrate each stage. The chapter further highlighted the influence of dignity and independence as factors that affect ageing populations disproportionately. This is due to the nature and complexity of the health conditions that they typically experience and when dignity and independence have not been considered adequately as influencing design inputs and outputs during technology development.

## References

1. Barnett, K., Mercer, S.W., Norbury, M., Watt, G., Wyke, S., Guthrie, B.: Epidemiology of multimorbidity and implications for health care, research, and medical education: a cross-sectional study. Lancet **380**(9836), 37–43 (2012)
2. Salive, M.E.: Multimorbidity in older adults. Epidemiol. Rev. **35**(1), 75–83 (2013)
3. Collerton, J., Jagger, C., Yadegarfar, M.E., Davies, K., Parker, S.G., Robinson, L., Kirkwood, T.B.L.: Deconstructing Complex multimorbidity in the very old: findings from the Newcastle 85+ study. BioMed Res. Int. **2016**(Article ID 8745670) (2016)
4. Department of Health: Report. Long-term Conditions Compendium of Information, 3rd edn. https://www.gov.uk/government/publications/long-term-conditions-compendium-of-information-third-edition (2012). Accessed: 03/04/2019
5. NHS England: Improving Care for Older People. https://www.england.nhs.uk/ourwork/clinical-policy/older-people/improving-care-for-older-people/ (2015). Accessed: 03/04/2019
6. World Population Review: http://worldpopulationreview.com/countries/united-kingdom-population/ (2019). Accessed 04/04/2019
7. Office for National Statistics: Disability-Free Life Expectancy (DFLE) and Life Expectancy (LE): At Age 65 by Region, England (2016)
8. Population Reference Bureau (PRB): 2018 World Population Data Sheet With Focus on Changing Age Structures. https://www.prb.org/2018-world-population-data-sheet-with-focus-on-changing-age-structures/ (2018). Accessed: 03/04/2019
9. The Kings Fund: Long Term Conditions and MultiMorbidities, from Time to Think Differently. https://www.kingsfund.org.uk/projects/time-think-differently/trends-disease-and-disability-long-term-conditions-multi-morbidity (2015). Accessed: 04/04/2019
10. Couser, W.G., Remuzzi, G., Mendis, S., Tonelli, M.: The contribution of chronic kidney disease to the global burden of major noncommunicable diseases. Kidney Int. **80**(12), 1258–1270 (2011)
11. Stroke Association: State of the Nation Stroke Statistics. www.stroke.org.uk (2018). Accessed 3 Apr 2019
12. NHS England: https://www.england.nhs.uk/ourwork/clinical-policy/ltc/our-work-on-long-term-conditions/neurological/ (2014). Accessed 04/04/2019
13. Buckley, B.S., Lapitan, M.C.M.: Prevalence of urinary and faecal incontinence and nocturnal enuresis and attitudes to treatment and help-seeking amongst a community-based representative sample of adults in the United Kingdom. Int. J. Clin. Pract. **63**(4), 568–573 (2009)
14. Boyd, C.M., Darer, J., Boult, C., Fried, L.P., Boult, L., Wu, A.W.: Clinical practice guidelines and quality of care for older patients with multiple comorbid diseases. JAMA **294**, 716–724 (2005)
15. Van Weel, C., Schellevis, F.G.: Comorbidity and guidelines: conflicting interests. Lancet **2006**(367), 550–551 (2006)
16. Bajorek, Z., Hind, A., Bevan, S.: The Impact of Long Term Conditions on Employment and the Wider UK Economy. http://www.theworkfoundation.com/wp-content/uploads/2016/11/397_The-impact-of-long-term-conditions-on-the-economy.pdf (2016). Accessed: 03/04/2019
17. Devices for Dignity: My Dignity Means: Patient Voices. https://devicesfordignity.org.uk/wp-content/uploads/2016/12/My-Dignity-Means-Patient-led-Event-Report-2-Dec-2015.pdf (2015). Accessed: 03/04/2019
18. Institute for Social and Economic Research and NatCen Social Research: Understanding Society: Waves 1–5, 2009–2014 [computer file], 7th edn. Colchester, Essex: UK Data Archive [distributor]. SN: 6614 (2015)
19. Holt-Lunstad, J., Smith, T.B., Baker, M., Harris, T., Stephenson, D.: Loneliness and social isolation as risk factors for mortality: a meta-analytic review. Perspect. Psychol. Sci. **10**, 227–237 (2015)
20. Valtorta, N.H., Kanaan, M., Gilbody, S., Ronzi, S., Hanratty, B.: Loneliness and social isolation as risk factors for coronary heart disease and stroke: systematic review and meta-analysis of longitudinal observational studies. BMJ **102**(13), 1009–1016 (2016)

21. Dixon, S., Palfreyman, S., Shackley, P., Brazier, J.: What is dignity? A Literature Review and Conceptual Mapping. HEDS Discussion Paper 11/13 (Unpublished). Available for download at: http://eprints.whiterose.ac.uk/43280/ (2013). Accessed 01/04/2019
22. Euroqol: EQ-5D Instruments | About EQ-5D. https://euroqol.org/eq-5d-instruments/ (2019). Accessed 01/04/2019
23. Sen, A.: Commodities and Capabilities. Oxford: Oxford University Press (1985). ISBN: 9780195650389
24. Potter, C.M., Kelly, L., Hunter, C., Fitzpatrick, R., Peters, M.: The context of coping: a qualitative exploration of underlying inequalities that influence health service support for people living with long-term conditions. Sociol. Health Illn. **40**(1), 130–145 (2018)
25. National Institute for Health Research (NIHR): Who We Are. https://www.nihr.ac.uk/about-us/who-we-are/ (2018). Accessed: 03/04/2019
26. Kryl, D., Marjanovic, S., Ni Chonaill, S., Ridsdale, H., Yaqub, O.: Healthcare Technology Co-operatives: Filling a Niche in the English R&D Landscape. Rand technical report TR932 (2011)
27. Department of Health: Innovation Health and Wealth, Accelerating Adoption and Diffusion in the NHS. https://webarchive.nationalarchives.gov.uk/20130107070708, http://www.dh.gov.uk/prod_consum_dh/groups/dh_digitalassets/documents/digitalasset/dh_134597.pdf (2011). Accessed: 03/04/2019
28. Heron, N.M., Tindale, W.B.: Healthcare technology co-operatives: innovative about innovation. J. Med. Eng. Technol. **39**(7), 378–381 (2015)
29. National Institute for Health and Care Excellence: Lower Urinary Tract Symptoms in Men: Management. Clinical guideline [CG97] Published date: May 2010. https://www.nice.org.uk/guidance/cg97/chapter/Introduction (2010, updated 2015). Accessed: 02/04/2019
30. National Institute for Health and Care Excellence (NICE): Urinary Incontinence and Pelvic Organ Prolapse in Women: Management. NICE guideline [NG123] (2019)
31. Singh, S., Hamdy, S.: Dysphagia in stroke patients. Postgrad. Med. J. **82**(968), 383–391 (2006)
32. Ekberg, O., Hamdy, S., Woisard, V., Wuttge-Hanning, A., Ortega, P.: Social and psychological burden of dysphagia: its impact on diagnosis and treatment. Dysphagia **17**(2), 139–146 (2002)
33. Farri, A., Accornero, A., BurdesS, C.: Social importance of dysphagia: its impact on diagnosis and therapy. Acta Otorhinolaryngol. Ital. **27**(2), 83–86 (2007)
34. Freed, M., Freed, L., Chatburn, R., Christian, M.: Electrical stimulation for swallowing disorders caused by stroke. Respir. Care **46**(5), 466–474 (2001)
35. Sproson, L., Pownall, S., Freeman, J., Enderby, P.: Combined electrical stimulation and exercise for swallow rehabilitation post-stroke: a pilot randomized control trial. Int. J Lang. Commun. Disord. **53**(2), 405–417 (2018)
36. Geeganage, C., Beavan, J., Ellender, S., Bath, P.M.: Interventions for dysphagia and nutritional support in acute and subacute stroke. Cochrane Database Syst. Rev. (10) (2012)
37. National Institute for Health and Care Excellence (NICE): Transcutaneous Neuromuscular Electrical Stimulation for Oropharyngeal Dysphagia. NICE Interventional Procedure Guidance No. IPG490 (2014)
38. Bebb, C., Breen, C., Day, C., Mitra, S., Milad, J.E., Grainger, J.: Implementation of a compact haemodialysis system in a real-world environment: assessment of user uptake, safety, performance and usability. In: Presented at EDTA Conference: Copenhagen, Denmark. http://www.quantadt.com/wp-content/uploads/2018/05/SM-00030-A-EDTA-Poster.pdf (2018). Accessed 02/04/2019
39. Harasemiw, O., Day, C., Milad, J.E., Grainger, J., Ferguson, T., Komenda, P.: Human factors testing of the Quanta SC+ Haemodialysis System: An Innovative System for Home and Clinic use. http://www.quantadt.com/wp-content/uploads/2018/11/2018-ASN-Poster.pdf (2018). Accessed 02/04/2019***
40. https://www.ons.gov.uk/file?uri=/peoplepopulationandcommunity/healthandsocialcare/healthandlifeexpectancies/datasets/disabilityfreelifeexpectancydfleandlifeexpectancyleatage65byregionengland/2006to20082007to20092008to20102009to20112010to20122011to2013and2012to2014/regionsat65.xls. Accessed: 04/04/2019

41. The World Bank: Disability Inclusion. https://www.worldbank.org/en/topic/disability (2018). Accessed: 03/04/2019

**Dr Katherine Jeays-Ward** is part of NIHR Devices for Dignity MIC, where she is NHS Innovations Manager, Renal Theme Manager, and Communications Lead. Devices for Dignity is a national MedTech and In vitro diagnostic Co-operative (MIC) funded by the NIHR to act as a catalyst within the NHS for the development of new medical devices and healthcare technologies. Katherine oversees and contributes to a portfolio of technology development projects including renal, paediatric and urinary continence technologies from initial concept through to evidence generation and adoption into practice. Katherine has over ten years' experience in managing collaborative multidisciplinary, cross-sector projects from within the NHS and Higher Education sector, ranging from research through to national policy and practice. She holds a PhD in developmental genetics from the Institute of Cancer Research, and spent a further four years as a post-doctoral research associate at the University of Sheffield.

**Dr. Vitaveska Lanfranchi** is a Senior Lecturer in Medical Computing at The University of Sheffield in the OAK Group and in Sheffield Teaching Hospital NHS Foundation Trust as part of the Medical Imaging and Medical Physics Group and affiliated with NIHR Devices for Dignity MedTech Co-operative with the mission of enhancing dignity and independence for people with prolonged ill-health and addressing health technology needs of children and adults. Her work has a strong applied focus, designing and implementing applications for real-world scenarios, at all technology readiness levels. She has a long experience in stakeholder analysis and translation of stakeholders' needs into technology, providing a useful link between technical innovation and user perspective. She has published over 50 papers in international conferences and journals.

**Dr. Avril D. McCarthy** is based at Sheffield Teaching Hospitals NHS Foundation Trust in the UK where she leads the innovation team within one of the largest NHS Clinical Engineering departments. A clinical scientist with a broad background across a number of clinical areas, she specializes in the development of medical technologies to support people living with long term conditions, particularly those that impact an individual's independence and dignity. She is the MedTech Theme lead for the NIHR Devices for Dignity MedTech Co-operative, having been one of its founders in 2008, and is an active member of the Medical Technologies Advisory Committee of the National Institute for Health and Care Excellence (NICE).

**Liz Pryde** Devices for Dignity is a national MedTech and In vitro diagnostic Co-operative (MIC) funded by the NIHR to act as a catalyst within the NHS for the development of new medical devices and healthcare technologies for people with one or more chronic health conditions. Liz works for NIHR Devices for Dignity MIC as Programme Manager and is responsible for the operational leadership of all the MIC activities. Liz is responsible for the management of the MICs annual objectives and overall effectiveness of the consortium by ensuring an effective framework for collaboration and continuous improvement is in place. She leads the implementation and ongoing coordination of a high quality project selection process and manages the delivery of the ongoing portfolio of projects.

**Lise Sproson** works for NIHR Devices for Dignity MIC as NHS Innovations Manager, Diabetes Theme Manager and PPI Lead.

Devices for Dignity is a national MedTech and In vitro diagnostic Co-operative (MIC) funded by the NIHR to act as a catalyst within the NHS for the development of new medical devices and healthcare technologies for people with one or more chronic health conditions.

Lise is also a Speech and Language Therapist with specialist expertise in the assessment and management of dysphagia (swallowing) and communication difficulties. She is a committee member of the Council for Allied Health Professions Research (CAHPR) and the UK Swallow Research Group (UKSRG) and Specialist Advisor on dysphagia for the Royal College of Speech and Language Therapists (RCSLT) and the National Institute for Health and Care Excellence (NICE).

**Wendy B. Tindale** (OBE, MSc, PhD, CSci, FIPEM, ARCP) is Consultant Clinical Scientist and Scientific & Innovation Director at Sheffield Teaching Hospitals. She holds a Professorship in Medical Physics at the University of Sheffield and nationally is Director of the NIHR 'Devices for Dignity' MedTech Co-operative. She has an interest in combining clinical academic excellence and patient benefit with business opportunities through the translation of innovative medical technologies and has worked with both public and private sectors to facilitate successful collaborations. Wendy has contributed to numerous national and international committees. She received a Beacon Award for outstanding contribution to healthcare and was honoured in the Health Service Journal's inaugural Inspirational Women awards for outstanding leadership in her field, becoming Healthcare Scientist of the Year in 2016.

# Part IV
# New Technologies

# Chapter 9
# Unlocking Service Flow—Fast and Frugal Digital Healthcare Design

**John Knight, Elliot Ross, Chris Gibbons and Tom McEwan**

**Abstract** This paper reports on the discovery phase of an e-consultation project that aimed to deliver a technology-enabled service for consultations between healthcare professionals and patients using digital communication tools. The project involved a mixed methods research phase that supported Agile service design work. A number of learnings were derived from the project and these are contextualised into a Flow Methodology for agile service design, delivery and operations.

**Keywords** Healthcare · Service design · Agile · User-centred design

## 9.1 Introduction

This paper reports on the discovery phase of a digital healthcare project that investigated the introduction of a new e-consultancy service to the UK National Health Service (NHS). The project defined a vision for a new digital service and created a proof of concept to enable face-to-face consultations between patients and General Practitioners (GPs) using a proprietary digital communication platform (Skype for Business). The project was funded and delivered by Accenture and aimed to explore virtual consultations as a way of meeting increasing demand from patients for digital access to health and care services. The resulting prototype was completed in two months, and enabled online consultations, appointment management and basic functionality centred on five core use cases:

J. Knight (✉)
Aalto University of Arts, Design and Architecture, Espoo, Finland
e-mail: john.knight@aalto.fi

E. Ross
Avanade, London, UK

C. Gibbons
Accenture, London, UK

T. McEwan
tgmcewan, London, UK



A. Woodcock et al. (eds.), *Design of Assistive Technology for Ageing Populations*, Intelligent Systems Reference Library 167,
https://doi.org/10.1007/978-3-030-26292-1_9

- Book an appointment
- Cancel an appointment
- Reschedule an appointment
- Edit an appointment
- Take part in a consultation

The future roll-out of the service would offer many benefits to patients, GPs and health professionals alike. These include increased and improved access to primary care, particularly for hard-to-reach patients and those with stigmatised conditions, including mental and sexual health issues. The service is also anticipated to reduce time and cost in consultancy delivery, streamline operations, and provide better and more secure data collection. Patients engaged during the discovery phase also noted the potential to reduce contagion by not having to travel in person to surgeries. Most importantly, online consultations offer the potential for better quality and more timely diagnosis, through broader knowledge sharing and improved, faster ways of triaging. Together, these relatively small incremental improvements contribute to better healthcare in the broadest sense, and specifically for:

- Follow up appointments and doctors
- Repeat prescriptions and doctors
- Sexual health advice and test results
- Mental health advice and check-ups
- Medication reviews for doctors
- Getting advice and results online

Despite the benefits, poor connectivity and unfamiliarity with technology (especially for older patients) were cited as negative aspects of the new service. Further evidence is currently being collected on these issues, as well as further insights into the value and impact of video consultations in healthcare more generally. In addition, work is continuing in testing the prototype and evaluating options to enhance functionality and run a broader pilot study.

Prior to a three-week immersive research discovery phase, a competitor analysis was conducted as part of project initiation. Several virtual consultation brands were reviewed, and this reinforced initial research findings that showed that GPs believed that some consultations, such as medication reviews, would be as effective over video as in person. Often the doctor is required to travel to perform these reviews, costing valuable time and money when being able to see and hear the patient would be adequate. Yet other appointments, such as a mother concerned about her young child, may require a face-to-face consultation. If the concerned mother had been able to request a video consultation without providing a reason, the patient may indeed have to have a repeat appointment with the doctor face-to-face, thus increasing the workload of the GP. Somewhat unusually the project was also informed by an academic literature review. This (see Sect. 9.2) informed how research was conducted and the methods for analysis as well as the design process and general direction of the work.

Understanding the concerns of clinicians and fellow researchers enabled designers to further explore these areas, find solutions, and thus design a more effective solution ready for evaluation using classic human-centred design tools and methods. This design work involved creating candidate solutions, making prototypes and testing them with the various user agent groups, including patients.

Development work was tightly aligned to design and having defined a successful solution the focus shifted to delivery and operations. The experience gained during the project was distilled into an amalgamated human-centred, agile, service design approach. This 'Flow' methodology is not only born out of the practical realities of delivering tangible positive change in healthcare, but also reflects limitations with all of those individual contributing influences. We believe adding focus on frugality and flow provides the necessary counterbalance to these shortcomings in order to arrive at a successful fusion.

## 9.2 Theoretical Foundations

E-consultation studies provide substantive individual studies, and meta-research findings report common benefits. Advantages span both Primary Care Providers (PCP) [23] and patients [26], including young people who might otherwise be disengaged with health services [15]. Positive effects of ECS are reported in increased access, efficiency and speed of delivery [18]. Similarly, greater patient satisfaction [24] and reduced costs were noted by Nijland et al. [26].

These benefits resonate with the broader potential of e-health [2] in providing a mediating platform between stakeholders [37] and specifically in facilitating progressive changes in the practitioner/patient relationship [10]. In this context, patients are becoming 'pre-informed' [35] about their condition and diagnosis and are therefore better placed to make decisions in a 'partnership' model [9] of healthcare. Likewise, the educational value of ECS to PCPs is also well documented [24].

Realising these advantages is not without problems [27]. A 'patient-centred' philosophy [34] can mitigate risks (e.g. low adoption) and ensure a good fit to stakeholder needs, through structured engagement [22]. This approach fits well, both philosophically and practically, with Participatory Design (PD) [7].

Methodologically this means integrating end users in the delivery process, which has proven to be effective [8] in a range of contexts and areas of health informatics. PD methods (e.g. co-design) are tried and tested in a broad range of domains and are characterised by active involvement of stakeholders, primarily in the early stages of design. Health related work includes early preliminary exploration by Patel and Kushniruk [28] through to cases by [4], Clemensen et al. [12] and Pilemalm and Timpka [29] up to the recent past: Donetto et al., [14] and 'Patient centredness'.

According to [3] 'traditional processes and services for the delivery of health and care are experiencing a drastic shift to meet people's present and future needs'. The shift is predicated on mass adoption of digital technologies, consumer demand for highly usable products and services and an economy characterised by 'disruptive' newcomers who can bring new offerings quickly to market through agile methods. This gap is understandable given that agility is naturally focused on delivering 'working software' at pace while healthcare understandably fixates on risk mitigation. An obvious way of balancing pace with risk mitigation is to augment agile delivery within a strategic framework for technology adoption. Such a framework would maintain agility's ability to deliver quickly together with an adoption framework that would ensure project direction fits with the broader possibilities of the technology.

Technology Acceptance Model (TAM) [13] offers such a balance. TAM is a commonly applied digital adoption framework. TAM measures a set of consistent variables (including 'Perceived Usefulness' (PU) and 'Perceived Ease of Use (PEU)) to assess how users adopt new technologies. [1] is one example of many that apply TAM to healthcare applications. Perceived ease of use is less likely to be a determinant of attitude and usage intention according to studies of telemedicine [17].

The apparent strength and relevance of TAM to health care studies is noted by Holden and Karsh [16] albeit with the caveat that the domain necessitates extensions to the basic model. Criticisms of TAM fixate on its lack of predictive power. These limitations have led to new extended models including UTAUT [11] that also address the predictive requirements of many health projects (e.g. Bennani and Oumlil [6]).

## 9.3 Approach

User research was carried out during a short (three week), immersive research phase. This work included in situ interviews, observation and generative workshops. Sessions took place in various locations, as well as some conducted remotely using video telephony. A hybrid approach was taken to research and design, as the new service related to three different domains. Firstly, the work focused on the development and adoption of a new technology (socio-technical focus) where adoption is critical. Secondly, the outcome is a public service (service design focus) where a holistic approach is needed to orchestrate the various actors and activities into an optimal whole. Lastly, this new service comprises screens, buttons and interfaces to a proprietary software product (human-centred design and technology focus) that were to be developed using Agile methods.

The hybrid approach was exemplified in using socio-technical methods (e.g. Technology Acceptance Model (TAM), [13]). Insights from TAM assisted and augmented the service design [32] work, specifically uncovering adoption issues and opportunities that would otherwise have been neglected in more traditional user studies. Davis's [13] original TAM predicates acceptance based on the end-user's *perceived usefulness* (PU) and *perceived ease of use* (PEOU) of the technology for a specific purpose. As part of the research, TAM was utilised to understand how staff in the NHS would adapt to new technology, and whether they feel supported using it. If staff believed that virtual consultations were useful (perceived usefulness), easy to use (perceived ease of use) and that senior management supported their endeavours, technology would likely be accepted and utilised fully. By using the TAM, areas of concern, and areas where staff feel more comfortable, were identified.

We found that, despite doctors and admin staff having a positive attitude towards virtual consultations, the perceived usefulness of the service was comparatively low. We also discovered that despite participants having high self-efficacy, they were anxious about having such a service rolled out. Moreover, staff felt that the value of consultation could be diminished if online consultations are either overused or misused. This may have contributed to user anxiety. Lastly, we learnt that staff felt that, while management support around the service would be limited, they believed they have the technological equipment required to launch the services as a kind of Minimum Viable Product.

As part of the design process, user research is foremost in structuring the application, prioritising components and features, and exploring and mitigating the issues that may arise. This user centric design methodology encompassed two research techniques, ethnographic research and co-design. The research took place through face-to-face individual interviews, group workshops and online interviews through Skype for Business. In total, the research phase had twenty-five (25) participants, with eight (8) doctors, eleven (11) patients and six (6) administrators. In choosing the participants, there was an important emphasis on reflecting the diverse group of people that may use this application. Personas were iteratively developed from the patient cohort rather than from a pre-defined profile to expedite design work.

A total of 5 individual patient interviews took place, primarily on Skype for Business due to distances between patients. The interviews were focussed on determining current processes for booking and attending appointments, and, once this had been established, moved onto how virtual consultations could enhance this process. The primary questions included:

- How do you usually book appointments?
- Would you be more likely to attend appointments if you had access to virtual consultations?
- What are the types of appointments that virtual consultations could be used for?
- What are the positives and negatives of virtual consultations?

Virtual consultations are not designed to completely replace face-to-face consultations, but rather to provide a more efficient and effective service to the patient. Thus, an understanding of what types of appointments can be served through an online service is imperative. The most popular appointment types for virtual consultations were:

- Follow up appointments
- Repeat prescriptions
- Sexual health advice and test results
- Mental health advice and check ups

The design thinking methodology of the classic 'Design Thinking' method 'Rose, Bud, Thorn' was also utilised during the interviews to understand what was good about this concept, but also the concerns and areas for enhancement. The results were consistent across all interviews:

**Positives**:

- No travel to surgery
- No contamination from surgery
- Saves me time
- More likely to attend

**Negatives**:

- Internet connection
- I'm not great with tech
- Tech equipment needed
- Less personal

Patients also took part in a design thinking workshop, where a variety of methods were used to determine user stories, validate personas, and help understand how this application could be used with patients. The workshop confirmed much of what was stated in the patient interviews, although there was a difference in how each age group perceived the negatives. Younger participants had much less concern regarding technology, instead seeing technology as the solution to some of their issues, rather than a constraint. The workshops built upon the concepts and features that were defined in the interviews, and a co-design session was utilised to help design the patient waiting room and the Skype for Business screens. These screens were later used to find common themes and to design the patient screens.

Eight (8) doctors were interviewed individually and in one small group to determine the requirements of the clinicians. The interviews were spread over multiple hospitals, doctor's surgeries and locations to ensure a diverse group of doctors. The interviews were focussed on user stories and the types of appointments that this technology could serve. The top five suggestions on appointments were similar to those of patients:

- Follow up appointments
- Repeat prescription appointments
- Medication reviews
- Mental health check ups
- Some results could be provided online (only minor)

The doctors also reported on the perceived benefits and negatives of such a technology, and the most popular responses are listed below:

**Positives**:

- No travel required to patients
- More personal, GP rapport (in contrast to phone consultations)

- Could save money from less travel
- Quality of diagnosis higher than phone consultation (especially mental health)
- Saves patients having to travel, and patients are therefore more likely to attend

**Negatives**:

- Internet connection
- Elderly patients may not feel this is appropriate for them
- Tech equipment needed
- If a replacement for face-to-face appointments, there could be a loss of rapport

In addition to the above, there was a focus on the functionality of both the doctor's dashboard and Skype screen. To discover these pieces of functionality, co-design was utilised with the doctors to draw out common themes.

The final user group identified were the administrators, who are responsible for booking appointments. Six (6) administrators participated in a workshop, focussing on the current appointment booking process, concepts, user stories, positives and constraints and then co-design. The current booking process was identified as a key research piece. By understanding the pain points of the current journey, and how new technology can provide a more effective booking process, we identified the following requirements:

- Calendar-based, similar to the current SystemOne view
- Ability to choose the GP, and thus view each GP's calendars
- Need to be able to input patient details, as well as any notes
- Requires confirmation that the appointment has been made.

Having completed designs based upon the interviews and workshops conducted with users, these were tested as wireframe prototypes. The aim of this was to ensure that the customer journey is efficient, and that users can progress through that journey within a sufficient time-frame. Different tasks were set for doctors, administrators and patients, as some of the tasks included booking an appointment (administrator), joining a session (patients) and opening an appointment (doctors). To test the usability, the expected way to complete the task was recorded by each participant, and the actual data after task completion. This was in addition to the time expected to complete the task, the perceived difficulty, and participants' perceptions of the necessary steps to complete the task. Additionally, the System Usability Scale (SUS) was utilised to provide a numerical figure by which we could benchmark the usability of the application. SUS suggests a minimum usability score of sixty-eight (68) out of one hundred (100); the virtual consultation application received an overall score of 93.5/100, deeming the application to have 'excellent' usability according to the SUS model.

## 9.4 Analysis

To conduct this analysis, a qualitative framework was used, where high, medium and low certainty facts for patients, GPs, and administrators were drawn out from the data and put into high level themes, in order to identify the key blockers to success and the most important attributes. The insights gained were analyzed using an Agile Coding and Framework approach, where data was reviewed with a research facilitator and an initial model of the data and framework produced (top down). Themes were explored using this top-down approach as well as a bottom up Framework [30] approach. Together, the data mapped to four themes comprising:

- Access to Technology
- Access to Healthcare
- Quality of Healthcare
- Quality of Experience

As every data point had been classified and counted, it was possible to quantify relative weights to the groupings and also the relative size of constituency (e.g. patients vs. GPs) among all participants. While the data pertained to a specific healthcare project, it can be generalised to the design and development of other technology-enabled healthcare services. The findings suggested that enabling access to healthcare through technology is both 'locked' in terms of excluding adoption by less technically savvy patients and providers. At the same time, it is 'unlocked' as it drives adoption generally and specifically among more technically astute users (e.g. younger, geographically dispersed groups etc.) who might otherwise be less willing to get treatment. The notion of lock also suggests that digital healthcare services must therefore transition through states of increasing sophistication through 'adoption thresholds'. This means that adoption needs to be built through incremental service improvement, with each step widening the potential audience for the service, but without risking developing expensive and/or risky functionality.

### *9.4.1 Results*

Despite doctors and admin staff having a positive attitude towards virtual consultations, the perceived usefulness of the service was comparatively low. This should be considered when rolling this out into the UK NHS—doctors and admin staff need to be educated on the usefulness of the service, which, in turn, will inevitably raise their views of how this service fits into their everyday life. Despite participants having high self-efficacy, they were anxious about having such a service rolled out in the NHS. Moreover, staff felt that their image around the practice would not be enhanced through using such a service, and this may have contributed to being anxious about using it. Lastly, NHS staff felt that management support around the service would be average, but that they believe they have the technological equipment required to succeed.

**Convenience And Efficiency**
A key theme that derived from both patients and clinicians was convenience. Patients (especially millennials and those who were reliant on public transport) viewed this as one of the most important features. They stated that they were more likely to attend online appointments, and so more patients would be treated sooner. For clinicians, the convenience of being able to conduct some home visits virtually would save approximately 30–40 min per day in travel time. These benefits were operationalised in the prototype through a GP and admin dashboard and appointment form.

**Consultation**
The rapport between patients and GPs was noted as a possible challenge to virtual consultations. Elderly patients feared the relationship, built up with their doctor over many years, might become less personal. GPs, in most cases, agreed with this assessment, but reported that everyday face-to-face consultations would not be replaced entirely and rather that only specific appointments could be served through this channel. For younger patients, the loss of rapport was not an issue, particularly because they did not feel the emotional connection with one particular doctor, and just wanted to be seen quickly.

**Better Patient Care**
Better patient care was a key theme across all patients and clinicians. Millennials noted that they were more likely to book an appointment and turn up if it was online, meaning their health concern would be looked at, and, if required, treated via this channel. For the younger generation, mental health was a key concern (also confirmed by GPs) and most check-ups are already done over the phone. Being able to see the body language of the patient online was seen as a positive attribute of the new service, as it would provide a greater insight into the patient's feelings, thus offering the potential to support greater patient care. The elderly had less faith that virtual consultations could improve health outcomes. These insights fed directly into the design work, through providing a patient waiting room and help and advice content relating to general medical issues.

**Technology**
From a patient's perspective, internet connectivity was one of the main areas of concern. Most believed that this would be the greatest barrier to this technology performing well. The user's choice of device was also a discussion point. GPs had similar worries on internet connectivity and felt that this might actually cause delays in their appointments. Most GPs noted that they already had most of the necessary equipment but would need to invest in webcams or microphones. All participants agreed that, by using technology, access to healthcare could be increased and hard to reach patients would receive more and better healthcare. A WIFI signal check was built into the prototype to help alleviate this issue.

**Adoption**

From the interviews and particularly from the TAM related data, we found that both patients and doctors, had concerns around the social groups that would be able to use virtual consultations. Millennials were the most enthusiastic about such a system, while parents with small children were sceptical and would rather their children be physically examined. Elderly patients were mixed in their response and were dependent upon their specific device (such as a tablet). Clinicians concurred that virtual consultations would make treating the hard to reach much easier, which in turn would result in better healthcare.

## 9.5 Agile Service Design

After the research phase, the project progressed into design and development. This was carried out using an Agile approach. While Agile has established itself as a tried and tested software development methodology, cases relating to healthcare applications are few and even fewer when wrapped up with a service design approach (e.g. Vink et al. [36] and Lee [21]). Agile manifests itself in developing a 'Minimum Viable Product' (MVP) rather than a fully provisioned end-to-end service. The service also utilised existing technology, which included much of the necessary functions, rather than developing a fully bespoke solution. On the negative side, this means that the resulting product or service is always only capable of providing a subset of the fully functioning end-state version.

Agile is also based on regular deliveries of 'working software' delivered in short sprints. This is counter to lengthy 'big bang' projects that are typically seen in waterfall. This means that releases can be trialed, and feedback gained early and often. In the case of this project, this data-driven approach was further enhanced by providing a patient feedback screen. From an operational perspective, Agile's common language of 'user stories' provides an accessible and easy way to ensure full and deep stakeholder involvement. For healthcare projects this clearly requires balance and a focus on reducing risk, driving adoption and data driven enhancements.

Service design's affinity with 'digital transformation' in general and public service optimisation specifically makes it a natural partner in healthcare technology. Evidencing services out of seemingly disparate service encounters (whether digital or human) would seem to be a positive differentiator to other approaches. Similarly, designing across multiple, individual actors, interactions and systems sets service design apart from the single interaction focus of user experience design. In addition, co-creation (designing with a broad set of active stakeholders) distinguishes it from more instrumental User-centred Design methodologies. Despite this potential, there are arguably a number of areas in which service design needs rebalancing in order to support technology-based healthcare applications.

Immersive research is ubiquitous within the service design domain. Augmenting 'ethnographism' with more socio-technical methods would make sense given contemporary everyday life is bound within non-visible networks, databases and intelligent automations. A different kind of data that goes beyond observing technology's end users' behaviour is needed for designing viable healthcare services. Naturally, service design privileges the kind of data produced through immersive research as inputs into design. The value of such data is unquestionable in its usefulness in ideating new service concepts. However, data pertaining to and generated from the real-time, actual use of service is arguably even more valuable in verifying concepts and optimising the experience for healthcare services.

Emphasising human activity (at the expense of technological capability) in the here and now (e.g. before service use) naturally follows through into the predisposition for mapping, blueprinting and modelling in service design. Having mapped out an optimal snapshot of human activity for a service, it is relatively easy to overlay the technological 'magic'. This 'design futurism' discounts the challenging minutiae of actual service delivery and optimisation which is, at best, vision setting and at worse misleading.

Blueprints also embody an architectural metaphor which lack relevance to living services where interactions track to fixed paths and journeys that extend over time. The notion of dynamically-created and evolving services is, to some extent, counter to service design orthodoxy, which focuses on mapping extant to future prescribed services. Instead this research and resulting framework suggest that, while mapping actors and activities is important to the design of a service, another lens is required (that of living service) to ensure the sustainability of the service.

The holistic but largely conceptual focus of service design makes sense when projects are design-led, timescales and budgets are generous and project goals are transformational. The lack of health-related cases is perhaps an indication that these conditions are relatively rare and perhaps more importantly out of step with best practice in delivering positive change through technologically based innovation. Agile is not only a reaction to the failure of delivering large-scale expensive software projects but a potentially valuable counterbalance to the 'blue-sky thinking' that underpins so much of service design. Instead, we should augment Agile's focus on delivering tangible, working systems but framed within a strategic design approach that delivers incremental as well as long-term value.

The rebalancing of service design is thus partially built on recasting the role of technology. Technology must shift from afterthought to primary importance in the research and design of a service, in order to fully realise the value of the approach. For health, this is a critical and relevant focus as healthcare providers' challenge is most often connecting legacy systems into usable services. This does not mean that service design should be constrained by technical limitations. What is needed is a more nuanced and sophisticated approach to services that recognises the centrality of the materials of their construction and the organic nature of their use and development, beyond design.

As with all of the project's design work, this was carried out in close collaboration with the developers and took account of the platform's limitations and capabilities. Designing for a specific platform is a commonplace approach in commercial design and requires creativity and technical ingenuity to deliver well-crafted living services. Design systems are a natural corollary of designing services to fit platform constraints and enablers. Unlike traditional 'blank sheet of paper' design, this approach utilises common components that integrate with the platform well, offering a consistent experience to users. At the same time, designers can focus their creativity on where elements are lacking.

### 9.5.1 Integrating Service Design, Delivery and Operations

The project underlined the need to deliver tangible positive change at pace, rather than expansive research, abstract blueprints and perfectionist solutions, seen in the service design orthodoxy. While aligning service to Agile is not unique, the data produced by the project research activities highlight a distinctively different approach to delivering health design projects. As well as highlighting practical design issues around accounting for barriers to adoption and the need to define an overarching vision and roadmap, the project led to a focus on frugality and flow.

We believe that the most important data in health service design relate to flow. This is information that can be used to benchmark current, proposed and newly operational services. While this data will be contextually specific, we think that there are likely to be commonalities in terms of service access and throughput, quality and experience of health provision that balance against cost and speed of delivery. Furthermore, we think that this information should be the starting point for new projects.

Structuring releases to adoption thresholds is not just a pragmatic delivery approach but also shifts focus to continuous improvement. In this context, practitioners should operate 'In Service' rather than separate from it, or via any form of abstracted knowledge. This would mean truly participatory design (with service agents), not before production but within it. There is a risk, evident in this project, that short, incremental change may not build to strategic positive service transformation. The research pointed to the need for a Service Roadmap that tackled three service adoption thresholds. These started at broadening access. Then the service would need to reach critical mass (and with the necessary features to do that) and then, finally, it ought to surpass the quality offered by the non-digital service.

Not only is design work a service production cost, but cost to value ought to be a factor in practice. Without this boundary, design becomes not only utopian but disconnected from its real-world context in the here-and-now. Any Agile/lean service operations project should deliver high Sprint Value to low Sprint Cost. Similarly, the impact of a release should be to increase Service Utilisation and lower Operational Cost. These rather dull factors should be both a spur for designers' creativity and critical areas to tackle through design thinking. All design work should be continuously estimated, and outcomes measured, in order to maximise value and to steer further sprints.

### 9.5.2 The Flow Methodology

The Flow Methodology is founded on three stages of service delivery, starting with a Minimum Viable Service (MVS). An MVS is focused on enabling access to a simplified, stripped-down version of the service. Such a proposition is basic, by nature, and supports a few use cases, but is built to ensure it enables core functions and proves the viability of the service. An MVS is, however, more than a conceptual proof of concept. Its adoption and use enables early service flow data collection that can help develop richer features. More importantly it enables mapping service supply and demand data. In this sense, a simple digitalised entry point for healthcare becomes the first manifestation of a Living Service, where data is used to drive features and touchpoints that build toward reaching a critical mass of users.

Having established a strong and sustainable user community, the next phase of the methodology focuses on delivering a Minimum Living Service (MLS). This is focused on delivering an optimised experience that can drive broad adoption. This means that accounting for a broader set of users, enabling service encounters on a wider variety of devices and providing consumer levels of usability. Finally, a Minimum Sustainable Service (MSS) focuses on facilitating value co-creation and operational excellence. This means not just replicating an existing service through digitalisation, but creating richer experiences that deliver comparably better healthcare on the basis of a strong cost-to-benefit ratio.

Having finalised the project, learnings and feedback were incorporated into a standardised service design, delivery and operations approach (Flow Methodology). This overarching process integrated insights from a range of Agile projects, addressed limitations in the service design orthodoxy (see above) and extended scale and scope of activities beyond blueprinting and into operations. Stories provide a common thread between agile and service design, and this narrative blending has been enriched by adding Muller's Pictive method [25] to the methodology. In Flow, stories are card-based (postcard-sized cardboard sheets) descriptions (story-cards) of key service attributes including the 6As:

- Agents—People, systems and technology involved in the service ecosystem
- Attitudes—Data pertaining to insights of different stakeholders and agents
- Activities and Actions—The atomic and sub-atomic (user stories) service interactions
- Artefacts—Tangible manifestations of the service, such as diaries and schedules
- Axioms—Rules (implicit and explicit) that underpin the living service

Unlike traditional user-centred design, the story-card approach is based on the assumption that there are significant similarities between different healthcare projects and that findings are never finalised. This means that rather than creating static research reports, initial story-cards are created before research, based on the team's hypotheses and current domain knowledge.

For example, a persona for a healthcare professional might be 'mashed together' from existing knowledge and then this initial version would be iterated through research sessions until a fixed and stable version is arrived at. Story-cards can utilise text or visuals and even sound or a combination of media. They can also be written or printed on cards or put in digital form. Digital story-cards can then be used in online research sessions that are less time-consuming and can survey a wider cohort than face-to-face sessions.

**Initiation**

Projects usually begin with an Initiation phase, which starts after concluding all commercial and contractual aspects of the work. Even in-house projects require agreement on scope, plans and resourcing. This material is usually documented in a Scope of Work. As well as detailing basic information about the project it usually includes profiles of the people working on the project, including the project pod. Depending on the size of the project this most often includes a service designer, developer and analyst who work according to Agile principles and team structures.

Having established core aspects of the project and the team that will deliver the work there is often a set of activities that take place 'pre-discovery'. These include internal workshops with the pod, project management and sponsor teams. This activity aims to understand the project context, and identify risks and opportunities, by getting to a deeper level of detail on the project. This helps foster collaborative work, and explore the scope, timing and direction of the work. This can be aided by the Shaping Game [20].

A key difference in focus and methods is the focus on service flow. Operationally, this means that all design, development and operation focuses on optimising flow, that is the quality, throughput and frugality of the service. This is developed through initial design concepts that attempt to maximise flow at minimum cost and interruption to existing service provision.

In order to do this, discovery work explores the technical context, looking for ways to reuse existing platforms and assets and map out the service architecture in terms of data and functionality. Cost benefits and their relationship to maximising flow are investigated through strategic business case building in this phase. Lastly, existing knowledge across the organisation, and from the technical and design viewpoints, is harvested in the first story-card harvest.

**Discovery**

Exploratory work follows, which closely aligns to traditional service and experience design projects. End user research is undertaken using interviews, observation and online shared work spaces. Concepts are developed as early prototypes and the story-cards are elaborated and aligned to development tasks and backlog. In parallel, a high-level service vision and roadmap is agreed that maps to the minimal viable, valuable and sustainable service levels.

Design starts with a collaborative activity to 'Design the Design', that is, to methodically check requirements and evaluate process options to ensure that the creative process used in the project provides the best fit in terms of goals, costs and agreed levels of quality. Also in the design phase, ideation is carried out to ensure that the best solution is developed from the standpoint of service flow. This can be done using 'Ideation Games' [20] that aim to unpick the underlying problem and solution in order to fully explore the domain and understand alternatives, including ones that might nullify the need for the project itself or even a technical solution.

Other work focuses on project flow, where existing patterns, assets, code and processes are checked for potential reuse. The story-cards and prototypes are elaborated too, through co-design sessions that incrementally add detail and validate decisions. This can be done face-to-face in facilitated workshops or online using tools such as Invision Freehand. Outcomes of this work focus on implementing a new or improved service as quickly as possible with the least resources. This is often done through running end-to-end service flow prototype sessions where tasks and encounters are tried out in-life to check the solution fit and make initial service flow measurements.

**Delivery**
The build phase utilises Agile methods including sprint-based development from an integrated design and development story backlog. Similar roles and ceremonies as those found in Scrum Development are adapted to service delivery, including an adapted service-owner role. The service-owner provides a single point of decision-making for the project and working with a scrum master ensures the smooth running of the project and its compliance to cost and quality levels agreed in the pre-discovery work.

**Operations**
As the service goes live, effort and focus shifts to operations. Here service flow measurements are augmented with triangulated data from user research, service analytics and customer satisfaction data. This then provides quality and value calculations that can be computed against cost of provision and operation in order to provide benchmarking data. Operational work is usually handed off to an internal team (if initially run externally) who take ownership of the service backlog, vision and roadmap and then go on to deliver incremental service improvements using Agile methods.

## 9.6 Conclusion

This project started as a relatively small e-consultancy project. Considerable learnings from applying service design in an Agile environment, as well as practical insights from delivering healthcare innovation at pace, emerged from the work. We believe that a significant shift is needed in service design, towards rapid, incremental delivery, where service flow and frugality is put at a premium. This change requires different kinds of data than those collected in traditional UCD projects, and also modifying attitudes to technology.

If we are serious about transforming healthcare, then we need to be in a position to harness rather than constrain our choices. This means radically moving from 'user-centric' agnosticism to a more nuanced position where technology is an asset in the designers' toolkit. Design is not enough, in itself, to make sustainable and progressive change either. Instead of blueprint, we need to embrace delivery and operations, with Agile's focus on continuous, incremental innovation. Together, these factors ought to put us in a strong position to improve healthcare in the future and here-and-now.

## References


1. Ahadzadeh, A.S., Pahlevan, S.S., Ong, F.S., Khong, K.W.: Integrating health belief model and technology acceptance model: an investigation of health-related Internet use. J. Med. Internet Res. **17**(2), e45 (2015)
2. Ball, M.J., Lillis, J.: E-Health: transforming the physician/patient relationship. Int. J. Med. Inform. **61**(1), 1–10 (2001)
3. Banos, O., Villalonga, C., Garcia, R., Saez, A., Damas, M., Holgado-Terriza, J.A., Lee, S., Pomares, H., Rojas, I.: Design, implementation and validation of a novel open framework for agile development of mobile health applications. Biomed. Eng. Online **14**(Suppl2). S6 (2015)
4. Bate, P., Robert, G.: Experience-based design: from redesigning the system around the patient to co-designing services with the patient. Qual. Saf. Healthc. **15**(5), 307–310 (2006)
5. Beck, K., Grenning, J., Martin, R.C., Beedle, M., Highsmith, J., Mellor, S., van Bennekum, A., Hunt, A., Schwaber, K., Cockburn, A., Jeffries, R., Sutherland, J., Cunningham, W., Kern, J., Thomas, D., Fowler, M., & Marick, B.: Manifesto for Agile Software Development. http://agilemanifesto.org (2001). Accessed on 27 Jan 2018
6. Bennani, A.E., Oumlil, R.: IT acceptance by nurses in Morocco: application of a modified unified theory of acceptance and use of technology. IBIMA Bus. Rev 1–10 (2014)
7. Bødker, S., Ehn, P., Kammersgaard, J., Kyng, M., Sundblad, Y.: A Utopian experience. In: Bjerknes, G., Ehn, P., Kyng, M. (eds.), Computers and democracy: a Scandinavian Challenge. Avebury (1987).
8. Bowen, S., McSeveny, K., Lockley, E., Wolstenholme, D., Cobb, M, Dearden, A.: How was it for you? Experiences of participatory design in the UK Health Service. Co-Design **9**(4) (2013)
9. Charles, C., Whelan, T., Gafni, A.: What do we mean by partnership in making decisions about treatment? BMJ: Brit. Med. J. **319**(7212), 780–782 (1999)
10. Christmann, S.: The impact of online health information on the doctor-patient relationship. In: Master's Thesis, London School of Economics. [Retrieved 14/08/2017] from Media@lse Electronic MSc Dissertation Series (2013)
11. Cimperman, M., Brenčič, M.M., Trkman, P.: Analyzing older users' home Telehealth services acceptance behavior—applying an extended UTAUT model. Int. J. Med. Informatics **90**, 22–31 (2016)
12. Clemensen, J., Larsen, S.B, Kirkevold, M, Kyng, M.: Participatory design in health sciences: using cooperative-experimental methods in developing health services and computer technology. Qual. Health Res. **17**(1) (2007)
13. Davis, F.: Perceived usefulness, perceived ease of use, and user acceptance of information technology. MIS Q. **13**(3), 319–340 (1989)
14. Donetto, S., Pierri, P., Tsianakas, V., Robert, G.: Experience-based co-design and healthcare improvement: realizing participatory design in the public sector. Des. J. **18**(2), 227–248 (2015)
15. Griffiths, F., Bryce, C., Cave, J., Dritsaki, M., Fraser, J., Hamilton, K., Huxley, C., Ignatowicz, A., Kim, S.W., Kimani, P.K., Madan, J., Slowther, A.M., Sujan, M., Sturt, J.: Timely digital patient-clinician communication in specialist clinical services for young people: a mixed-methods study (The LYNC study). J. Med. Internet Res. **19**(4), e102 (2017)
16. Holden, R.J., Karsh, B.T.: The technology acceptance model: its past and its future in health care. J. Biomed. Inf. **43**, 159–172 (2010)
17. Hu, P.J., Chau, P.Y.K., Liu Sheng, O.R., Yan Tam, K.: Examining thetechnology acceptance model using physician acceptance of telemedicine technology. J. Manage Inf. Syst. **16**(2), 91–112 (1999)

18. Keely, E., Liddy, C., Afkham, A.: Utilization, benefits, and impact of an e-consultation service across diverse specialties and primary care providers. Telemedicine J. e-health Off. J. Am. Telemedicine Assoc. **19**, 733–738 (2013)
19. Knight, J.: The experience design framework: supporting design thinking in the sevice domain. In: Miettinen, S, Valtonen, A (eds.) Service Design with Theory. Discussion on Value, Societal Change and Methods. Lapland University Press (2012)
20. Knight, J.: Ideation Games. https://www.slideshare.net/WorldofKnight/radical-innovation-33889361 (2014). Accessed 2nd Jan 2018
21. Lee, S.: Evaluating serviceability of healthcare servicescapes: service design perspective. Int. J. Des. **5**(2), 61–71 (2011)
22. Liddy, C., Rowan, M.S., Afkham, A., Maranger, J., Keely, E.: Building access to specialist care through e-consultation. Open Med. **7**(1), e1–e8 (2013)
23. Liddy, C., Afkham, A., Drosinis, P., Joschko, J., Keely, E.: Impact of and Satisfaction with a new eConsult service: a mixed methods study of primary care providers. J. Am. Board Fam. Med. **28**(3), 394–403 (2015)
24. Liddy, C., Drosinis, P., Keely, E.: Electronic consultation systems: worldwide prevalence and their impact on patient care-a systematic review. Fam. Pract. **33**(3), 274–285 (2016)
25. Muller, M.J.: Participatory design: the third space in HCI. In: Jacko, J.A., Sears, A. (eds.), The Human Computer Interaction Handbook, L. Erlbaum Associates Inc., Hillsdale, NJ, USA, pp. 1051–1068 (2002)
26. Nijland, N., van Gemert-Pijnen, J.E.W.C., Kelders, S.M., Brandenburg B. J., Seydel, E.R.: Evaluation of the use of an ask-the-expert e-consultation service for support on health-related requests. In: Proceedings, 2010 Second International Conference on eHealth, Telemedicine, and Social Medicine, St. Maarten, pp. 72–76 (2010)
27. Nijland, N., van Gemert-Pijnen, J.E.W.C., Boer, H. Steehouder, M.F., Seydel, E.R.: Increasing the use of e-consultation in primary care: results of an online survey among non-users of e-consultation. Int. J. Med. Inf. **78**(10), 688–703 (2009)
28. Patel, V.L., Kushniruk, A.W.: Medical informatics and the Communications. Graphics and Web design based science of cognition. J. Am. Med. Inf. Assoc. **5**(6), 493–502 (1998)
29. Pilemalm, S., & Timpka, T.: Third generation participatory design in health informatics-Making user participation applicable to large-scale information system projects. J. Biomed. Inf. **41**(2), 327–339 (2008)
30. Ritchie, J., Lewis, J.: Qualitative Research Practice: A Guide for Social Science Students and Researchers. Sage, London (2003)
31. Robert, G., Cornwell, J., Locock, L., Purushotham, L,. Sturmey, G., Gager, M.: Patients and staff as codesigners of healthcare services. BMJ **350**, g7714 (2015)
32. Shostack, L.: How to design a service. Eur. J. Mark. **16**(1), 49–63 (1982)
33. Taleb-Bendiab, A., England, D., Randles, M., Miseldine, P., Murphy, K.: A principled approach to the design of healthcare systems: autonomy vs governance. Reliab. Eng. Syst. Saf. **91**(12), 1576–1585 (2006)
34. Tsianakas, V., Robert, G., Maben, J., Richardson, A., Dale, C., Wiseman, T.: Implementing patient-centred cancer care: using experience-based co-design to improve patient experience in breast and lung cancer services. Support. Care Cancer **20**(11), 2639–2647 (2012)
35. Wald, H.S., Dube, C.E., Anthony, D.C.: Untangling the web—the impact of internet use on health care and the physician-patient relationship. Patient Educ. Couns. **68**, 218–224 (2007)
36. Vink, J., Mahr, D., Odekerken-Schröder, G.: Shifting the balance of power in healthcare through service design. In: Mager, B. (ed.) Service Design Impact Report—Health Sector Service Design Network (2017)
37. Weiner, J.P.: Doctor-patient communication in the e-health era. Israel J. Health Policy Res. **1**(1), 33 (2012)

# Chapter 10
# MATUROLIFE: Using Advanced Material Science to Develop the Future of Assistive Technology

**Louise Moody and Andrew J. Cobley**

**Abstract** Faced with an ageing and increasingly dependent population, novel approaches are needed to support independent and healthy living for longer. An increased focus on the design of effective,usable, desirable and stigma free solutions is required to ensure that technological innovation benefits both the individual and wider society. Too often solutions are developed that whilst functionally effective, fail to meet user needs and achieve their full purported therapeutic or health benefits. This chapter outlines the MATUROLIFE project, and its aim to utilize smart materials in the development of novel assistive technology that is accpeted by the end-user. The project combines design with electrochemistry and material science innovation. It employs selective metallization to coat the fibres within a textile with a thin layer of copper to create a multi-functional material. These multi-functional materials enable the embedding of 'smart' technology within textile-based products, such as clothing, footwear, upholstery and furniture. The chapter will outline how these new functional textiles will provide the opportunity to design more discrete and subtle assistive solutions to support independent ageing.

**Keywords** Assistive technology · Smart textiles · Selective metallisation · Co-creation

## 10.1 Introduction

Urban areas in Europe are seeing an increasing population of older adults and existing approaches to care for them are becoming unsustainable, creating a European-wide

---

L. Moody (✉)
Centre for Arts, Memory and Communities, Faculty of Arts and Humanities, Coventry University, Coventry, UK
e-mail: l.moody@coventry.ac.uk

A. J. Cobley
The Functional Materials Research Group, Coventry University, Priory Street, Coventry CV1 5FB, UK
e-mail: andrew.cobley@coventry.ac.uk



A. Woodcock et al. (eds.), *Design of Assistive Technology for Ageing Populations*, Intelligent Systems Reference Library 167,
https://doi.org/10.1007/978-3-030-26292-1_10

societal challenge [1, 2]. Older adults typically want to stay in their homes independently for as long as possible [3]. As well as the benefits for the individual and those that support them, there is a need to reduce the health and social care burden of an ageing population [4]. There is a growing market therefore for products that support healthy ageing and independent living. Assistive Technology (AT) can help provide older adults with the security that will enable them to live independently, and without the need for care for longer. Despite assistive aids being available, AT is often regarded as undesirable and stigmatizes the user. Research has shown high abandonment rates and a lack of user acceptance and continued use [5, 6]. There is a drive therefore for products that can help people in more subtle and discrete ways, and that are attractive and desirable to own. The 3 year EU-Horizon 2020 funded MATUROLIFE (**Meta**llisation of **T**extiles to make **Ur**ban living for **Ol**der people more **I**ndependent and **F**ashionabl**e**) project (2018–2020) is developing products to meet this need through use of smart materials and E-Textiles to embed assistive functionality discreetly in everyday items.

### 10.1.1 Smart Functionality and Assistive Technology

Fabrics able to sense or respond to stimuli, and embed functionality are often referred to as 'smart' or 'e-textiles' [7–10]. The extent of intelligence offered through materials can vary from 'passive' solutions that sense the environment; to active smart textiles that can sense and react to the environment; to very smart textiles that have the ability to adapt their behaviour to the circumstances [11, 12]. Increasingly, the miniaturisation of circuits and micro-components make them almost invisible and easier to embed in flexible substrates [7, 13]. Example applications include material- based smart phone controls [14]; heating systems (i.e. outerwear, gloves and footwear) [15–17]; and the use of sensors for fitness and performance monitoring [12, 14, 18]. It is envisaged that smart material technology will increasingly be used to help older adults for example by providing health-monitoring capabilities, built-in communications and safety features, and alarms and falls detectors [19].

### 10.1.2 MATUROLIFE

The MATUROLIFE project will make use of innovations in smart materials, and techniques developed to metallise fabrics, to embed assistive functionality in a range of products. The project involves a range of expertise including electronics, printing, electrochemistry, material science, design, clothing, footwear and furniture manufacturing. Partners represent 20 organizations from 9 European countries (France, Italy, Poland, Spain, Turkey, Belgium, Germany, Slovenia, United Kingdom).

The funding stream (H2020-EU.2.1.3. NMBP-05-2017 [20]) targets 'Advanced materials and innovative design for improved functionality and aesthetics in high added value consumer goods' and the application of innovative advanced material solutions in the creative industries. The MATUROLIFE project seeks to address the quality of life of older adults through innovation in assistive products enhanced with advanced or 'smart materials'. The involvement of design expertise in the project enables a focus on improved desired functionality, usability and aesthetics to ensure that desirable products are developed. The emerging product range will include a piece of furniture, clothing and footwear that offer assistive functions by embedding metallised fabrics. The assistance offered will be subtle, with well-designed items adapting and responding to individual or environmental changes to provide the user with support. The development and use of metallised textiles will enable discrete addition of electronic components without adversely affecting the appearance, comfort or the weight of the products.

## 10.2 Smart Material Innovation

To make textiles 'smart' and perform a useful function generally requires the integration of some form of electronic component e.g. sensors that might detect body heat, vital signs etc. Rather than have such sensors as separate 'wearable' devices the MATUROLIFE project team aims to integrate them into textiles for use in the production of material-based products.

Electrochemistry and material science innovation is taking place through the selective metallization of textiles. To develop smart solutions there must be interconnections between the sensors and other essential components e.g. the battery, the aerial etc. These interconnections are typically created using copper 'tracks'. Copper is relatively cheap and a highly electrically conductive metal. An example of these copper tracks can be seen on printed circuit boards (PCBs) which are ubiquitous in electronic devices [21]. Figure 10.1 shows a simulated image of a PCB electroplated

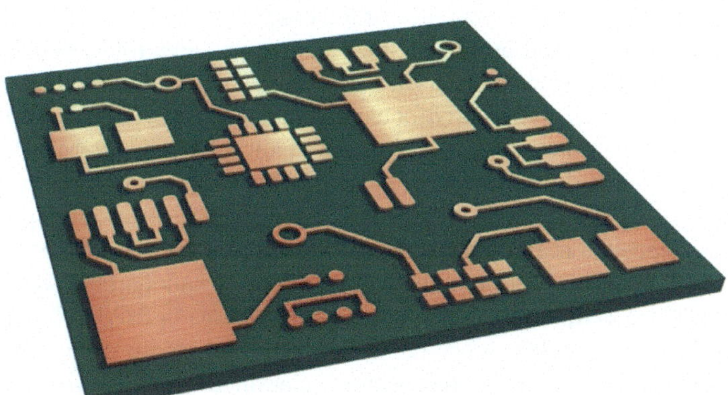

**Fig. 10.1** An example of a copper electroplated PCB

with copper. These copper tracks are subsequently coated with solderable, protective coatings and will be used to connect the various components that will be attached to the PCB.

Figure 10.1 illustrates that a traditional PCB is produced on a rigid base which is often an epoxy based dielectric (i.e. non-conductive) material. The use of a dielectric is essential to prevent shorting between the copper tracks. However, the use of such rigid substrates for smart textiles would make the textile feel rigid and add bulk and weight. To improve the integration of electronics into textiles there has been a move in recent years to use the textile itself as the base for the copper interconnections. Like the dielectric materials used in PCBs, textiles are non-conductive and therefore, as long as the can be selectively metallised, make a suitable base onto which to place circuitry.

There are several ways to introduce electronic connectivity to a textile. Solid metal wires can be knitted or weaved into the textile [22] but this tends to make the textile feel more rigid and can add significant weight. In addition, some wires can be brittle and, with frequent flexing, might break. Conductive inks can be printed onto the surface of a textile [23] to produce connectivity but these are often printed onto a polyurethane film [24], which is then adhered to the textile. The inks used require curing adding additional process stages and cost to manufacture. In addition, the conductivity can fail over time as the printed pattern deteriorates with use, washing etc.

The approach being applied in MATUROLIFE is an additive process that aims to coat the fibres within the textile with a thin layer of copper. By coating the fibres within the textile a multi-functional material can be created. The addition of a thin coating of copper means that there is little change to the feel and weight of the textile whilst those fibres that have been copper coated are now highly conductive and can be used to connect electronic components. In addition, the properties of the textile are maintained i.e. flexibility, washability etc. Figure 10.2 shows an enhanced high magnification Scanning Electron Microscopy (SEM) image of fibres within a polyester textile that have been coated with copper. Close observation of the fibres reveals the slightly granular appearance of the copper deposit.

The coating process used in MATUROLIFE is called electroless copper plating. Electroless copper plating has been used for many years to metallise non-conductive materials such as textiles particularly for electro-magnetic interference (EMI) shielding [25]. Indeed, in a project that laid much of the platform for MATUROLIFE Coventry University and the National Physical Laboratory (NPL) developed such a process for the selective metallisation of textiles [26]. Electroless plating requires a catalyst to be deposited to initiate the deposition reaction (see Fig. 10.3).

Typically, this catalyst is a precious metal colloid or nanoparticle such as Palladium [27] or Silver [28]. One of the material innovations in MATUROLIFE is the use of a copper nanoparticle catalyst [29]. This is not only more sustainable (copper is an earth abundant metal, whilst palladium is classified by the EU as a critical raw materials (CRM)) but, as the catalyst is the most expensive stage of the metallisation process, will also significantly reduce manufacturing costs. The catalyst will be selectively applied to the textile as a catalytic ink [30] in the pattern required to create the necessary electronic circuitry. Subsequent electroless copper plating

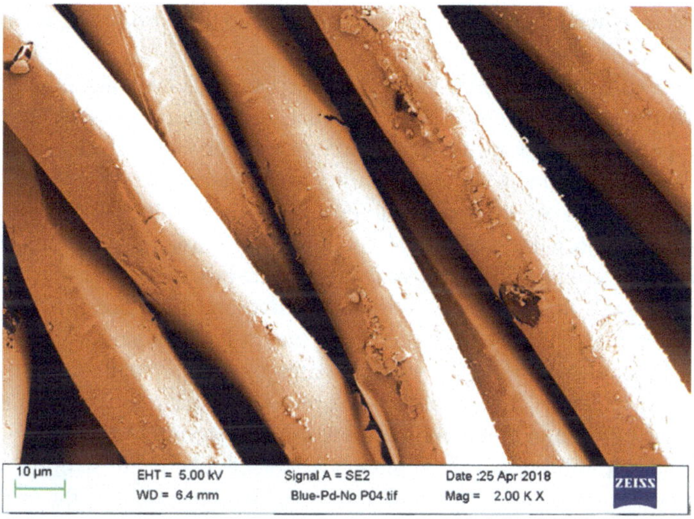


**Fig. 10.2** Enhanced/coloured SEM image of copper coated fibres in a polyester textile

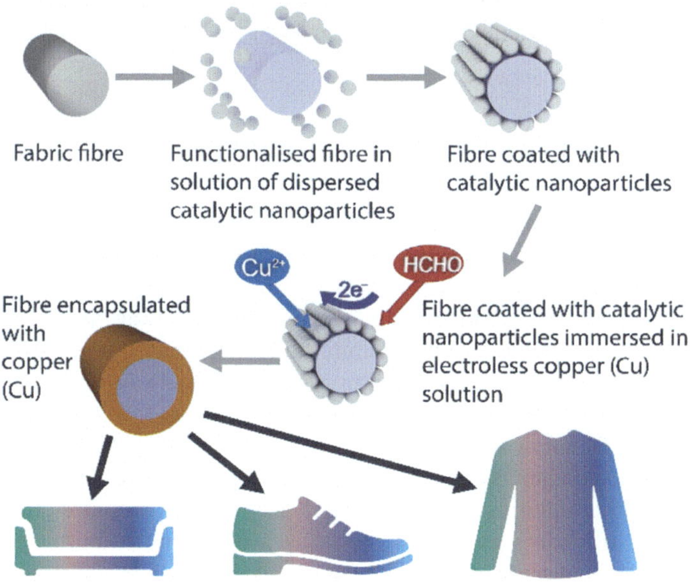


**Fig. 10.3** Illustrated process

will only occur where the catalyst has been deposited thus producing the electronic connectivity.

Once the textile has been copper plated the copper tracks will need to be protected from the environment. This could be from sweat, atmospheric moisture or the effects of washing powders. Without this protection the copper will corrode/oxidise and its properties such as conductivity and ductility will deteriorate. The MATUROLIFE team are investigating two approaches to protecting the conductive tracks which can be categorised as inorganic and organic coatings. Inorganic coatings might include the deposition of a very thin (a few nm) layer of metals such as gold [31] or silver (Long and Toscano 2013) which are very passive and resistant to oxidation and corrosion. These metals can be deposited using electroless deposition and/or a process known as immersion plating. The organic coatings encapsulate the copper tracks in a polymer type material [32] which forms a physical barrier between the metal and any aggressive environment the copper track might encounter.

In summary, an additive process is planned to selectively metallise fibres within the textile to produce a truly multi-functional material which maintains the properties of the textile (feel, drape, weight) whilst adding electronic connectivity. This will enable better integration of electronics and more discreet and functional AT. It is intended that the versatility of the resulting technology will be wide with applicability to many industries (textiles, electronics, surface engineering, etc.).

## 10.3 Design and Development of Prototypes

The MATUROLIFE project is complex in the level of scientific innovation and technical development required. In order to bring together the material innovation and the development of functional AT products a design-led approach is being adopted. The focus is on furniture, clothing and footwear solutions. Many smart independent living solutions are restricted to use within the home where the available systems and connectivity (e.g. telecare, panic alarms) are available. Development of clothing and footwear enables use outside of the home, whilst 'smart furniture' is a relatively unexplored area to exploit.

A design management approach is guiding the development pathway and seeking to reconcile and integrate the material science research alongside the technical development of the prototypes whilst ensuring a user-centric approach [33]. Too often assistive technology is developed without the direct involvement of the users in the end-to-end design process, and is driven by functional requirements rather than taking into account the emotional requirements of the end user. The outcome does not provide a meaningful, valuable service as a result [34, 35].

There is a desire to create a new market around desirable AT that embeds innovative technology, but where the products are not driven by the technology alone but centred on user need and requirements. User-centred design (UCD) is both a philosophy and variety of methods by which end-users influence and are involved in design [36, 37]. It typically involves identifying the intended users of a product, then ascertaining and prioritizing their needs and requirements, as well as the task requirements; developing and testing prototypes; evaluating design alternatives; analysing and resolving usability problems; and testing the design and its features with users in an iterative manner [36]. UCD can involve consulting users about their needs and involving them at specific points during the design process; or it can involve users being involved as partners and co-designing throughout the development process.

In the MATUROLIFE project user and stakeholder engagement is being achieved through a number of design research methods that have included interviews, scenario mapping, personas and co-creation activities, involvement of a stakeholder panel, and iterative testing of prototypes. By working with older adults across a number of European countries, it is intended that the views, cultural values and ideas of older people will become central to the development process and support innovation and creation of new meaning around the core values of aesthetics, desirability, independence and security [34].

### *10.3.1 Forming a Design Direction*

Our initial understanding of the needs of older adults needs was informed by a literature review and a series of semi-structured interviews. Thirty-seven older adults (aged over 65 years of age) from 6 European countries (i.e. France, Italy, Poland, Spain, Turkey, and United Kingdom) were interviewed. The questions explored current and future health concerns, digital literacy, use of technologies promoting independence and well-being, and views about the acceptability and accessibility of technology. The interviews also explored contextual issues such as their living environment, typical day, and the activities performed to ascertain an understanding of overlooked requirements and needs.

Acknowledging age-related health conditions, the participants' recognized physical change such as a tendency to complete tasks and activities more slowly, perceived changes in body temperature and a change in sleep patterns. Falls were one of their main fears. Factors affecting their emotional wellbeing included loneliness, perceived loss of social value, feeling unable to fully contribute as a member of their community and in some cases depression. Overall, a fear of becoming more dependent on others was expressed. Different strategies were adopted by the participants to improve well-being and independence. These include maintaining an active life (both physically and mentally); taking care of their body by eating healthily; exploring new hobbies and interests; and engaging in social activities and volunteering.

Despite the adoption of strategies to remain independent, a number of factors were perceived as potential barriers to well-being. Economic and financial concerns were raised in terms of sustaining long-term independence, as well as ease of use of transport to go out independently. The attitude towards technology amongst interview participants varied. Some use the internet, tablet and smartphones regularly. Others used more basic phone models. Most participants recognized the value of technology adoption for a purpose that would improve their lives, but indicated concerns about data sharing and privacy.

These interviews provided a good contextual grounding for the project and the qualitative views guided the design of a number of design tools, specifically personas, design insights and the structure and focus of co-creation workshops [38, 39]. Ten co-creation workshops were then run in 9 European countries (Spain, Italy, Belgium, UK, France, Slovenia, Germany, Poland and Turkey). These workshops built on the insights and functional requirements collected during the interviews outlined above, and aimed to specify and develop the MATUROLIFE products. The premise being that the products would be more attractive and representative of user needs because they are designed in collaboration with the people who are to use them.

A total of 94 participants (63 female and 31 male) took part across the 10 workshops. During the workshops, participants worked collaboratively to develop ideas for new products. The first 4 workshops were used to scope initial requirements and agree 3 design briefs—one for a smart item of furniture, one for footwear and one for clothing. With both Maturolife team members and participants coming from diverse subject backgrounds and different cultures the workshops were developed iteratively to refine the co-creation approach and adapt it to local needs. The design brief was then translated by the designers into a series of concepts which were further developed and iterated upon during the following 6 workshops. This second set to workshops aimed to prioritise and agree the functionalities that users wanted to see incorporated in the clothing, footwear and furniture products.

### *10.3.2 Design Development*

Bringing together the design research and output from the co-creation workshops, 3 multidisciplinary design teams were formed that involve designers, manufacturers, electronics specialists as well as material scientists to further develop 3 design concepts:

1. Assistive footwear to support balance and reduce falls
2. Assistive clothing to help regulate temperature
3. Assistive furniture to enable improved sleep and mobility

A series of workshops have been run between project partners to share ideas and review the concepts as they develop. Service design maps have been generated for each of the proposed concepts to guide the development of the prototypes. A service blueprint visualises the relationships between different components of the product and the underlying resources and processes that make the product or service function. These are supported by a use case scenarios and user journeys which explain and illustrate how the products will be used from set-up, use, data collection and feedback, and alerting help if required.

Each concept will be supported by a software/mobile application that will enable the set-up of the system, choice of parameters and characteristics and provide informative data to both the user and wider support network. The data acquired about the user via the products will be uploaded to a big data platform. This information flow will enable 'intelligence' and advanced algorithms for detecting patterns, making predictions and generating reports for the user.

2D sketches, 3D CAD and 3D mock-ups are being developed to provide the low-resolution 'look and feel' and indicate functionality, appearance and interaction elements. Indicative aesthetic designs and materials are provided to illustrate how the technology will potentially be embedded without compromising the appearance of the product.

### *10.3.3 Technical Development and Integration*

The project will achieve the development of system prototypes that can be demonstrated in an operational environment (TRL7) by the end of the three-year funded period (EC) [40]. Preliminary system architecture diagrams have been developed which illustrate the system components and information flow. The current expectation of how metallised fabric will be utilised in the 3 emerging prototypes is indicated.

Ongoing development activity will focus on the integration of sensors and electronic components and using the selectively metallised fabrics to produce usable prototypes that are supported by appropriate software applications. The fabrics will be used to interconnect the core electronics and potentially as a means of integrating controls and displays. They are still under design and development at this stage (18 months into the 3 year project) and will continue to be further specified as the project continues. Provision of power and charging, algorithm development and sensor calibration are also key elements of the development pathway. The prototypes will collect data on user behaviour and characteristics to support service provision. The data will be captured and compiled in a big data platform, and therefore behaviour patterns, user profiling and alerts and recommendation can be provided. Associated user interfaces are also under development for reporting and visualising the captured information to both the direct user (i.e. the older adult) and those supporting them (e.g. carers, family members). Partner companies have expertise in the design and manufacture of clothing, footwear and furniture to enable development and exploitation of solutions in these areas.

## 10.4 Testing and Evaluation

The testing of the functional prototypes will be undertaken at a number of the levels.

### 10.4.1 Iterative Testing

Iterative development and testing is essential to ensure our products remain user-centric [37] as well as being crucial in a complex multi-disciplinary project where the design work is evolving alongside scientific developments. As well as the interviews and co-creation activity outlined above, user and stakeholder engagement is planned through the development and testing process. Relationships have been built and maintained with representative users as well as wider stakeholders including designers, healthcare professionals, manufacturers and carers etc. across partner countries via a Stakeholder Representative Panel (SRP), a stakeholder database and an External Expert Advisory Board (EEAB). These groups will guide and shape product development as well as market impact. It is aimed that this involvement will increase take-up of the resulting products, ensure usability and accessibility as well as the long-term viability of the material development and market implementation.

Both the EEAB and SRP were consulted on the conclusions from the user research and early design concepts. Progress and early design concepts were presented to the groups and feedback gained. It is intended that both groups will be regularly consulted through the remainder of the project on our progress and feedback sought on the various stages of design development.

More informal iterative feedback will be collected from a wider group through local user testing groups. This feedback will be more frequent and relate to key design decisions and stages of development. It will be undertaken locally to the partners leading on the development of the prototypes rather than across all 9 countries for practical reasons (e.g. the movability of a sofa).

### 10.4.2 Technical Testing

Once fully functional prototypes have been developed testing will be undertaken. Different technical features (mechanical, physical and chemical) as well as functionality, electrical/electronic compatibility, safety and behaviour will be checked for the conductive textiles and the 3 prototypes. In terms of the textiles, the surface properties and conductivity of the metallised textiles will be considered. Their response to mechanical stress and washing will be tested, alongside their response to sweat. The electrostatic properties of the textiles and the electrical functioning of the prototypes will be considered through characterization tests. Each prototype will then be validated in terms of a series of physical, mechanical, chemical, comfort, safety and aging tests against relevant industry standards (e.g. against ISO standards). The technical testing will ensure the feasibility of each of the prototypes and the results will inform future development and scaling of MATUROLIFE processes and production.

### 10.4.3 Final Evaluation

In the latter stages of the project, the resulting prototypes will be tested with end-users. It is envisaged that representative older adults will make use of the prototypes either in the home setting or a usability laboratory. Product walkthroughs will be employed to identify remaining barriers to use and participants will be interviewed to give final feedback. The SRP and EEAB will also review the final prototypes in terms of the potential impact on the independence of older adults and the extent the project has achieved the intended aims.

## 10.5 Conclusions

MATUROLIFE partners have developed a novel method to selectively metallize textiles and fabrics. Technological advances have produced a highly innovative selective metallization process that utilizes nanotechnology, electrochemistry and materials science to encapsulate the fibres in textiles with metal. It is also aimed to develop a novel copper nanoparticle catalyst to replace the use of palladium in the metallisation process leading to significant impact on the cost and sustainability of E-textiles. The metallised textiles will provide conductivity and electronic connectivity allowing better integration of electronics and sensors into the fabrics and textiles, as additional wires are not required. By coating fibres within the textile with a thin layer of copper a truly multi-functional material can be produced with the benefits of a highly conductivity metal (electronic connectivity) with those of a textile (light, flexible, washable etc.). These multi-functional materials will enable the embedding of 'smart' technology without significant increase in weight, and can be utilized in the development of textile-based products, such as clothing, footwear, upholstery and furniture. This new way of developing functional textiles can be more discrete and subtle, and affect the overall aesthetic design of the solution compared to the use of wiring that might otherwise be required.

To actually integrate these materials in products that are acceptable to and utilised by older adults is the end-goal of the MATUROLIFE project. This requires multidisciplinary collaboration from partners in a number of specialist areas as well as ongoing involvement of older adults. An ongoing iterative testing process will be critical in ensuring the projects ambitious gaols are achieved.

**Acknowledgements** The authors wish to acknowledge and thank the participating older adults who gave their time for the interviews and the co-creation activity. This publication is based on work undertaken and planned as part of the MATUROLIFE (Metallisation of Textiles to make Urban Living for Older People more Independent and Fashionable), project which has received funding from the European Union's Horizon 2020 research and innovation programme under Grant Agreement No. 760789. Any dissemination reflects the authors' view only and the European Commission is not responsible for any use that may be made of the information it contains. The views and opinions expressed in this paper are those of the authors and are not intended to represent the position or opinions of the MATUROLIFE consortium or any of the individual partner organisations.

## References


1. Beard, J.R., Officer, A., de Carvalho, I.A., Sadana, R., Pot, A.M., Michel, J.-P., Chatterji, S.: The World report on ageing and health: a policy framework for healthy ageing. Lancet **387**(10033), 2145–2154 (2016). https://doi.org/10.1016/S0140-6736(15)00516-4
2. Bloom, D.E., Chatterji, S., Kowal, P., Lloyd-Sherlock, P., McKee, M., Rechel, B., Smith, J.P.: Macroeconomic implications of population ageing and selected policy responses. Lancet **385**(9968), 649–657 (2015). https://doi.org/10.1016/S0140-6736(14)61464-1
3. Gaßner, K., Conrad, M.: ICT enabled independent living for elderly. In: A Status-quo Analysis on Products and the Research Landscape in the Field of Ambient Assisted Living (AAL) in EU-27. VDI (2010)
4. Mostaghel, R.: Innovation and technology for the elderly: Systematic literature review. J. Bus. Res. **69**(11), 4896–4900 (2016)
5. Chaiwoo, L.: Adoption of smart technology among older adults: challenges and issues. Public Policy Aging Rep. **24**(1), 14–17 (2013)
6. Yusif, S., Soar, J., Hafeez-Baig, A.: Older people, assistive technologies, and the barriers to adoption: a systematic review. Int. J. Med. Inf. **94**, 112–116 (2016). https://doi.org/10.1016/j.ijmedinf.2016.07.004
7. Jones, D.C.: The role of wearable electronics in meeting the needs of the active ageing population. In: McCann, J., Bryson, D. (eds.) Textile-led Design for the Active Ageing Population. Elsevier Woodhead Publishing, Cambridge (2015)
8. Lee, J., Kwon, H., Seo, J., Shin, S., Koo, J.H., Pang, C., Lee, T.: Conductive fiber-based ultrasensitive textile pressure sensor for wearable electronics. Adv. Mater. (2015)
9. Stoppa, M., Chiolerio, A.: Wearable electronics and smart textiles: a critical review. Sensors **14**(7), 11957 (2014)
10. Stylios, G.K., Lam Po Tang, S.: An overview of smart technologies for clothing design and engineering. Int. J. Clothing Sci. Technol. **18**(2), 108–128 (2006). https://doi.org/10.1108/09556220610645766
11. Tao, X.-M.: Sensor in garments. Text. Asia 38–41 (2002)
12. Van Langenhove, L., Hertleer, C.: Smart clothing: a new life. Int. J. Clothing Sci. Technol. **16**(1/2), 63–72 (2004). https://doi.org/10.1108/09556220410520360
13. Zeng, W., Shu, L., Li, Q., Chen, S., Wang, F., Tao, X.-M.: Fiber-based wearable electronics: a review of materials. Fab. Devices Appl. **26**(31), 5310–5336 (2014). https://onlinelibrary.wiley.com/doi/abs/10.1002/adma.201400633
14. Chan, M., Estève, D., Fourniols, J.-Y., Escriba, C., Campo, E.: Smart wearable systems: current status and future challenges. Artif. Intell. Med. **56**(3), 137–156 (2012). https://doi.org/10.1016/j.artmed.2012.09.003
15. Hu, J., Mondal, S.: Temperature sensitive shape memory polymers for smart textile applications. In: Mattila, H.R. (ed.) Intelligent Textiles and Clothing. CRC—Woodhead Publishing Limited, Cambridge (2006)
16. Mondal, S.: Phase change materials for smart textiles—an overview. Appl. Thermal Eng. **28**(11), 1536–1550. https://doi.org/10.1016/j.applthermaleng.2007.08.009
17. Rantanen, J., Alfthan, N., Impio, J., Karinsalo, T., Malmivaara, M., Matala, R., Vanhala, J.: Smart clothing for the arctic environment. Paper presented at the digest of papers. fourth international symposium on wearable computers (2000)
18. Solaz, J., Belda-Lois, J.M., Garcia, A.C., Barbera, R., Dora, T.V., Gomez, J.A., Prat, J.M.: Intelligent textiles for medical and monitoring applications. In: Mattila, H.R. (ed.) Intelligent Textiles and Clothing. CRC—Woodhead Publishing Limited, Cambridge (2006)
19. Eskofier, M.B., Lee, I.S., Baron, M., Simon, A., Martindale, F.C., Gaßner, H., Klucken, J.: An overview of smart shoes in the internet of health things: gait and mobility assessment in health promotion and disease monitoring. Appl. Sci. **7**(10) (2017). https://doi.org/10.3390/app710098
20. European Commission: NMBP-05–2017—Advanced Materials and Innovative Design for Improved Functionality and Aesthetics in High Added Value Consumer Goods. Available from: https://cordis.europa.eu/programme/rcn/701801/en (2017). Last accessed 09/01/2019

21. LaDou, J.: Printed circuit board industry. Int. J. Hyg. Environ. Health **209**, 211–219 (2006)
22. Cherenack, K., Van Pieterson L.: Smart textiles: challenges and opportunities. J. Appl. Phys. **112** (2012)
23. Kao, H., Chuang, C., Chang, L., Cho C., Chiu, H.: Inkjet-printed silver films on textiles for wearable electronics applications. Surf. Coat. Technol. (2019)
24. Gysling, H.J.: Nanoinks in inkjet metallization—evolution of simple additive-type metal patterning. Curr. Opin. Colloid Interface Sci. **19**, 155–162 (2014)
25. Han, E.G., Kim, E.A., Oh, K.W.: Electromagnetic interference shielding effectiveness of electroless Cu-plated PET fabrics. Synth. Met. **123**, 469–476 (2001)
26. Wills, K.A., Krzyzak, K., Bush, J., Ashayer-Soltani, R., Graves, J.E., Hunt, C., et al.: Additive process for patterned metallized conductive tracks on cotton with applications in smart textiles. J. Text. Inst. **2017**, 1–10 (2017)
27. Osaka, T., Takematsu, H., Nihei K.: Study on activation and acceleration by mixed PDCL//2/SNCL//2 catalysts for electroless metal deposition. J. Electrochem. Soc. **127**, 1021–1029 (1980)
28. Chao, Y., Liu, C., Pan L., Lee, C.: Activity of small silver nanocubes as activators for electroless copper deposition. Electrochim. Acta (2015)
29. Graves, J.E., Sugden, M., Litchfield, R.E., Hutt, D.A., Mason, T.J., Cobley, A.J.: Ultrasound assisted dispersal of a copper nanopowder for electroless copper activation. Ultrason. Sonochem. **29**, 428–438 (2016)
30. Krykpayev, B., Farooqui, M.F., Bilal, R.M., Vaseem, M., Shamim, A.: A wearable tracking device inkjet-printed on textile. Microelectron. J. **65**, 40–48 (2017)
31. Bui, Q.V., Nam, N.D., Choi, D.H., Lee, J.B., Lee, C.Y., Kar, A., et al.: Corrosion protection of ENIG surface finishing using electrochemical methods. Mater. Res. Bull. **45**, 305–308 (2010)
32. Jablonski, M., Bossuyt, F., Vanfleteren, J., Vervust, T., de Vries, H.: Reliability of a stretchable interconnect utilizing terminated, in-plane meandered copper conductor. Microelectron. Reliab. **53**, 956–963 (2013)
33. Moody, L., York, N., Ozkan, G., Cobley, A.: Bringing Assistive Technology Innovation and Material Science together through Design. In: Howlett, R.J., Jain, L.C. (eds.) Smart Innovation, Systems and Technologies. Springer, Berlin (2019)
34. Norman, D.A., Verganti, R.: Incremental and radical innovation: design research vs. technology and meaning change. Des. Issues **30**(1), 78–96 (2014)
35. Verganti, R.: Design Driven Innovation: Changing the Rules of Competition by Radically Innovating What Things Mean. Harvard Business Press (2009)
36. Moody, L.: User-centred health design: reflections on D4D's experiences and challenges. J. Med. Eng. Technol. **39**(7), 395–403 (2015)
37. Giacomin, J.: What is human centred design? Des. J. **17**(4), 606–623 (2014)
38. Ramaswamy, V., Ozcan, K.: What is co-creation? An interactional creation framework and its implications for value creation. J. Bus. Res. **84**, 196–205 (2018). https://doi.org/10.1016/j.jbusres.2017.11.027
39. Sanders, E.B.N., Stappers, P.J.: Co-creation and the new landscapes of design, codesign. Int. J. CoCreation Des. Arts **4**, 5–18 (2008)
40. European Commission: Technology Readiness Levels (TRL). Available from: https://ec.europa.eu/research/participants/data/ref/h2020/wp/2014_2015/annexes/h2020-wp1415-annex-g-trl_en.pdf. Last accessed 09/01/2019
41. Long, E., Toscano, L.: Electroless Nickel/Immersion Silver—A new surface finish PCB applications. Met. Finish. **111**, 12–19 (2013)

**Louise Moody** is Professor of health design and human factors in the Centre for Arts, Memory and Communities at Coventry University. She is also Human Factors Theme Lead in the NIHR Devices for Dignity MedTech Co-operative. She has a BSc in psychology and sports science, and a PhD in electronic and electrical engineering from the University of Birmingham. After working as Human Factors Researcher and Consultant, she joined Coventry University in 2006, initially as Senior Lecturer in intelligent product design in the Department of Industrial Design. Since then, she has held a number of management roles within the School of Art and Design as well as leading a research theme related to arts and well-being. Her research is focused on the design and development of products, interventions and services to benefit health and well-being. She employs a range of research methods as well as art-based approaches to ensure that new solutions are functional as well as being desirable and acceptable to end-users and stakeholders. Her focus is on ensuring the involvement and representation of diverse user needs and requirements from the early stages of user need assessment, through to the design, development and evaluation of solutions. Her current projects are particularly focused on assistive and self-management technologies.

**Prof. Andrew J. Cobley** began working in printed circuit board (PCB) manufacture in 1986 for British Aerospace Dynamics (Stevenage). Subsequently, in 1989, he joined Shipley Europe Ltd. where he installed PCB manufacturing processes at some of the largest electronic manufacturers in Europe including Siemens, Sony and Ericsson. Andrew obtained his Ph.D. from Loughborough University in 2002 and joined Coventry University in 2006 where he is now a Professor of Electrochemical Deposition and leader of the Functional Materials Research Group. During his academic career he has lead several externally funded projects including, most recently, the 6 million EU H2020 project MATUROLIFE. He has over 40 academic publications, is a named inventor on over 30 patents and has presented at more than 50 International conferences. Prof. Cobley is an evaluator for Innovate UK and the European Commission and is Chairman of the Institute of Circuit Technology. He is a Chartered Chemist and Fellow of the Royal Society of Chemistry.

# Chapter 11
# Domestic Robots for Older Adults: Design Approaches and Recommendations

**Rebecca Wiczorek, Megan A. Bayles and Wendy A. Rogers**

**Abstract** This chapter provides an overview of design considerations for domestic robots to be used by older adults. We describe characteristics of older adults such as sensory, cognitive and motoric functions, as well relevant context factors of the home environment regarding motion and navigation or installation and maintenance. Methods relevant for the design of domestic robots are described including user-centered design as well as ethical considerations such as safety, security, privacy, and autonomy. Relevant tasks of daily living as well as functionalities and features of robots are presented. We describe examples of robots that have been evaluated in research for use by older adults. The robot designs are discussed with regard to benefits and limitations in the development of domestic robots for supporting daily activities of older adults in their home environment.

**Keywords** Domestic robots · Older adults · Human robot interaction (HRI) · Design recommendations

## 11.1 Introduction

The current chapter provides an overview of human factors aspects relevant to the design of domestic robots for older users. Older adults are a heterogeneous group and therefore the definition of who to include in this group is sometimes difficult. Older people can be defined based on their biological age, functional age, sociological age and some more [1]. The easiest and therefore most common definition refers to the chronological age with people of 65 years and more being described as old and people

---

R. Wiczorek (✉)
Department of Psychology and Ergonomics, Technische Universität Berlin, Berlin, Germany
e-mail: wiczorek@tu-berlin.de

M. A. Bayles · W. A. Rogers
College of Applied Health Sciences, University of Illinois, Urbana, Champaign, USA
e-mail: mbayles2@illinois.edu

W. A. Rogers
e-mail: wendyr@illinois.edu

 
A. Woodcock et al. (eds.), *Design of Assistive Technology for Ageing Populations*, Intelligent Systems Reference Library 167,
https://doi.org/10.1007/978-3-030-26292-1_11

above 85 being very old [2]. Of course it is not the chronological age that matters when it comes to human robot interaction (HRI), but rather certain characteristics that often correlate with age. The chapter gives an overview of the most important of these characteristics and how they should be considered in the context of robot design.

The advantage of robots over other technical systems is that they can carry out all the four stages of automation [3]. That means they can perceive information, analyze it, make decisions, and carry out actions. Therefore, they can assist in a variety of tasks and on different levels. In the best case their functionalities or amount of support can even be adapted to the users' (decreasing) abilities. The chapter explains the types of tasks for which robots can be designed to support older adults. We provide examples of current robots of three classes: assistive, social and social-assistive, that have been assessed for use by older adults.

Furthermore, the most relevant factors of the home use context will be described to create an awareness of the environment domestic robots have to fit in. Design methods and essential ethical considerations relevant for the development of domestic robots are also discussed. Finally, current gaps are identified and future developmental needs are addressed.

## 11.2 Characteristics of the Target Group of Older Adults

When designing technical systems, functionalities should meet the users' needs. Usability is important, and intuitive design can help facilitate the HRI (i.e., human-machine interaction (HMI) in general). Of course that holds true for all user groups, but may be more challenging in the design for older adults. Not only do they have special needs, but they also have diverse capabilities and limitations. To be familiar with the different possible age-related declines can help in developing solutions that optimize the HRI.

### *11.2.1 Functional Decline with Age*

The decline of some functions is a normal part of the aging process. As aging is a general and gradual process, nearly every older person will experience some declines. However, this process is characterized by inter-individual differences regarding the type of declines as well as the degree of decline at a certain point in life. That results in a very heterogeneous target group of users [1]. It is not possible to present reliable numbers of how many older people are experiencing specific losses, as this varies according to the underlying definitions of 'loss' and 'older'. Stuart-Hamilton [1] provided an example for hearing loss. According to several different studies, either a third of people in their seventies and eighties or half of the people in their eighties and older suffer from severe hearing loss. Considering a different criterion, half of the people older than 65 suffer from age-related hearing loss (ARHL) according to one author versus 70–80% according to another. The question whether or not people correct their hearing problems and to which degree by using hearing aids renders the picture even more complex.

Regarding HRI, the most relevant aspects are potential regressions of sensory functions, cognitive functions, and motor functions. People need these functions to perceive information given by a robot, to understand this information, to decide how to respond, and to communicate their intentions.

#### 11.2.1.1 Sensory Functions

Sensory functions decline with age for most individuals, including vision, hearing, haptics, smell, and taste (for an overview see [1]). Vision and hearing are especially important when it comes to HRI, because a lot of information given by the robot is presented visually or auditory. Very common age-related declines in vision are for example the reduction in acuity, lower contrast sensitivity, or changes in color perception (i.e., yellowing of the lens). The most relevant age-related hearing decline is the reduction of sensitivity to high frequencies, which affects both sound and speech.

When using classical display interfaces, there are certain suggestions that can be followed, for an overview see [4, 5]. To improve visual perception for example, symbols and letters must be presented not only in sufficient size, but also with a high contrast; that is, the background and color schemes should be based on complementary colors. To accommodate for hearing limitation, lower frequencies should be preferred. It also helps to reduce the speed of the speech and to use short sentences with simple syntax. Whenever possible the option to repeat the auditory information can be provided to further improve comprehension. Also whenever possible, the use of multi-modal information should be preferred over single modality information, because the redundancy increases the probability of successful perception [6].

The design of robots however, offers a lot more options for interface design than simply using a display. Due to its embodiment, the robot can make use of more natural ways of communication. That can be the use of gestures and facial expressions, as well as pointing to items or handing them over directly. Combined with a verbal output, these characteristics may render HRI much more intuitive for older users compared to interacting with other technical devices.

#### 11.2.1.2 Cognitive Functions

To understand the information given by the robot and to choose the proper response or action, cognitive abilities are needed that also tend to decline with age [1, 5]. Two of the most crucial aspects are attention and working memory, which can both decline with age, leading to several complications. For example, selective attention, which involves focusing on relevant information while actively suppressing distracting elements, becomes more difficult for older adults. The same holds true for division of attention, which is needed for multi-tasking requirements. Working memory, responsible for further processing the perceived information also declines with age. It represents the short-term storage of pieces of information that should be memorized or manipulated (e.g., interpretation, mathematical operations, mental rotation), and it is limited in capacity (number of items that can be stored) and time (duration of storage). With increasing age, both capacity and time get reduced. Thus, parallel processing, memory of many items at once, or time consuming operations become more difficult or even impossible. Regarding the higher order intellectual abilities, age-related declines are also almost always found. Reasoning ability, numerical abilities, and spatial-figural abilities decline with age. The intellectual properties that remain most stable during age are verbal ability, semantic memory, and metacognitive ability.

Thus, to improve older users' comprehension on a cognitive level, the design of interfaces should respect certain rules. First, irrelevant and distracting items should be avoided when presenting information to older users to improve attention. In addition, the necessity of focusing on more than one source of information, as required for multi-tasking, should be eliminated. To relieve the working memory, requirements to memorize or manipulate items should be reduced to a minimum. Making breaks and repeating information can also help. To support higher order intellectual abilities, there are some features such as consistency or grouping of items, as well as the use of verbal (written) information that can facilitate comprehension or using metaphors that are familiar to older adults (e.g., a filing system).

HRI offers more possibilities over other technical devices, such as interacting in a more natural way, for example, through verbal communication. Of course consistency is important also in speech and gestures. That means, for example, using the same words and same gestures. With a more human-to-human like communication, it is easier to integrate a feedback loop. In this way the user can indicate whether the previous information was understood or not. For instance, this can be done using simple short pieces of communication like "Ok?", "All right?" and "Yes!" or "Ok!" that are not very time consuming or annoying.

#### 11.2.1.3 Motor Functions

Motor functions are relevant when giving physical input to the robot, which can be in the way of feedback or of commands. Movement of older adults is generally slower, less precise, and less accurate, for an overview see [1, 4, 5]. Consequently, the timely coordination of movements become more challenging. In addition, some older adults experience different types of hand tremors that further impairs precise motor actions. Not only manual motor actions are affected but also overall stability and balance is impaired, which results affects walking ability and makes older adults more prone to falls [7, 8].

For conventional input devices, recommendations are to increase the size of buttons, to avoid very precise and timely coordinated motions and to use speech interfaces, when possible. As was mentioned, speech is a very natural way of interacting with a robot. In addition, a robot has a physical presence and is able to move. Thus, it can support older users by carrying out tasks that are motorically challenging for them such as opening bottles, inserting a thread into a needle eye, and many more. Of course it is important to make sure the robot does not become an additional fall risk for the users. On the contrary, it may rather be designed to support safe walking by offering support such as handles.

### *11.2.2 Other Characteristics of Older Adults*

Younger adults are often surrounded by a lot of different devices, and therefore familiar with the specific communication styles of input- and output-devices. However, some older people may have a mostly analogue life. What may be 'intuitive' for a digital native, may be completely new to an older adult and must be learned prior to communicate with a robot. However, learning becomes more difficult in later life and should be minimized [9]. Therefore, given the widespread unfamiliarity of older adults with robots, the choice of more natural and human like communication styles may be the easiest way to make HRI effective for them.

Finally, it should be mentioned that everyday tasks are often found to be less impaired than one would expect based on the pure execution of sensory, cognitive or motor tests [1]. Humans are creative and have the ability to develop alternative strategies to carry out tasks. This can be through an emphasis in planning before execution, the increase of cognitive resources invested in tasks or the use of prior knowledge. The design of robots should consider these compensatory strategies and try not to undermine them.

### *11.2.3 Summary*

Successful robot designs will consider the characteristics of the older user group into account. Most important are declines in sensory, cognitive, and motoric functions, but also prior experience with technology in general. The base for a suitable robot solution is a successful HRI. That includes a comprehensive communication robot-to-human and a facilitated communication human-to-robot. Both can be achieved by following design recommendations and by taking advantage of the specific robot characteristics. These allow for a more natural communication style using speech interfaces as well as recognition of facial expression and gestures on both sides [10].

## 11.3 Context Factors of the Home Environment

When designing domestic robots, the specific home environment has to be taken into account. Certain features are very distinct from the outdoors or in non-private buildings such as nursing homes. In fact, older peoples' homes do impose some challenges to the safe operation of a domestic robot, but they also offer some advantages as well.

### *11.3.1 Motion and Navigation*

People's homes may be very different in size, but are always restricted in space compared to the outdoor world. Certain rooms such as the bathroom are often fairly small and the remaining space for a robot to pass is rather narrow. Furniture is placed all over the apartments or houses, representing a parkour of obstacles for a robot. In addition, fragile items such as vases may be placed on tables and sideboards and can be destroyed easily when motions of the robot are too far-reaching. Unlike outdoor or other environments, home floors are often covered by carpeting and equipped with doorsills, which can represent a challenge for the safe motion of the robot. The handling of doors can also be an issue if the robot navigates through the entire place. There might also be animals living in the space with the older person such as cats or dogs. It is important that the robot does not imply a danger for them nor do the animals for the robot.

On the positive side, the layout of the apartment and the placement of furniture will usually not change very often. Thus, navigation becomes easier once the robot knows its way. There also may be some variation in how to place the single pieces of furniture, allowing for creation of suitable robot paths. Another advantage is that the robot does not have to face weather conditions such as bright sunlight or rain that may interfere with its sensory functions or requires special protection.

### *11.3.2 Interaction with Older People in the Home*

Even though many older people are living alone in their homes, others live as couples or together with other family members. Additionally, friends or relatives may visit. Thus, the robot must be able to identify the relevant person or also distinguish between different inhabitants. That is especially important as they may have different needs in assistance such as medication necessities or diverse levels of impairments requiring the corresponding amount of support. Certainly, it would be useful if robots are able to adapt to different communication situations, where users are not obligated to position themselves in an optimal way for the robot. Instead, it should be possible for them to stand, sit, or even lie in bed with the robot adapting its communication interface on the required height, position, and angle to guarantee an optimal HRI.

### *11.3.3 Installation and Maintenance*

Installation and maintenance should be considered when designing the robot. Proper instruction and training of the users should be part of the procedure, as well as explanation of additional features that might be necessary to install in the home for proper functioning. Information about functions, privacy, and security should be a standardized part of this process to not scare the older users.

Maintenance should put the least burden on the older users as possible. One reason is that they might forget important procedures or maybe not even understand them in the first place. Also, they might be afraid of getting injured or destroying the robot. Thus, the best way would be, if the robot was able to take care of most of the maintenance issues by itself. That would include automatic recharging and storing as well as the indication of malfunctions. In case of a failure it should give advice what to do or whom to call. In addition, the older adults must be reassured that someone will be there soon to take care of the problem. While in some cases this might be a relative, in other cases a technician or other specialist may be needed. Thus, whenever developing a commercial robot, designers must reassure the users that there will be maintenance staff working at a hotline and coming to peoples' homes to fix problems.

### *11.3.4 Summary*

Domestic robots have to face specific challenges to properly function in older peoples' homes. That regards the navigation and motion, interaction with the inhabitants as well as installation and maintenance. Thus, finding adequate solutions for these challenges should be considered in the design phase.

## 11.4 Methods Relevant for the Design of Domestic Robots for Older Adults

When developing domestic robots for older users, the use of appropriate methods can facilitate the design process and help to create a robot that fits the needs of older users. Engaging in user-centered design as well as considering relevant ethical issues will help designing products that are useful for the target group as well as accepted.

### *11.4.1 User-Centered Development*

To develop a domestic robot that meets the special needs of the older user group and to consider their characteristics, users should be integrated in the design process from the beginning. The idea of user-centered design is to follow an iterative way of development where analysis, development and evaluation follow each other in circles until an optimal or pleasing state of prototype has been reached. There are several methods suitable for the user-centered design process. For the analysis phase, qualitative methods such as focus groups, interviews, cognitive walk-through, etc. can be used. To facilitate the development especially in the early phases, methods of rapid prototyping can be used. These include 3D printing, mock ups, paper prototypes, etc., that do not take long to be created and can easily be adjusted [11, 12]. Certain techniques such as Wizard of Oz may be used to simulate functions, especially related to communication to test whether they are suitable before doing an elaborated implementation [13]. Evaluation can either be qualitative, quantitative, or a combination of both. Qualitative methods are observation, thinking out loud-techniques, or focus groups, whereas experiments or expert ratings are common quantitative methods. For more information regarding user-centered design for older users we suggest reading [4, 5].

### *11.4.2 Ethical Considerations*

An important, but often neglected part of the design process is the consideration of ethical aspects such as safety, security, privacy, autonomy, and others. Sorell and Draper [14] suggested an ethical framework for the development of domestic robots for older users. As they have indirect and direct implications for HRI, use and purchase intention, ethical considerations are an essential part of the development of domestic robots for older adults.

#### 11.4.2.1 Safety and Security

The handling of the robot should not impose any danger to the user or other people (or domestic animals). This includes not presenting obstacles that could make the older adult stumble, not to squeeze them, prevention from falling on the user, rolling over their feet, etc. Also in case of malfunction, fail-safe methods should be applied to avoid hazards.

Furthermore, the presence of the robot should not compromise security of the older people. That includes to not open doors to strangers, giving private information to unauthorized people as well as not giving obvious signs of absence of the inhabitants. In addition, the use of a domestic robot should not compromise using security systems such as alarm systems with movement sensors for instance.

#### 11.4.2.2 Privacy

The matter of privacy is a very crucial point when introducing technologies in people's homes. For navigation and proper treatment, the robot needs a lot of information about the environment and the user, both online and stored. Thus, it is important to develop secure ways of handling and storing this information. The least amount necessary should be stored. Transferring data via internet is always a weak point where strangers can get access to private information and should thereby be limited to the absolute minimum.

Moreover, the type of technology used implies different levels of invasion of privacy. While several sensors (e.g., infrared) only generate an abstract "image" of the environment, camera pictures convey a lot of private information. If not used for the purpose of surveillance it should be questioned whether camera images are really needed and if so whether it is necessary to transfer and/or store them. Furthermore, it is important to inform the older users about the current situation of privacy and its violation and to explain to them what information is collected and how it is handled, transferred, and stored. Sometimes it is not easy for older users to understand all the involved technical terms. Thus, it is the obligation of the designers to find a way and a language to make the users understand what happens to their data.

#### 11.4.2.3 Autonomy

One aim for the use of domestic robots is to help older people staying longer at home and live their life as independently and autonomously as possible. Autonomy includes the execution of everyday tasks as independently as possible as well as taking their own decisions without having to explain themselves to others. It is important for older people to carry out the tasks they are still able to do by themselves to prevent skill degeneration and boredom. Sometimes there may be a trade-off between safety and autonomy. In this case it is important to remember that mentally healthy older adults must be given the full responsibility over their own life decisions, even if these decisions can seem wrong to designers (and relatives) from time to time. The value of autonomy increases with the decline of physical and cognitive abilities and the progressing need to rely on others.

### *11.4.3 Summary*

The design of domestic robots should follow the approach of user-centered development. In line with that ethical considerations are an essential part of the developmental process. That includes the consideration and sometimes the trade-off of aspects such as safety, security, privacy and autonomy.

## 11.5 Tasks

When designing technologies for older adults, one needs to first determine how older adults' needs differ compared to other age groups. For example, a younger adult might need help remembering to engage in the activity of brushing their teeth, whereas an older adult with a tremor may need help physically getting the toothpaste onto the toothbrush. Three general categories describe older adults' needs: Activities of Daily Living (ADLs), Instrumental Activities of Daily Living (IADLs), and Enhanced Activities of Daily Living (EADLs). In the preliminary processes of designing technologies, the designer should distinguish the type of activity the older adults will be engaging in as this can influence the factors that mediate acceptance among older adults. Alongside determining the activities, the designer will also need to consider how these activities may require different assistance for older adult users compared to other cohorts.

### *11.5.1 Activities of Daily Living (ADLs)*

ADLs were labeled as such in the 1960s [15] to describe everyday tasks that someone would do that are involved with self-care [16]. An ADL can be anything along the lines of bathing, eating, transferring, or getting dressed. These activities often have the greatest influence on the ability of an older adult to be able to live independently. Older adults who need assistance with ADLs may include, but are not limited to, those who have mobility impairments, cognitive impairments, or are recovering from surgery or illness.

A variety of environmental factors in the home can pose potential barriers for older adults' ability in performing specific ADLs. These environmental barriers are one focal point for the goals of robotics aiding in ADLs because they may increase the potential for older adults to fall. "… fall rates are expected to increase, leading to a rise in accidental injury and injury-related deaths, and placing an escalating burden on health care systems" [1, p. 5990]. Using robotics to assist with fall avoidance could not only keep older adults living independently longer, but also minimize the number of medical visits for older adults.

Potential fall threats in the home are rugs, wires, limited lighting, and lack of hand rails. These environmental factors could be fixed in other ways besides robots. However, the value of robots may depend on the degree to which their introduction provides the most expedient and effective solution. Consider transferring (e.g., from a wheelchair to a bed or toilet). Transferring is a cause for falls that can be mended well with the help of a robot by supporting the older adults' weight and compensating for loss of balance.

When designing robots to assist with ADLs, designers should consider environmental barriers that may intrude on the older adults' ability to continue living independently (tripping hazards, transferring assistance, eating assistance, etc.). Robots might assist in detecting misplaced objects within a walking path, provide weight-bearing assistance for transferring around the home, lighting pathways, or generally assisting with physical tasks that may be difficult for an older adult to perform (bending down to grab something off the floor, reaching the top shelf). However, it is also important that older adults perceive the robots as both useful and easy to use, as those are predictors of technology acceptance [4]. To illustrate, consider an individual who has mobility impairments and lives alone. This person will likely perceive a robot that assists with transferring to be very useful (compared to someone with a live-in caregiver, for example). Moreover, if the person has experience with a range of technologies, they might be likely to perceive the robot as easy to use. Both of these factors will increase acceptance of the robot in the home.

### *11.5.2 Instrumental Activities of Daily Living (IADLs)*

Instrumental Activities of Daily Living (IADL) emerged as a term to further classify the activities older adults engage in that enable them to continue living independently. IADLs include medication management, managing finances, shopping, cooking, and doing housework [16]. Non-adherence to a medication regimen, for example, is one factor as to why older adults' are admitted to hospitals or transitioned to nursing homes [17]. Older adults may face challenges performing IADLs due to age-related changes in memory, vision, hearing, or mobility. Robots could be designed to assist in IADLs such as helping with medication management, assisting with vacuuming or other house chores, and reminding the user that bills are due.

Successful robot design for IADL support will depend on the degree to which the older adult trusts the technology to manage these activities. 'Trust' was defined by the Almere Model of robotic acceptance as "… the belief the robot performs with personal integrity and reliability" [18, p. 364]. Older adults reported factors important to trusting a robot performing ADLs and IADLs included the robot's professional skill, if the robot was personable, and it communicated clearly [19]. Another key factor in improving trust for older adults in relation to robotics is to show them that they have control of the robot [6].

### *11.5.3 Enhanced Activities of Daily Living (EADLs)*

EADLs can be considered broadly as activities that contribute to continuous development and engagement of oneself in aspects like social interaction and emotional support. We originally developed the concept of defined EADLs to capture the broader range of activities that older adults engage in—beyond ADLs and IADLs: "… existence as an independently living, active older adult … requires the ability to adapt to a changing environment. The willingness to accept these new challenges and to learn may be key to staying fully functional in a changing environment. We have labeled these additional behaviors of active elders' enhanced activities of daily living (EADLs)" [20, p. 111]. Examples of EADLs are staying in touch with friends and family, game playing, keeping up with technology, and continuing to implement improvements for healthy living (both physically and mentally). We are unaware of current research that primarily focuses on robots that assist in EADLs. However, robots designed to assist with ADLs and IADLs may also assist in EADLs in some ways.

Perhaps the most critical characteristic for robots designed to assist with EADLs will be enjoyment and engagement. Given that EADLs are optional (i.e., not required for independence), the choice of an older adult to use a robot in this context is likely to be dependent on these hedonistic characteristics, more so than ease of use and usefulness.

## 11.6 Functionalities and Corresponding Features

As robots become increasingly prevalent within the consumer population [21], it becomes more important to determine aspects of acceptance among different demographic populations. Robotics is a broad field, and designers are developing robots with different goals in mind. Consequently, there are a multitude of ways to classify robots, especially those that are intended to assist older adults with everyday activities: "While much attention has been paid to robots that provide assistance to people through physical contact (which we call contact assistive robotics), and to robots that entertain through social interaction (social interactive robotics), so far there is no clear definition of socially assistive robotics" [22, p. 465].

When designing robots for older adults, it is important to identify a goal that is relevant to the activity the robot is trying to assist. Will the robots aim be to help the older adult eat breakfast (ADL)? Or will it help the older adult with medication management (IADLs)? Perhaps it will be a combination, assisting the older adult with eating breakfast (ADL) while simultaneously encouraging the older adult to talk about their day (EADL). Defining the specific robot's roles will inform the design and ultimately enhance successful HRI and acceptance by older adults.

The definition of an assistive robot was broadly stated by Mataric [22] as a robot that gives aid to a human user. Assistive robots are generally going to be tasked with helping older adults preform ADLs—supporting the older adult user with physical aid more so than other kinds of aid.

Social robots differ from assistive robots; as they are intended to provide or assist with social interaction and emotional support [23]. Social interaction can consist of holding a conversation, playing games, or providing emotional engagement (this can be verbal or non-verbal). Social intelligence in this context can span from responding to touch to being able to detect emotion from facial expression.

Social-Assistive robots bring these two definitions together and can be defined as: "A system that employs hands-off interaction strategies, including the use of speech, facial expressions, and communicative gestures, to provide assistance in accordance with the particular healthcare context" [24, p. 3]. Social-assistive robots may be deployed in more fields than healthcare; thus we define them more broadly as socially intelligent entities that encourage engagement while simultaneously helping the user with ADLs, IADLs, or EADLs.

Below are examples of robots in these categories that are currently on the market that might be suitable for older adults. They are classified according to the definitions above. The following list is not meant to be comprehensive, but to illustrate examples of each type of robot (Table 11.1).

Older individuals acceptance of these various robots will likely vary depending on their predisposed conception of the robots' function [25]. To elaborate, the older adult users purchasing a robot to assist with an IADL may not care what it looks like as long as it functions properly. However, if the same older adult were to purchase a robot to assist in ADLs, they may prefer that the robot's appearance be comforting. Robots should be designed with the older adults preferences, capabilities, and limitations in mind. To support that effort, we provide design suggestions and robot limitations for social assistive robots intended for use by older adults.

In the preliminary stages of designing a social assistive robot for older adults, there are consideration that will influence the likelihood that the robot will be accepted. The first thing a designer should do is to familiarize themselves with general human factors design tips that are specific for older adults (e.g., [4, 5]). The next step is to focus on identifying the range of considerations for robots being designed for older adults to use in their home. We have illuminated many of the issues in the present chapter (see also [26]).

Specific factors to consider will include proxemics preferences (how the robot approaches the user), display font (if there is a digitized aspect on the robot), appearance preferences, and abiding to the older adults' historical concept of technology to avoid confusion and ensure their constructs of technology are not challenged. For social assistive robots in particular, it will be critical to consider the specific tasks that

**Table 11.1** Classification of robots along with the activity type they assist with

| Classification | Robot name | Assists with ... | References |
|---|---|---|---|
| Assistive | Healthbot | ADL | University of Auckland, New Zealand |
| Assistive | Obi | ADL | John Dekar (Desin) |
| Assistive | PR 2 | ADL | Willow Garage |
| Social | Paro | EADL | Takanori Shibata—Intelligent System Research Institute (Intelligent System Co.) |
| Social | Aibo | EADL | Sony's Digital Creatures Lab and Toshitada Doi |
| Social | Nao | EADL, IADL | Aldebaran Robotics (acquired by Softbank in 2015) |
| Social-assistive | Pillo | IADL, EADL | Pillo Health |
| Social-assistive | Jibo | IADL, EADL | Cynthia Breazeal; MIT |
| Social-assistive | Pepper | ADL, IADL | Softbank Robotics |

will be supported as well as the unique needs of older adults and their home environment. Usability testing from formative to summative evaluations should involve older adults, ideally in the target context of use (actual or simulated home environments) and performing realistic daily tasks. Usability testing is always a necessary step in the design process. However, it may be especially important for emerging technologies that are unfamiliar to the target users as well as for technologies intended to be integrated into one's home environments. Robots in the homes of older adults are unique and there may be unintended consequences or misperceptions that can be remedied during the design process with direct engagement with a diversity of older adults. This will help avoid putting a product on the market that falls short due to a design flaw that could have been detected and diminished prior to implementation.

When considering social assistive robots, there are some limitations that should be addressed so they have a higher chance of being accepted among the older adult cohort. One limitation that may reduce acceptance is the lack of instruction the robot provides to the older adult. Providing the older adults with easy to follow instructions on how to work with the robot might increase their confidence, therefor improving acceptance (by improving perceived ease of use and therefor affecting perceived usefulness) [27]. Next, as discussed earlier in the chapter, it should be clear to the older adult user that the social assistive robot is useful in their life; increasing perceived usefulness increases acceptance. A final thought on limitations for social assistive robots is the stigmatization older adults' have about the technology itself. An older adults' views about robots can affect their acceptance (either good or bad) [25, 28]. For example, if they have seen a movie where the robot is nice and happy, they will have higher acceptance. If the robot is bad in a movie they have seen, they may have more negative acceptance towards robots. The field should work on decreasing the stereotype of robots that provide consumers with negative affect towards the field, increase the perceived usefulness of the robots, and finally, increase perceptions of ease of use.

## 11.7 Discussion and Conclusion

Robots have the potential to support the everyday activities of older adults. The characteristics of the robots need to match the needs, goals, and capabilities of older adults. First, there are well-documented general age-related changes in perceptual, motor, and cognitive capabilities that may influence an older adult's ability to interact with a robot. Second, there are broad task categories (ADLs, IADLs, EADLs) with which older adults may need or want assistance. Different types of robots will be better suited for different types of task goals. Third, if the robot is to be used in the home environment, that specific context of use must be considered in the design process. Fourth, a variety of robot types are being explored in the context of supporting older adults such as assistive, social, or social-assistive. The specifications for these robots and their success in supporting older adults' needs are still being explored in the literature. The potential is there but the broad availability of robots remains in the future while technical as well as HRI issues are being specified.

We provide guidance from techniques used in human factors that will increase the likelihood of robot that will be useful to and usable by older adults. The degree to which older adults will accept robots into their home to support their activities and independence will be determined by the attention paid to design processes that engage older adults from conceptualization to implementation. Moreover, considerations of ethics, privacy, trust, and support for long-term use will be paramount to successful deployment of domestic robots.

## References


1. Stuart-Hamilton, I.: The Psychology of Ageing: An Introduction. J. Kingsley Publishers, London (2012)
2. Erber, J.T.: Aging and Older Adulthood. Wiley-Blackwell, Chichester, West Sussex (2013)
3. Parasuraman, R., Sheridan, T.B., Wickens, C.D.: A model for types and levels of human interaction with automation. IEEE Trans. Syst. Man Cybern. Part A Syst. Hum. **30**(3), 286–297 (2000)
4. Czaja, S.J., Boot, W.R., Charness, N., Rogers, W.A.: Designing for Older Adults: Principles and Creative Human Factors Approaches, 3rd edn. CRC Press, Boca Raton, FL (2019)
5. Pak, R., McLaughlin, A.: Designing Displays for Older Adults. CRC Press, Boca Raton, FL (2010)
6. Lee, J.H., Poliakoff, E., Spence, C.: The effect of multimodal feedback presented via a touch screen on the performance of older adults. In: International Conference on Haptic and Audio Interaction Design, pp. 128–135. Springer, Berlin, Heidelberg (2009)
7. Dietz, V.: Proprioception and locomotor disorders. Nat. Rev. Neurosci. **3**(10), 781–790 (2002)
8. Tinetti, M.E., De Leon, C.F.M., Doucette, J.T., Baker, D.I.: Fear of falling and fall-related efficacy in relationship to functioning among community-living elders. J. Gerontol. **49**(3), M140–M147 (1994)
9. Barnard, Y., Bradley, M.D., Hodgson, F., Lloyd, A.D.: Learning to use new technologies by older adults: perceived difficulties, experimentation behaviour and usability. Comput. Hum. Behav. **29**(4), 1715–1724 (2013)
10. Beer, J.M., Rogers, W.A.: Enabling older adults to interact with robots: why input methods are critical for usability. In: Markowitz, J., Neustein, A. (eds.) Speech and Automata in Healthcare, pp. 163–181. DeGruyter (2014)
11. Beaudouin-Lafon, M., Mackay, W.E.: Prototyping development and tools. In: Jacko, J.A., Sears, A. (eds.) Handbook of Human-Computer Interaction, pp. 1006–1031. Lawrence Erlbaum Associates, New York (2002)
12. Snyder, C.: Paper Prototyping: The Fast an Easy Way to Design and Refine User Interfaces. Morgan Kaufmann, San Diego (2003)
13. Portet, F., Vacher, M., Golanski, C., Roux, C., Meillon, B.: Design and evaluation of a smart home voice interface for the elderly: acceptability and objection aspects. Pers. Ubiquit. Comput. **17**(1), 127–144 (2013)
14. Sorell, T., Draper, H.: Robot carers, ethics and older people. Ethics Inf. Technol. **16**, 183–195 (2014)
15. Katz, S., Ford, A.B., Moskowitz, R.W., Jackson, B.A., Jaffe, M.W.: Studies of illness in the aged: the index of ADL, a standardized measure of biological and psychosocial function. J. Am. Med. Assoc. **185**, 914–919 (1963)
16. Activities of Daily Living (ADLs): Retrieved 10 Feb 2019, from https://www.seniorliving.org/caregiving/activities-of-daily-living/ (n.d.)
17. Scarlett, W., Young, S.: Medical noncompliance: the most ignored national epidemic. J. Am. Osteopath. Assoc. **116**(8), 554 (2016). https://doi.org/10.7556/jaoa.2016.111

18. Heerink, M., Kröse, B., Evers, V., Wielinga, B.: Assessing acceptance of assistive social agent technology by older adults: the Almere model. Int. J. Soc. Robot. **2**(4), 361–375 (2010). https://doi.org/10.1007/s12369-010-0068-5
19. Stuck, R.E., Rogers, W.A.: Older adults' perceptions of supporting factors of trust in a robot care provider. J. Robot. **2018**, 1–11 (2018)
20. Rogers, W.A., Meyer, B., Walker, N., Fisk, A.D.: Functional limitations to daily living tasks in the aged: a focus group analysis. Hum. Factors **40**(1), 111–125 (1998). https://doi.org/10.1518/001872098779480613
21. Tobe, F.: Robotics industry growing faster than expected—the robot report. The Robot Report. Retrieved 14 Feb 2019, from https://www.therobotreport.com/robotics-industry-growing-faster-than-expected/ (2017)
22. Feil-Seifer, D., Mataric, M.J.: Defining socially assistive robotics. In: 9th International Conference on Rehabilitation Robotics, pp. 465–468. https://doi.org/10.1109/icorr.2005.1501143 (2005)
23. Dautenhahn, K.: Socially intelligent robots: dimensions of human-robot interaction. Philos. Trans. R. Soc. B Biol. Sci. **362**(1480), 679–704 (2007). https://doi.org/10.1098/rstb.2006.2004
24. Fasola, J., Matarić, M.J.: A socially assistive robot exercise coach for the elderly. J. Hum. Robot Interaction **2**, 3–32 (2013). https://doi.org/10.5898/JHRI.2.2.Fasola
25. Prakash, A., Rogers, W.: Why some humanoid faces are perceived more positively than others: effects of human-likeness and task. Int. J. Soc. Robot. **7**(2), 309–331 (2014). https://doi.org/10.1007/s12369-014-0269-4
26. Rogers, W.A., Mitzner, T.L.: Human-robot interaction for older adults. In: Encyclopedia of Computer Science and Technology, 2nd edn, pp. 1–11. Taylor and Francis (2017)
27. Lin, C.-S., Hung, P.-H., Hsiao, H.-P., Chen, H.-P.: A study on technology acceptance model with the service of the E-gate: a case of mini three links in Kinmen. In: The International Academic Forum. The Asian Conference on Asian Studies 2016 (2016)
28. Sundar, S.S., Waddell, T.F., Jung, E.H.: The hollywood robot syndrome media effects on older adults' attitudes toward robots and adoption intentions. In: 11th ACM/IEEE International Conference on Human-Robot Interaction (HRI), pp. 343–350. https://doi.org/10.1109/hri.2016.7451771 (2016)

**Rebecca Wiczorek, Ph.D.** is the leader of the interdisciplinary research group FANS (Fußgänger-Assistenzsystem für ältere Nutzerinnen und Nutzer im Straßenverkehr—Pedestrian assistance for older road users) at the Technische Universität Berlin in Germany. She received her Ph.D. in Human Factors in 2012. Her research interests include design for aging, assistance systems, human–robot interaction, and traffic psychology with focus on pedestrians.

**Megan A. Bayles, B.S.** is a graduate research assistant in the Human Factors and Aging Laboratory at the University of Illinois Urbana–Champaign. She received her Bachelor of Science in Psychology from Florida State University in 2017. Her research interests include human factors, technology acceptance, social robotics, and design for aging. Current projects include telewellness, computer support for older adults at-risk for social isolation, attitudes towards socially assistive robots, and digital accessibility. She is currently pursuing her Master of Science in Community Health.

**Wendy A. Rogers, Ph.D.** is the Khan Professor of Applied Health Sciences at the University of Illinois Urbana–Champaign. She is the Director of the Health Technology Education Program; the Program Director of CHART: Collaborations in Health, Aging, Research, and Technology; and the Director of the Human Factors and Aging Laboratory. Her research interests include design for aging; technology acceptance; human–automation interaction; aging-in-place; human–robot interaction; and aging with disabilities.

# Part V
# Supporting the Mobility Needs of Older People

# Chapter 12
# Strategies for Mitigating the Injury Risks of Using Motorised Mobility Scooters

**Selby Coxon and Jennie Oxley**

**Abstract** Transport is an important arbiter of engagement in society. Disability and functional limitations (cognitive and physical) associated with older age can make such participation challenging. The design and engineering community have been strident in recent decades to use technology to mitigate, as far as possible, the limitations imposed upon individuals as a consequence of the effects of ageing. This chapter explores one such assistive technology; the motorised mobility scooter (MMS) and in particular examines some contradictions and ambiguities associated with this increasingly popular form of mobility. For example, MMS are classified as pedestrians yet can travel at speeds of 10 km/h. They can look and perform similar to bigger motor scooters but are not permitted to travel in traffic lanes. These issues have significant implications upon the safety of the user and other pedestrians. From a review of the scientific literature and examination of injury data from the Australian experience, some implications are drawn for systems-based strategies engaging design, policy and regulation approaches to enhance the safety of this form of assistive technology. These approaches have potential benefits that may extent to other parts of the world where this form of mobility is also growing in popularity.

**Keywords** Mobility · Disability · Ageing · Motorised mobility scooter · Pedestrians · Assistive technology · Injury data

## 12.1 Background

With ageing, comes physiological, sensory and cognitive impairments to a varying degree and alacrity. While there are many individual differences in the ageing process, even relatively healthy older adults are likely to experience some level of functional

S. Coxon (✉)
Faculty of Art, Design and Architecture, Monash University, Melbourne, VIC 3800, Australia
e-mail: selby.coxon@monash.edu

J. Oxley
Monash University Accident Research Centre, Monash University, Melbourne, VIC 3800, Australia



A. Woodcock et al. (eds.), *Design of Assistive Technology for Ageing Populations*, Intelligent Systems Reference Library 167,
https://doi.org/10.1007/978-3-030-26292-1_12

decline in sensory, cognitive and physical abilities. These may include: decline in visual acuity and/or contrast sensitivity; visual field loss; reduced dark adaptation and glare recovery; loss of auditory capacity; reduced perceptual performance; reductions in motion perception; a decline in attentional and/or cognitive processing ability; reduced memory functions; neuromuscular and strength loss; postural control and gait changes, and slowed reaction time [1–3]. There is also evidence suggesting that functional limitations related to specific medical conditions impacts safe mobility [4, 5].

Safe participation in traffic, especially making and executing decisions in complex environments and during complex manoeuvres, requires adequate functioning of a range of sensory, perceptual, cognitive, executive and physical abilities, particularly attention, perception of speed and distance, processing of sensory input, judgement, decision-making and memory.

Inflexibility in the limbs requires an individual to take more time to initiate and complete movements. A decline in posture and stability, restricted gait and balance make even the most automated of human movements such as walking more difficult. Moving with a small gait and experiencing difficulty surmounting small steps or obstacles impede the most basic of mobility. Indeed, there is evidence that older adults are at increased risk of falling while walking [6, 7]. Walking sticks provide stability and support for a time, but any increased physical deterioration leads eventually to the need for wheeled transport either self-powered or pushed by way of a carer. These impairments mean that public transport with its sudden stops and starts, urgency and fellow passengers are not always a long-term solution and indeed for some even the most innocuous of short trips becomes a mobility challenge. Older passengers of public transport are significantly over represented in emergency departments from falls using public transport. With an increase in the elderly population [8] and high expectations among older adult with regard to maintaining personal mobility [9], this undesirable trend is set to continue.

So, for the ambulant impaired who have stopped driving, independent wheeled mobility can mean a new lease of life, enabling the most basic of mobility and freedom. In response, the availability and design of all manner of assisted personal wheeled devices has developed significantly. Wheelchairs, especially those powered by the arms of the user require sustained upper body strength, a potency not easily found amongst the elderly, and only a viable transport option for very short trips. In contrast, self-powered devices can afford transport for longer trips. Indeed, since the 1950s the notion of an electric motorised wheeled conveyance or scooter has been a viable option to meet the mobility needs of the infirm.

The Motorised Mobility Scooter (MMS), has developed considerably since this advertising image in Popular Mechanics from 1954 (Fig. 12.1). These assisted technology devices vary greatly from the motorised wheel chair. They are far more vehicular in form (enclosed mechanical components) and the rider can typical get on and off the MMS in ways not expected of occupants of motorised wheelchairs. Contemporary designs take the visual form language of a more substantial and sophisticated vehicles; for example, the front tiller occasionally reflecting the geometry and appearance of a motorcycle (Fig. 12.2). Their performance too, in terms of speed,

**Fig. 12.1** An advertising image from Popular Mechanics, November 1954

**Fig. 12.2** A variation on the contemporary visual form language of a MMS. Authors photograph

belies the intention of conveying someone in a slow ponderous and sedately manner. It is precisely these qualities of mass, speed, appearance and the manner in which they are sometimes operated that has led to some safety concerns. Notwithstanding their popularity and potential to provide an alternative transport option for many older adults [10], these safety challenges include aspects of the vehicle itself, lack of regulations (purchase, design), and operator issues.

## 12.2 Vehicle Design and Operation

Mobility scooters come in a variety of shapes and sizes. Many manufacturers strive to present an appealing option for independent mobility through the design of the MMS, however, in many cases these design features may also suggests more robust and sophisticated qualities not actually relevant to the intentions of the vehicle. The handle bar and tiller controls are often styled in sweeping upward curves and adorned with lighting and wing mirrors which evoke the appearance of a motorcycle or larger two wheeled road worthy scooters. The foot plate or floor of a MMS often contains integrated wheel arches, to prevent water and dirt splashing the rider. This parallels the appearance of more substantial cars and other road based vehicles. Manufacturers consider these visual affects to create authenticity and confidence in the efficacy of the device but there is also the danger that they belie the true character of the scooters performance and location in the panoply of mobility options.

The physical accoutrements of a contemporary mobility scooter also contribute to both the visual and actual mass of the object. Larger sized MMS can exceed the maximum un-laden mass of 110 kg excluding the rider. This is a heavy object moving at relative speed amongst pedestrians. Mobility scooters can also be used in conjunction with public transport, boarding trains and where physically possible buses. Unlike wheelchairs the mobility scooter places more load on ramps and takes up more space within the larger vehicle. The interiors of many public spaces have requirements for accessibility compliance in which the turning circles of wheelchairs are prescribed. However, mobility scooters struggle to make an elegant 180° turn, i.e. without making a multiple-point turn or the like, within the relevant AS (Australian Standard 1428.2) dimensions of 2250 mm by 2250 mm.

In Australia, as many countries, MMS are classified as pedestrians and so are permitted to use the pavement. Despite their appearance and geometry, they are not defined as a vehicle and not permitted to occupy any traffic lanes along with cars and other larger vehicles. Their maximum speed is around 10 km/h powered by an electric motor with a rechargeable battery.

## 12.3 Regulations

There are very few regulations or standards pertaining to MMS. For example, there is generally no requirement to take a competency test before purchasing one as one might other forms of motorised mobility. An exception to this exists in the State of Victoria where a qualified health professional is required to authenticate the validity of an application for financial assistance from a prospective user [11]. A MMS is not required to be registered with any overseeing authority, with the exception of Queensland, where there is such a legislative mechanism. Any purchase of a scooter in that State requires a certificate from a health practitioner and compulsory third-party insurance [12]. The absence of any formal assessment of a new owner to operate a MMS leaves open the risk of having users with significant functional impairments (visual, cognitive, physical) permitted to attempt to operate their MMS within the transport environment, and often in complex situations and amongst other pedestrians.

## 12.4 The Impact of Urban Environmental Design

The physical impairments to active mobility brought on with age are exacerbated by the environment in which they move, rendering even more disadvantage [7, 9]. Wheeled mobility encounter obstacles that stepping can overcome. Ground surfaces may be smooth or rough, sloped or stepped, paved or covered with grass, and each of these characteristics may be further altered by the weather. What may be easy to move through on one occasion can be rendered impassable during conditions of extreme cold, rain or soft snow. Even within clement conditions riders will need to negotiate changes in level and surface when crossing a road from pavement to pavement. Such manoeuvres can bring a heightened level of risk of overturning. Wider usage of MMS would suggest a need for greater provision in the design of infrastructure. This includes wider and even/smoother pathways, parking space and re-charging points. MMS are larger than bicycles but less flexible in where they may be left safely. Retail and hospitality venues, especially with step entranceways make accessibility difficult. Public transport infrastructure where a scooter rider is required to bridge the gap between the vehicle and the platform has to be carefully negotiated so that wheels don't become stuck. Again, if the heights between vehicle and platform are at variance then boarding can become insurmountable.

An indication of a progressive society is the provision of and indeed insistence upon universal accessibility to mobility. The Disability Discrimination Act (DDA) 1992 set out an overarching agenda for equitable treatment and access to public places and infrastructure. The specific standards created to meet compliance in the public transport arena are laid out in the Disability Standards for Accessible Public Transport (DSAPT) which sets requirements amongst others for wheeled personal transport to move unimpeded about stations, trains and buses. Full compliance in Australia is not yet in place and is undergoing a roll out process to a binding legal deadline in 2022. Accessibility to infrastructure on MMS are broadly accommodated with the exception of minimum turning spaces (AS1428.1). Curiously the MMS themselves have no standards as yet concerning their design or manufacture. Few scooters are built in Australia, most are manufactured in Asia with some models coming from the

US and Europe. Due to the vagueness around the defining of a MMS some standards are interchangeably described from electric wheelchairs, an altogether different type of device.

## 12.5 Risk of Injury

A recent analysis of injury data in Victoria[1] showed that, over a 10-year period from 2007/08 to 2016/17, there were over 1000 Emergency Department (ED) presentations for MMS-related injuries. The number of ED presentations increased over time from 84 in 2007/08 to 131 in 2016/17. The majority of injured adults (66.1%) were aged 70 years and above, with the oldest adults (85+ years) accounting for 24.6% of adults injured in a MMS-related collision.

The types of injuries sustained in these collisions included: fracture (28.8%), open wound (16.9%), dislocation, sprain and strain (13.2%), injury to muscle and tendons (5.4%), intracranial injury (3.4%), and superficial injury (15%). The most common body regions injured included: head (16.2%), knee and lower leg (13.8%), hip and thigh (11.8%), and multiple body regions (13.6%).

The majority of injuries were associated with a fall from the MMS (55%), with a further 15% the result of a collision with a vehicle (Fig. 12.3).

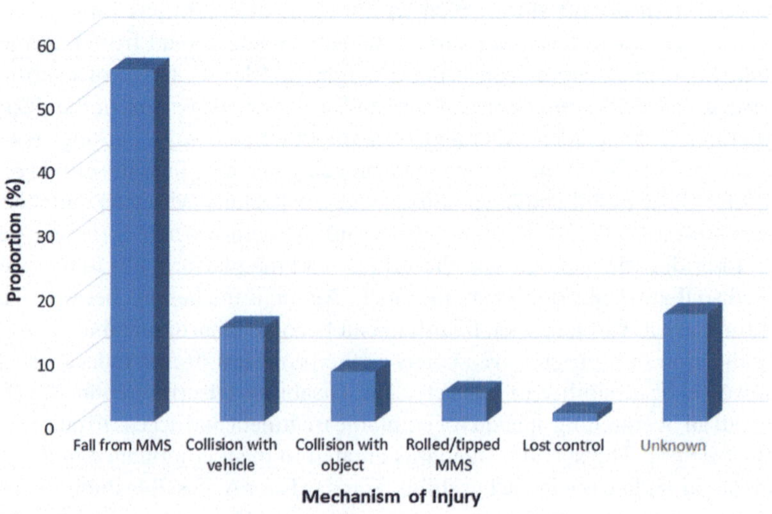


**Fig. 12.3** Mechanisms of injuries for MMS-related ED presentations. *Source* VISU, Monash University

[1]Data and data analysis provided by the Victorian Injury Surveillance Unit (VISU), Accident Research Centre, Monash University.

## 12.6 Mitigating Injury Risks and Enhancing Safe Mobility

The principles of various complementary concepts and approaches including the Safe System [13], Age-Friendly Cities [14], Smart, Accessible and Resilient Cities [15], Healthy Ageing, and Active Ageing are gaining increasing traction in the ageing research and design community and can be used to provide safe alternative transport options. Together, these approaches offer a framework in which effective measures to achieve reductions in the incidence and severity of MMS-related injuries whilst maintaining community access and participation can be considered.

There are three ways of conceptualising future developments around the design of safe MMS use, as indicated in the following figure.

At the base of this hierarchy lies the foundations of an MMS-friendly built environment, the scooter to ground surface materials and finish. A built environment that is sympathetic to the needs of older adults is by default sympathetic to a wide range of the population of any age. Attention to all weather smooth surfaces and continuous paving, sympathetic inclines and access ramps where impediments exist, wide paths with shoulders designed to prevent rollover in the event that a vehicle's wheels leave the pathway, the creation of parking spaces and re-charge points, graphical wayfinding, warning and other signage indictors for MMS as well as pedestrians on footpaths to alert users to navigation and maintain safety can benefit all road users.

The second-tier concerns the mobility scooter itself. It is clear that some confused perceptions exist around the intentions of the scooter when they have been aesthetically designed to resemble more substantial vehicles. This does not negate the value of creating an appropriate design of scooter that engenders a sense of quality, safety, efficacy and even desirability. Few things demean the disabled more than having to use ugly unappreciated objects that accentuate disability rather than celebrate their utility.

In the design referred to above the contemporary approach to motorised mobility changes the form language from the pseudo motorcycle more to that of an elegant chair. A number of design experiments have looked at the notion of articulating the seated chair posture to one that might fully articulate into an upright stance. The consideration here is the opportunity for the user to have greater dignity afforded by standing at full height amongst others.

Sophisticated visual treatments of a seated scooter can reinforce a sense of worth and wellbeing by association. And in particular reinforce the notion of independence through choice, taste and interaction with features that enable engagement with life activities. The emphasis being on ability rather than "look at me I have a disability". Equally a chair centric approach to scooters removes the perception that they should be equally at home trundling along the highway.

The top tier of Fig. 12.4 concerns the legislative and standards framework that has oversight of the modality, and includes fitness of the user to operate the scooter, scooter design standards and purchasing regulations. With regard to functional abilities to operate the vehicle, advocacy of a raft of measures is essential which will ensure an alignment between the capabilities and requirements of the user and the

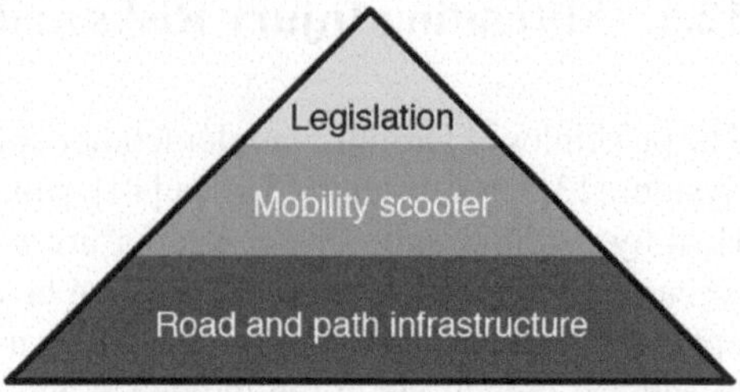


**Fig. 12.4** Building the layers for a safer mode. Authors diagram

mobility scooter are met. While it has been identified that about half (51%) of MMS users seek advice or assessment from mobility specialists such as occupational therapists when buying a scooter, only a quarter have safety training or tuition on their scooter [10]. Development of best-practice guidelines for occupational therapists or other health professionals conducting assessment or training for users of MMS is required to determine fitness to drive is critical. Appropriate training in the purchase and safe use of MMS [16] can complement assessment guidelines. In addition, consideration of revising eligibility of MMS ownership may improve the availability and accessibility of MMS to a wider population of older adults [12]. Since it is likely that the population of MMS riders will increase in the coming years then measures to address these issues for the wider public are likely to be helpful. Finally, development and implementation of manufacturing standards of performance and quality especially around safety in line with other supportive compliance legislation would be to the general good (Fig. 12.5).

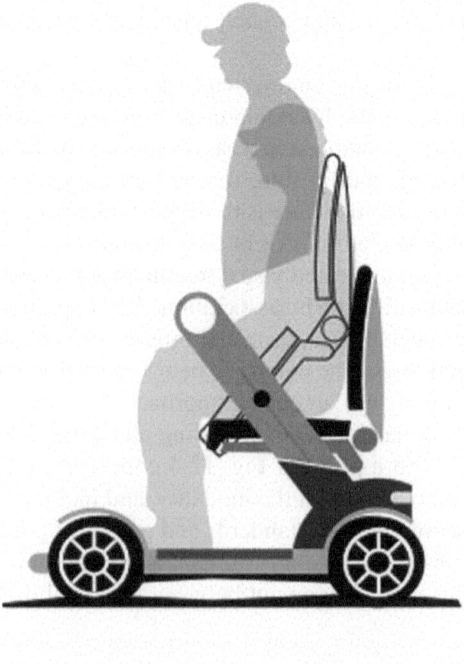

**Fig. 12.5** Future designs of an MMS might reflect a more human and utilitarian response than aping other vehicular norms. Authors drawing

## 12.7 Conclusion

The implications of a growing ageing population are widely documented [9, 14]. Amongst these concerns are the maintenance of key quality of life indicators as exemplified by personal mobility. Mobility is critical to carry out life's activities and to maintain independence, well-being and social/community engagement. In contrast, poor mobility places a substantial burden on the individual, family, community, and society. It is evident that private vehicle travel affords high mobility for older adults, however, for those unable to drive, provision of safe alternative travel options that allow easy access to services and amenities is a vital factor in maintaining mobility. While safe travel remains an essential goal for any society, recognition of the benefits of continued mobility and, conversely, serious consequences of loss of mobility should be considered [7].

To mitigate potential negative outcomes of driving reduction/cessation, it will be critical for governments and community organisations to recognise (i) the need to support older adult to full and active lives and remain healthy and vital members of the community, and (ii) the benefits of providing safe access to viable transport options as a key method to achieving a positive impact on healthy ageing. MMS certainly overcome many of the impediments of a non-ambulant, non-driving aged population. However, with this new mobility comes exposure to new risks and challenges for the design environment which hitherto remain under examined and researched. The Authors have shown, that in the Australian context, there are significant risks emerging from this form of mobility and that a raft of inconsistencies in legislation and design standards suggest that they are contributing to this suboptimal outcome.

A three-tiered design strategy is advocated where the environment, the MMS itself, and legislation are brought into alignment to address the underlying risk profile of this form of mobility. At the base level the development of 'age friendly cities' expanding upon the provisions of the Disability Discrimination Act includes wider environmental planning for a growing take up MMS. This goes beyond the provision of ramps but includes parking re-charging and wayfinding navigation.

The design of MMS should comply to a comprehensive set of standards about the engineering and design that protects both the safety of the rider and other pedestrians as well as support uphold the dignity of the user. Great assistive technology developments point to the potential for greater sophistication in amenity, but also described in a visual form language that is pertinent to the rider such as the provision to stand supported and does not attempt to replicate bigger and more powerful road based vehicles.

While scooters may benefit mobility, there is concern that many older adults who are deemed unfit to drive may choose to use a MMS as their main travel mode and therefore may not have the functional abilities to operate a MMS in a safe manner, particularly if they are driving on the road. Currently, there are many gaps in our understanding of the profile of MMS users, particularly regarding their functional abilities to operate MMS, and how they may be assessed for 'fitness to ride'. There is no licensing requirement for the use of a scooter and, unlike driving assessments, there are no known best practice guidelines for occupational therapists or other health professionals conducting assessment or training for users of MMS. Lack of consistent and comprehensive assessment may result in a lack of identification of cognitive, sensory or motor impairments of the potential user, and lack of appropriate training in the safe use of the MMS [16].

MMS users have the same desire and right to explore and engage in the wide range of spaces as ambulant people. The ability to remain seated and move about daily tasks safely with all the inherent benefits therein should be a realistic aspiration and right. There remains research work to be done to address the current shortcomings and develop a stronger more robust form of mobility scooter operating in a scooter friendly infrastructure framework [17].

## References


1. Andersen, G.: Aging and vision: changes in function and performance from optics to perception. Wiley Interdisc. Rev. Cogn. Sci. **3**(3), 403–410 (2015). https://doi.org/10.1002/wcs.1167
2. Murman, D.: The impact of age on cognition. Semin. Hear. **36**(3), 111–121 (2015). https://doi.org/10.1055/s-0035-1555115
3. Nigam, Y., Knight, J., Bhattacharya, S., Bayer, A.: Physiological changes associated with aging and immobility. J. Aging Res. **2012**. https://doi.org/10.1155/2012/468469 (2012)
4. Charlton, J., Koppel, S., Odell, M., et al.: Influence of Chronic Illness on Crash Involvement of Motor Vehicle Drivers. Monash University Accident Research Centre, Melbourne, Australia (2010)
5. Marshall, S.: The role of reduced fitness to drive due to medical impairments in explaining crashes involving older drivers. Traffic Inj. Prev. **9**(4), 291–298 (2008)
6. Naumann, R., Dellinger, A., Haileyesus, T., Ryan, G.: Older adult pedestrian injuries in the United States: causes and contributing circumstances. Int. J. Inj. Control Saf. Promot. **18**(1), 65–73 (2011). https://doi.org/10.1080/17457300.2010.517321
7. Oxley, J., O'Hern, S., Burtt, D., Rossiter, B.: Falling while walking: a hidden contributor to pedestrian injury. Accid. Anal. Prev. **114**, 77–82 (2018)
8. Australian Bureau of Statistics: Population Projections, Australia, 2012 (base) to 2101. Australian Bureau of Statistics, Canberra, ACT Australia (2013)
9. OECD: Ageing and Transport. Mobility Needs and Safety Issues. OECD Pubs., Paris, France (2001)
10. Australian Competition and Consumer Commission (ACCC), N. M. Services, CHOICE, EnableNSW, F. University: Mobility Scooter Usage and Safety Survey Report (2012)
11. Oxley, J., Whelan, M.: It cannot be all about road safety: the benefits of prolonged mobility. Traffic Inj. Prev. Spec. Issue Older Road Users **9**(4), 367–378 (2008)
12. Johnson, M., Rose, G., Oxley, J.: Motorised mobility scooters—understanding a growing transport mode for older Australians. In: Proceedings, Australasian Transport Research Forum, Brisbane, Australia, 2–4 Oct 2013
13. Austroads: Operational impacts of alternative private passenger vehicles. Austroads Publication No. AP-R351/10. A. Ltd., Sydney (2010)
14. VicRoads: Website accessed 8 Jan 2019. https://services.dhhs.vic.gov.au/eligibility-and-applications
15. City of Melbourne: Resilient Melbourne Strategy. City of Melbourne's Future Melbourne Committee (2016) https://resilientmelbourne.com.au/strategy/
16. Mortenson, B., Hoag, E., Higgins, R., et al.: Stakeholders' perspectives related to the development of a scooter training program. Disabil. Rehabil. Assist. Technol. **11**(4), 289–294 (2016)

17. World Health Organisation: World Report on Ageing and Health. WHO Press, Geneva, Switzerland (2015)

**Selby Coxon** Associate Professor, the founding director of Monash University Mobility Design Lab. His career spans thirty years in professional industrial design, research and higher education, most of which have been in leadership positions. During his time in the commercial sector Selby worked for leading consultancies and brands with a wide range of products and services in Europe, including Philips Electronics and Schlumberger. Selby has published widely on a range of design responses to challenging problems, but in recent years the focus of his research has been in the pursuit of improved public and private transport, particularly in new and emerging modalities.

**Jennie Oxley** Associate Professor, a leading internationally renowned academic in injury prevention research, having made a significant contribution to the reduction of injuries and safe mobility of vulnerable populations. Her main research interests are in the roles of health promotion, education and behaviour in injury prevention and trauma and investigating the role of human behaviour and performance in injury occurrence, focusing on the assessment and management of risk and a human-factors approach to understanding the human machine interface in safety systems. Jennie provides expert advice to government, key stakeholders and community organizations. Some of her research findings have made a substantial contribution to licensing and educational policy and guidelines/standards for enhanced design and infrastructure both in Australia and overseas, and include the development of innovative and practical assessment, management and education strategies to improve safety.

# Chapter 13
# Development of a Wheelchair Stability Assessment System: Design Tools and Approaches

**Louise Moody, Paul Magee and Dimitar Stefanov**

**Abstract** This chapter describes how design has been applied to the development of a system for supporting the prescription of wheelchairs. With an ageing population there is likely to be a continued rise in wheelchair usage, as well as wheelchair modifications for specific needs such as specialist seating and the addition of assistive devices. Ensuring the ease of use, stability, safety and performance of wheelchairs both occupied by, and attended to by older adults is an important consideration. This chapter describes the design methods employed in the development of WheelSense®, a system for use by wheelchair prescribers to support the assessment, adaptation and tuning of wheelchairs to meet individual needs. The system development has required a multidisciplinary approach bringing together designers, engineers, human factors specialists, clinical specialists alongside end-users and stakeholders. The resulting WheelSense® system combines electronics and a weighing system in a folding platform. It is supported by a handheld device and graphic user interface (GUI) for guiding the prescription process, enabling data entry and to support education of the wheelchair user chair.

**Keywords** Interdisciplinary design · Stability assessment · Wheelchair prescription · Load-cell · Optimising wheelchair performance · User-centred design

---

L. Moody (✉)
Centre for Arts, Memory and Communities, Faculty of Arts and Humanities, Coventry University, Coventry, UK
e-mail: l.moody@coventry.ac.uk

P. Magee
Faculty of Health and Life Sciences, Coventry University, Priory Street, Coventry CV1 5FB, UK
e-mail: p.magee@coventry.ac.uk

D. Stefanov
School of Science and Technology, Middlesex University, The Burroughs, Hendon, London NW4 4BT, UK
e-mail: D.Stefanov@mdx.ac.uk



A. Woodcock et al. (eds.), *Design of Assistive Technology for Ageing Populations*, Intelligent Systems Reference Library 167,
https://doi.org/10.1007/978-3-030-26292-1_13

## 13.1 Introduction

Mobility difficulties are a significant cause of disability among older adults [1]. With an ageing population there is increasing need for wheelchair usage, alongside requirements for specialist seating and other assistive devices [2–5]. Wheelchair stability, safety and performance are important considerations for the estimated 65 million people requiring the use of a wheelchair [6]. Not only should users feel safe when static in their wheelchair, they should be able to move around their environment, and engage confidently in daily activities with a clear understanding of the capabilities and performance of their wheelchair [7]. For older users balancing the ease of maneuverability against stability, catering for additional equipment being added to the wheelchair, and enabling control by both the occupant and potentially an attendant are important to ensure individual needs are met.

### *13.1.1 Wheelchair Prescription*

Wheelchair prescription involves selecting and reconfiguring the wheelchair to optimise it for specific user characteristics taking into account ability, age, competence, and behaviour (e.g. body movements and behavioural choices); as well as environmental features and conditions [7]. Wheelchair modification, special seating systems and units, and accessories, for example the addition of ventilators, oxygen cylinders and communication devices, will affect performance. Wheelchairs that are not appropriately modified to meet client requirements, lifestyle and environments can be prone to tipping, sliding and loss of traction, as well as propulsion and manoeuvrability difficulties [8]. Incidents can occur on ramps, kerbs, cambers, soft ground, or when modifications have been made to the chair which alter the centre of gravity (such as the addition of medical or assistive equipment). Loss of wheelchair stability can lead to a chair tipping and potential injury to the user. Poor wheelchair performance can lead to loss of confidence, falls and potentially serious injury [9, 10].

The wheelchair can be adjusted through the positioning of the seat, the rear wheels and front castors [7]. The International Standards Organisation [11, 12] recommends the stability testing of wheelchairs after modification to ensure the chair is unlikely to tip. However, there is currently a lack of a standardised assessment method for stability and performance. The prescription process within the NHS (UK) typically makes use of fixed-incline, ramp-based tests to demonstrate compliance to ISO 7176 [11, 12]. The testing process involves positioning the occupant and wheelchair on a fixed incline ramp (at 12° for push chairs and 16° for powered and self-propelled wheelchairs). This offer an assessment of static stability by demonstrating whether the wheelchair and occupant will tip at the specific angles, giving a pass/fail result. Variable angle ramps in contrast can test a range of angles, to find the exact point of tipping. In both cases, the wheelchair user is present in the chair during these tests, which can cause anxiety for the client and present manual-handling risks for the prescriber. Whilst the prescriber is informed when the wheelchair configuration has failed (because the chair tips), they are not guided as to the adjustments needed to achieve optimum stability and performance.

### *13.1.2 Design for Wheelchair Services*

To improve the reliability and accuracy of stability testing and inform the prescription and modification process, there have been attempts to develop more sophisticated stability assessment systems. These systems have been based on load cell technology which is widely used in vehicle stability measurement. Exemplar systems have been adopted by some wheelchair services to measure the weight distribution of the wheelchair and occupant [13] and calculate the centre of gravity. This can potentially be used to predict the performance of the wheelchair following modifications to the system. An example is shown in Fig. 13.1.

Though load cells have been developed in selected Wheelchair Services this has typically been through an engineering led approach driven by mechanical systems and mathematical modelling. The example system pictured above, utilised four standard weighing scales used in motorsport for tuning go-karts (Intercomp SW650).

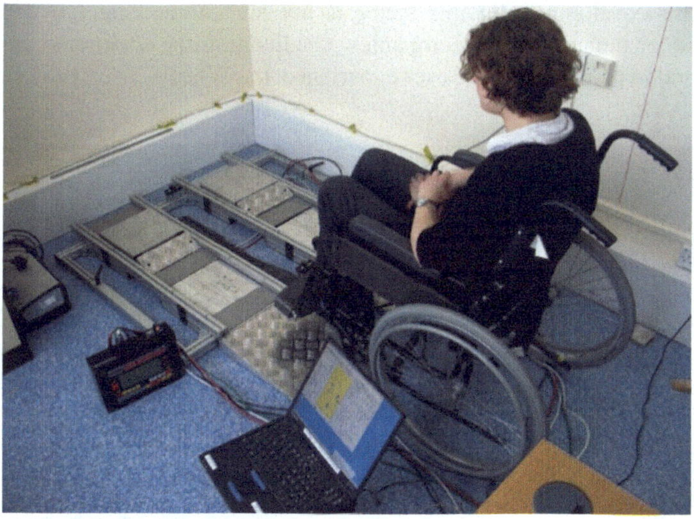

**Fig. 13.1** A load-cell based wheelchair assessment system

### 13.1.3 WheelSense®

WheelSense® [14] was developed through a NIHR i4i funded project between Coventry University, Birmingham Community Healthcare NHS Trust, Kings College Hospital NHS Foundation Trust, and Abertawe Bro Morgannwg, South Wales. The project sought to design, prototype and evaluate a stability assessment system for use by wheelchair prescribers, manufacturers and suppliers. This included the design and development of the weighing system, underlying mathematical modelling, the custom made platform rig and electronics, and accompanying graphical user interface. WheelSense® calculates the centre of gravity of the client-wheelchair system (i.e. the occupied wheelchair) and maps the stability points. This information is presented to the wheelchair prescriber on a Wi-Fi connected tablet computer so that it can be used to inform their adjustment of the wheelchair, and to model the impact of any intended modifications or additions. The system should benefit the wheelchair client and carer by giving clearer boundaries of wheelchair use and guide accurate adjustment by the prescriber to optimize performance. The resulting system is pictured in Fig. 13.2.

## 13.2 Design Approach and Methods

Traditionally development in this space is not are engineering led within NHS departments or manufacturing companies. Challenging the traditional engineering approach and to help address the user experience, the WheelSense® project aimed to

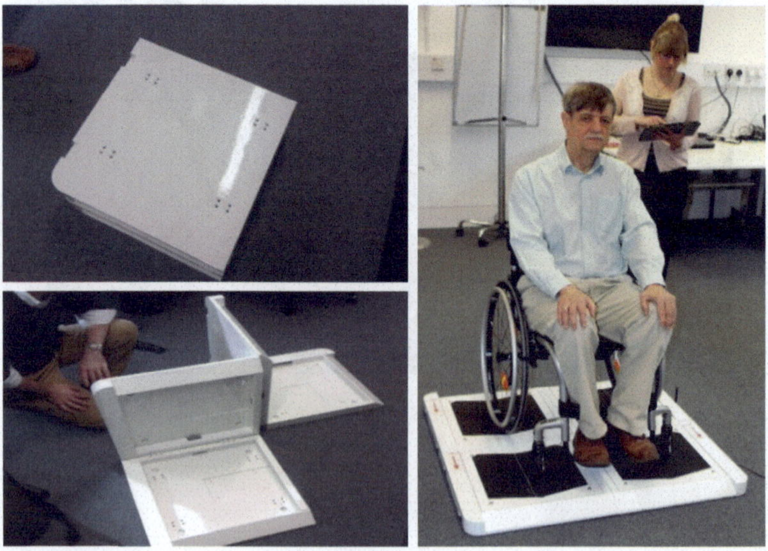

**Fig. 13.2** WheelSense® system—folded and laid out for use

employ a design led approach. Design approaches are increasingly documented in healthcare [15–18] with growing attention being paid to how they can be embedded within working practices to achieve innovation [19, 20].

### *13.2.1 Multidisciplinary Approach*

The WheelSense® research team was multidisciplinary involving rehabilitation engineers and wheelchair prescribers employed by three different NHS Trusts, and researchers in the fields of engineering, mathematical modelling, software development and gaming, human factors and product design. The team approach recognised the need to prioritise design and human factors to shift the experience of the prescription process and enable innovation [21]. The involvement of product designers was considered important to address the visual aesthetic and integration of the experiential elements of using the system either as a prescriber or wheelchair client. The team sought to package the electronic components and complex mathematical modelling within an accessible and easy to use product that would minimize anxiety and reduce the risks associated with ramp-based testing.

The team members worked on different elements of the development with level of involvement varying through the user needs research, development of the platform and user interface, technical development and evaluation stages. The wide range of professional and academic disciplines, at times made progress and communication challenging. The team were distributed across the UK as well as a later stage of the project involving collaborators in the USA and Japan. A range of communication tools were important including workshop activities and use of collaboration tools such as the CUbe (discussed further below).

### *13.2.2 User and Stakeholder Centered Design*

A user and stakeholder-centred design process was adopted [22]. The primary user for WheelSense® was identified as the prescriber (typically a rehabilitation engineer or occupational therapist), and the secondary users or beneficiaries as the wheelchair clients and carers. The scenario of use involves the prescriber positioning the wheelchair and occupant (client) on the WheelSense® platform; entering some of the key dimensions of the wheelchair into the system while other key dimensions (such as distance between contact points of wheels with the measurement surface) were sensed and measured automatically by the system. WheelSense® provides data to guide adjustment of the wheelchair. The wheelchair user/client does not operate or control the system. They may self-propel onto the platform, and can view visual information about the stability of the wheelchair with the prescriber via the tablet interface.

A stakeholder group was brought together at the outset of the project to guide the research, design and evaluation. Three of the project team had an active clinical role in wheelchair services, which facilitated active involvement of additional clinical staff, patients and carers as well as contacts working in relevant companies and organisations. The stakeholder group was made up of two wheelchair users, a wheelchair service representative, representatives of a manufacturer of wheelchairs, a manufacturer of special seating systems, and of commercial wheelchair providers. The group met every 6 months to review progress and provide critical input throughout the project.

The development process was informed by a scoping study to understand current wheelchair provision and needs [22, 23]. An online survey of 98 participants working in wheelchair provision, as well as 10 interviews and a focus group with professionals working in wheelchair provision in three NHS Trusts in the UK were undertaken. The project was advertised in the three Trusts and participation was voluntary. The results provided a broad picture of the current UK practice in stability testing and helped generate system requirements. Issues with the reliability and usefulness of the existing methods were highlighted. A large proportion of respondents (78.9%) reported that there were limitations to the stability assessment method that they currently used. The strongest recurring theme was that ramp tests only indicate static stability and are not sophisticated enough to mimic the real-world use of wheelchairs. The tests are performed indoors on a simple ramp, which is not necessarily reflective of the outdoor environment where slopes are often multi-planar. Interview participants were also concerned about the limited information provided by ramp-based tests and translating the output into a meaningful measure of stability.

Concern was expressed for the clients during the ramp test. Survey feedback indicated that the ramp tests can leave the wheelchair client feeling vulnerable as they are positioned at an angle on a ramp. Some prescribers expressed concern that on the steeper inclines (e.g. 16°) clients would lean or brace themselves to feel more secure, or need to be supported by the prescriber, which could further undermine the accuracy of the test. It was highlighted however, that the ramp test was useful in illustrating to the client how a particular angle of incline feels, and the limits of the chair.

The existing systems were not regarded as portable, yet ideally they would be used in field tests and domiciliary visits at patient homes. Some participants reported having sustained injury or causing damage to surroundings when moving the ramp. Others felt it was impractical to take the ramp on a home visit due to the size and weight. Interviewed prescribers reported manual handling issues with positioning wheelchairs on the ramp and conducting stability tests. It was indicated that stability tests would typically be undertaken by two people to reduce the injury risk and to reduce the test time. In addition, participants indicated that the ramp was out of keeping with the look and feel of modern healthcare equipment and did not inspire confidence.

The user needs research along with collaborative team working resulted in a list of requirements for the development of a new load cell-based wheelchair stability assessment system (see Table 13.1). This list included improved accuracy of tipping

**Table 13.1** Emerging requirements [22]

| |
|---|
| *Functionality of the system*: |
| • Enhanced accuracy of stability measurement<br>• Determine the maximum inclines at which the chair can be considered as stable (and at which point it will tip) without inclining the measured wheelchair to these extreme angles, preferably both in a static and dynamic test, and in each direction<br>• Ability to model or predict the effects of various configurations of the chair on its stability<br>• Ability to use in patients' home environments or environment where chair is most often used<br>• Be portable and lightweight to allow sharing between services in order to reduce cost barriers and facilitate wheelchair tuning at users homes<br>• Be simple to use but with the option of more complex features and functions to cater for novice and expert users<br>• Be easy to set up and reduce preparation times compared to existing load cell solutions in order to allow timely assessments<br>• Look attractive to avoid client apprehension<br>• Enable prescribers to demonstrate to patients the stability limits of the chair and the physical angle of tilt |
| *Outcomes and outputs*: |
| • Enhanced optimisation of chair stability for individual needs<br>• Record the assessment process and outcomes for clinical use<br>• Hints and guidance on interpreting the measurements whilst retaining clinical judgement and ability to adapt for different clients<br>• Enhanced posture and comfort for the client<br>• Reduced manual handling for the prescriber compared to the existing ramps<br>• Have the capability to show clients some visual representation of their chair capabilities<br>• Support education of the clients and carers in order to reduce misuse of chairs or client error. |
| *Purchasing and training*: |
| • Training and support in how to use the tool and maximise value—particularly for professionals who may be put off by the technicality of such a product.<br>• Low cost—less than £4 k for unit and training.<br>• The system should move away from traditional views on 'stability' and testing to encourage more consistent and frequent use to maximise chair performance and safety. |

angles, features to support record keeping, improved client/carer education support and the capacity to model or predict client-wheelchair system performance in different configurations.

### *13.2.3 Developing a Shared Design Direction*

Building upon a set of requirements a Product Design Specification (PDS) and brief is typically drafted at the outset of a project to inform the design and engineering tasks. As significant technical development and research was required, it was challenging to develop a clear PDS at the start of this project. Instead the specification emerged as the technology developed, with a focus on several key elements to the system:

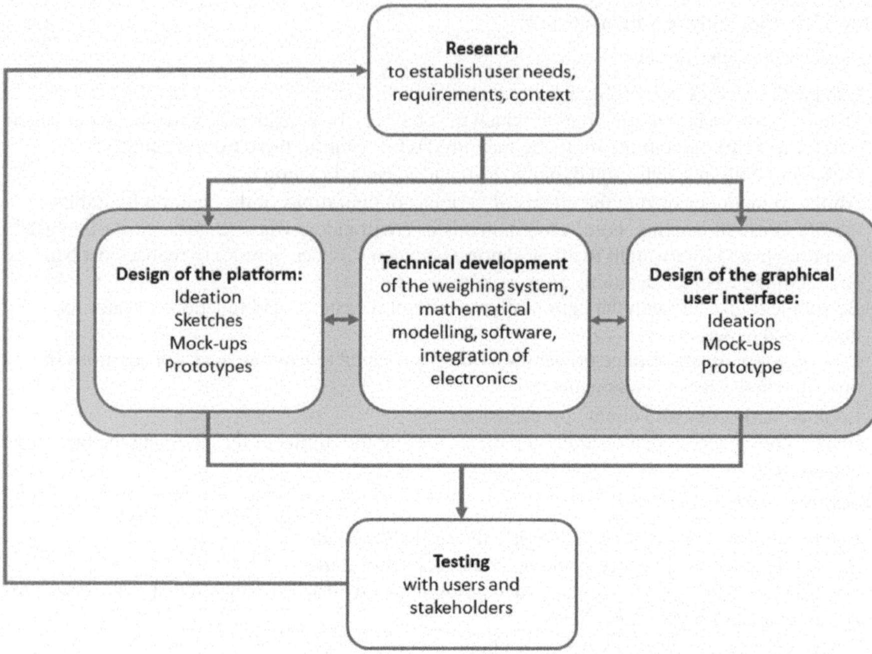


**Fig. 13.3** Integration of development strands

1. A platform on which the client and wheelchair system is positioned
2. The underlying electronics within the platform
3. The graphical user-interface to operate the system

The industrial design occurred iteratively and to some extent in parallel, with the technical and electronic system development (represented in Fig. 13.3).

### *13.2.4 The CUbe Ideas Generation Tool: Developing a Shared Vision for the Product*

To establish a vision and shared ethos for the design, the team used the CUbe ideas generation tool [24]. The CUbe was developed to support design collaboration between a variety of disciplines and to facilitate team communication and ideas generation [24]. It had been created and used previously by the one of the WheelSense® design team. It was used here to kick-start the creative process and create a shared product vision that would align the contributing partners and disciplines. It was chosen over other tools, as it was a demonstration of the design lead's own approach and methodology, as their discipline was introduced to the rest of the team.

The CUbe is an easily handled (30 $cm^3$) box covered in removable paper. The surface of the CUbe is used for writing, noting or illustration during a facilitated session (see example in Fig. 13.4). A question is posed by the facilitator. The CUbe is then passed around the group to add thoughts and ideas to the surface as the discussion continues. Connections are made from one idea to another regardless of which side of the cube they sit. Once the session is complete, the cube, constructed in a net format is opened up. Scanning the surface or digitally re-creating each side produces a graphic that can be stored, shared and developed further.

To guide the early concept development, an initial question was set: "What would the system look like if it was designed by Apple?" This question was posed because the design team recognised the impact the Apple aesthetic has on user experience due to visual simplicity and high perceptual value. In contrast, the existing ramps were recognised as heavily engineered tools, for which styling had not been considered. The question challenged the engineering approach familiar to a number of the partners. Expertise in design, human factors, rehabilitation engineering, mechanical engineering, electrical engineering, occupational therapy and robotics were involved in the exercise. The sessions were run outside of the usual office setting with participants standing around a table to heighten involvement in a driven, active experience. The sessions were limited to 30 minutes to maintain energy and focus.

The result was a series of concept ideas that were further explored by three more 30-minute sessions during the same day. The resulting notes were shared with the group; informing the agenda for follow-on meetings and exploration across the team into areas that would otherwise not have been considered without the use of this abstract idea generation technique.

By using this approach, we were able to agree a design direction, the main system requirements and then specific tasks for individuals or smaller groups to work on. It was agreed that the system design should not impose itself on the user, and should almost disappear into the environment so the focus of the prescription experience was low stress and could be driven by conversations rather than a test.

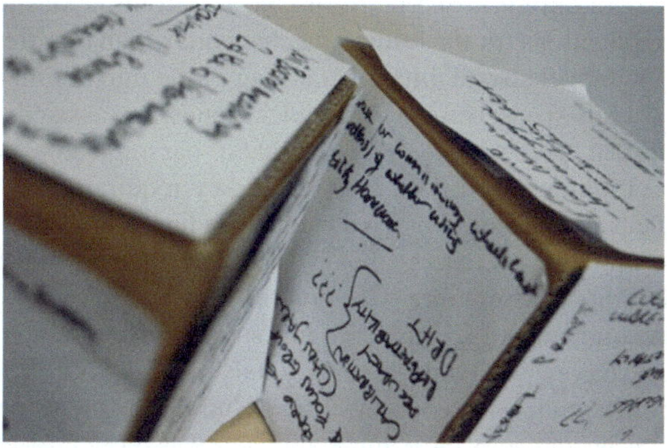

**Fig. 13.4** The Cube

### 13.2.5 Design Evolution and the Use of Sketch Models

Having established system requirements and an agreed design vision, the sketch and early modelling phase began. Sketch models are low-cost, rapidly produced demonstration models used to provide the team and stakeholders with a better understanding of development factors such as size, touch point position, component spacing etc., that may not be as effectively resolved by on-screen means alone, for example through Computer Aided Design (CAD). Sketch modelling allowed the team to propose design ideas and gain early feedback through multiple iteration cycles (in development circles referred to as failing fast [25]).

A series of sketch models were produced in different materials to fulfill different purposes as summarised in Fig. 13.5. These enabled the team to identify and address both design and electronic challenges. When establishing the external package size of the platform, a foam board model was produced. It was fast to create, easily adjusted and had low material cost. This basic model was used to convey the proportions of the design and led to the introduction of small details such as rounded edges, the approach ramp angles, and theoretical cable routing. A basic card tri-fold model with a carry handle helped consideration of platform portability and the size of a portable case (Fig. 13.5A).

A Medium Density Fibreboard (MDF) sketch model was used to explore the package size and agree the locations of various electronic parts. A further model of MDF and steel piano hinges allowed demonstration of the folding function (see Fig. 13.5B). Whilst the materials incurred some cost and production time they allowed experimentation, and the team to handle and apply loads to the model. Through the modelling the potential risk of electronic cross-talk and interference across the measurement platforms related to the hinge position was identified. In the next iteration this was addressed by the measurement platforms operating independently.

A foam board sketch model (see Fig. 13.5C) was used to demonstrate a triple hinged system and the first conduit locations to envelope electronics and cables. This model illustrated a risk of cross talk between each sensitive top plate leading to a re-design in the platform and it becoming an inverted form. This in turn had implications for the folding mechanism, which required a different kind of structure. The model also facilitated planning of the integration of the electronics through use of a full-size schematic of the electronics. It was used again later in the development to plan cable routing and a method of passing cables safely through the conduit.

Lightness and portability were key requirements for the system. The previous measurement systems used lengths of aluminium extrusions to form a rigid frame. As a result they were large, heavy and hard to move (30.6 kg + (4 × 3.4 kg)). To address weight, Carbon Fibre (CF) composites used in motorsport were considered. Due to the high cost of CF, material investigation focussed on Aluminium composites such as Alucore®. Alucore® (ASD Metals Ltd, UK) is a strong but lightweight aluminium honeycomb covered by 1–2 mm aluminium sheets and often used in architecture, marine and aerospace applications. It was more cost-effective and we experimented with Alucore® to understand how it may behave when modified (i.e. drilled, cut,

| | Image | Material | Purpose |
|---|---|---|---|
| A |  | A basic, card tri-fold model with a carry handle | To aid the discussion and begin to understand emerging portability implications |
| B |  | Medium Density Fibreboard (MDF) | To understand package size and to agree the locations of various electronic parts |
| C |  | Foam board sketch model | To demonstrate a triple hinged system and the first conduit locations to envelope electronics and cables |
| D | 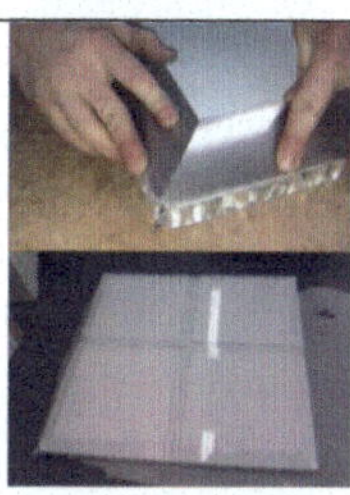 | Alucore® sheets, aluminium edging, wrapped in vinyl | To demonstrate the folding action and a method to hide the hinges into a flat surface. To help the production of a draft specification for the internal electronics. |
| E |  | Alucore® sheets, aluminium edging and aluminium box section, in automotive white spray paint finish. | To allow internal sensor attachment via a custom designed CNC machined aluminium mount, improving the repeatability of each sensor installation. Provision of a surface treatment to provide grip. |

**Fig. 13.5** Stages of sketch modelling

formed). The material was also used in a preliminary test model, in flat-sheet form to establish performance.

Grand Design Systems (GDS, Brackley), had extensive experience of using Alucore® to produce rigid, portable race team pit systems and were commissioned to produce multiple prototypes. An Alucore® model (Fig. 13.5D) was produced to demonstrate the folding action and a method to hide the hinges into a flat surface once fully opened or closed, to provide a smooth exterior. Due to the nature of this hinging system, the plates performed poorly in a lateral motion, suffering some twisting which required adjustment and use of a more typical laterally rotating hinge for the follow-up model. This model helped the production of a draft specification for the internal electronics.

The next version of the sketch model was finished in automotive white spray paint (Fig. 13.5E). In this final model, 16 internal sensors were attached to the base plate via a custom designed CNC machined aluminium mount, improving the repeatability of each sensor installation. Each top plate was supplied in white paint finish with an additional surface treatment to provide grip. This treatment was formed from a chevron shaped, self-adhesive, abrasive sheet, illustrating the direction of travel required to complete the stability test.

### *13.2.6 CAD Methods*

As well as traditional, analogue and fabricated (non-computer based) models, digital techniques were also used. SolidWorks CAD (Dassault Systèmes SolidWorks Corporation, USA) was used to test out form and interactions between components. A SolidWorks CAD model of each component was built virtually, as the part features were identified, and presentations to the stakeholder team involved computer generated, rendered images of the prototype virtual assembly, combined with iterative sketch models. A simple process was used to carve out material from a virtual 3D solid allowing for each part to be incorporated in a specific location. Electronic components were integrated at a late stage in the project to the CAD model, but had already been bench tested in a more basic form to ensure fit and access. All of the CAD data generated was easily supplied to external fabricators for production review and manufacture.

## 13.3 Integration of Components into the Final Functional Prototype

A semi-finalised batch of 4 WheelSense® fully functional pre-production prototypes were produced through fabrication and high-resolution 3D printing.

### 13.3.1 The Platform and Frame

The final prototype system consists of a platform containing a number of force-sensitive weighing that contain arrays of planar beam load sensors mounted in an aluminium frame linked mechanically with suitable hinges. The folding frame is constructed with Alucore® honeycomb aluminium sheets, CNC milled and then bonded to extruded aluminium. The resulting frame is strong, yet more lightweight and portable than comparative systems. The platform has a non-threatening white finish with extra grip details in a high colour contrast grey (and high friction texture). This combination has similarities to sports equipment (e.g. mountain bikes). To prevent any unwanted movement of the platform once deployed on a smooth floor, the base featured a rubber (Dycem, UK) sheet.

### 13.3.2 Internal Architecture

The platform incorporates 16 sensors installed within 4 sensing plates dimensioned 500 mm × 500 mm, 4 amplifiers (1 per sensing plate), 1 PSU (power supply unit) and power switch, and 1 WiFi antenna (see Fig. 13.6). The sensitive parts (sensors and signal amplifiers) are housed in commercially available rapid manufactured enclosures that were additionally modified with customised lids by using a high resolution Fused Deposition Modelling (FDM) 3D Printer with the use of Acrylonitrile Butadeine Styrene (ABSplus) material (Stratasys, USA). This provides specific enclosures with integrated mechanical fixing points for individual parts and attachment to the frame.

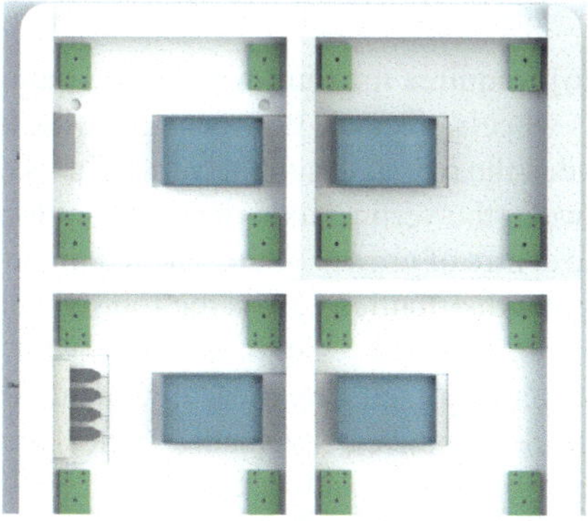

**Initial sketch of the topology of measurement boards**

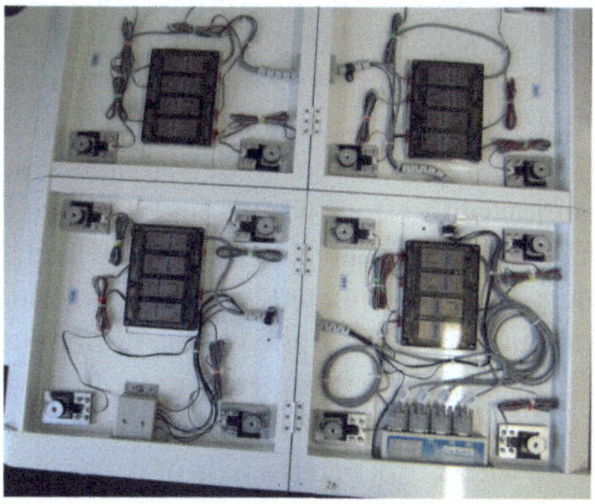

**The internal architecture**

**Fig. 13.6** Layout of the internal architecture of the product

Connecting cables are routed through 3D printed ports between each quadrant of the platform. They are grouped internally to prevent excessive movement whilst allowing the mechanical frame to fold without restriction. Sensor signals from the load cells are filtered and digitised. RS485 is used as a communication protocol between the sensor processing channels and a wireless portable router that transfers sensor signals to a portable tablet with Windows 8.1 OS. The sensors are of sufficient resolution to sense the load imposed by each wheel of the wheelchair and to detect the position of the centre of pressure of each wheel.

A communication interference issue was discovered once the components had been assembled. Signals were blocked as a result of housing the antenna within an all-aluminium structure (creating a faraday cage effect). In order to overcome this, the antenna is re-positioned on the outside edge of the mechanical frame where it can be partly concealed when not in use. Whilst this has resolved the interference issue, the externally mounted antenna remains vulnerable and needs further consideration.

### *13.3.3 Measurement System*

The measurement platform is comprised of four measurement segments that capture the force through each wheel. During the measurement process, each wheel is located on a separate measurement segment of the platform. For calculation of static wheelchair stability, at least two sets of load measurements are required. One of these measurement sets presents the load distribution measurements underneath each wheel when all contact points of the wheels lie in the horizontal plane and included measurement of the wheel reaction forces for 2 different swivel positions of the castors (front wheels oriented to the rear and front wheels oriented to the front). The second set of load measurements was collected when castor wheels are risen on a known angle and oriented to the rear.

The approach [14] (detailed further in [26]) requires measurement of the radius of the rear wheels, radius of the front (castor) wheels, distance between rear wheels, distance between front wheels, distance between the rear axle and the vertical axis of rotation of the front wheels and distance between the vertical axis of the wheel fork and the axis of rotation of the front wheel. The centre of gravity of the wheelchair and client is then calculated by measuring the weight distribution across the four wheels via four load cells aligned in the horizontal plane. To enable calculation of the centre of gravity, certain parameters regarding wheelchair geometry are measured manually (measuring the sizes of the rear and castor wheels, distance between the vertical axis of the wheel fork and the axis of rotation of the front wheel, and distance between the rear axle and the vertical axis of rotation of the front wheels). The developed system does not require manual measuring of the distances between wheel contact points, i.e. it was developed and applied an original algorithm where the distance between rear wheels and the distance between front wheels are determined automatically and do not require manual measurements. It was also considered that the evaluation of wheelchairs from known models can be facilitated if their dimensions are entered directly by referencing to a wheelchair database instead measuring them manually every time. Such a wheelchair database could be created centrally and distributed as internet updates to owners of measurement systems and considering options for customising the wheelchair data library additionally by the end user.

### 13.3.4 Deployment and Transportation

For portability, the platform was designed as a folding 4 quadrant frame. Deploying the platform is a manual process and the correct opening sequence is illustrated on the surface using a number system (i.e. 1, 2, 3) and direction arrows. The overall folded package is 500 mm × 500 mm × 240 mm. The system will fit into the boot of an average hatchback car, e.g. Ford Fiesta. When deployed it presents a floor mounted 1000 mm square platform, 50 mm in depth with a 45° ramp along two opposing sides. Using geometry that originates from the ISO stability standards [11, 12] the platform was optimised to a depth of 50 mm. This depth is required to perform the virtual tilt tests and minimises the physical change in angle (tilt forwards, backwards and side to side) that is experienced by the Wheelchair client during assessment. For transportation, the folded unit is contained within a padded, wheeled fabric case (i.e. when travelling between clinic and a patient's home or preferred test location). For transportation over greater distance, there is also a rigid case.

### 13.3.5 Software and the Graphic User Interface (GUI)

The wheelchair stability software is installed on the tablet and comprises of three main elements, the calculation algorithm software, the signal processing software [14, 26] and the graphical user interface (GUI). A web-based format was chosen in order to be platform agnostic and suitable for a wide range of devices. The system has an internal router that creates short-range Wi-Fi network signal to communicate with a tablet (more than one tablet could be connected to the same network where necessary).

The platform was designed as a fully calibrated system. The measurement procedure involves 3 easy steps where the wheelchair and occupant are moved sequentially into 3 positions on the platform and sensor data is automatically generated and recorded. The distance between each pair of wheels is calculated automatically from the positions detected by the sensors. The signal processing software continuously scans the sensors to detect wheelchair positioning and to prevent communication errors and loss of information. The load cell signals of the separate plates are preprocessed and sent via Wi-Fi link to a computer (typically a tablet) with signal processing and visualization software. Sensor signals are analysed by the software on the tablet, and the position of the centre of gravity and theoretical tipping angles are calculated and presented to the user.

The Graphical User Interface (GUI) can be accessed via a web browser (see GUI illustrated in Fig. 13.7). It was designed to be simple and intuitive to facilitate system set-up and present data in an accessible format and meaningful. It was intended to provide not just a clinical tool for the prescriber but a system that would educate the wheelchair client on the limits of their wheelchair. The GUI designs were developed

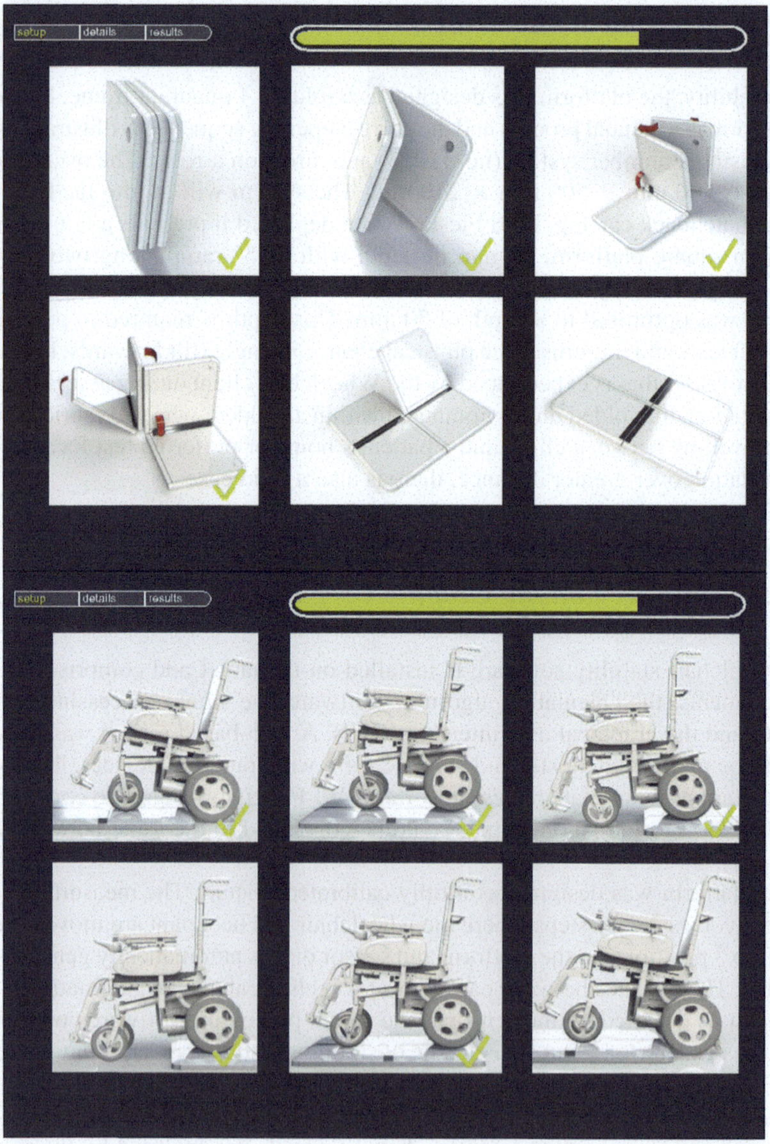


**Fig. 13.7** GUI screen shots of the tuning toolkit and the user guidance given

iteratively between the team and representative users. Initially mock-ups were created (e.g. hyperlinked PowerPoint documents) and features were added, or descriptive language modified before moving onto semi-functional prototypes.

The final interface has an icon driven menu offering a series of instructions and prompts. It offers different levels of functionality depending on the clinical expertise of the user. It takes the user through the stability assessment task by a series of prompts and guided steps to achieve the tipping angles. A graphic illustration of the measured wheelchair stability parameters is generated and can be saved. A 'Tuning Toolkit' mode is also included offering a virtual simulation intended to allow the user to test and visualise the potential effect on stability of any alterations to the wheelchair (Fig. 13.7).

## 13.4 Iterative Testing and Evaluation

As well as iterative design and technical testing throughout the project lifetime, final technical tests with both manual and powered wheelchairs were performed and the system demonstrated high measurement accuracy with both test users and 100 kg test dummy. WheelSense® was then evaluated in a clinical setting. It was tested within 3 NHS Trusts over a 3-month period by wheelchair prescribers including clinical engineering and physiotherapy specialists. It was intended that the prescribers would use the system in a range of scenarios including use for powered as well as manually propelled wheelchairs, with adaptations such as specialist seating, tray tables and grab bars, and with clients with a range of characteristics and conditions (including tetraplegia, cerebral palsy, amputation).

Effectiveness and usability were considered through observations of the system in use (n[1] = 7) with both powered and manually propelled wheelchairs and through usability assessments (n = 3). To understand the user experience, semi-structured interviews with engineers and occupational therapists (n = 10) and staff experience log and surveys were completed (n = 27). To understand the patient/wheelchair client experience surveys of patient and carer experience were collected (n = 8).

The results suggested that the system was regarded as a useful tool. Prescribers felt that the system was generally easy to use. The manual exertion required to use WheelSense® was considered more manageable than traditional inclined ramp systems. The prescription process was found to be efficient and the displayed results useful for patient education. The results showed that the prescribers felt safe when using the system, felt it was suitable for their clients, and had not observed anxiety from their clients during use. In comparison to ramp-based stability testing the following feedback was given:

> 'It's a more comfortable experience for the user just being put on a flat surface.' KCHS08

[1] n = number of participants.

> 'I've had a lot of comments that it's less scary. Because obviously I've been doing it on the static rig and then putting them on the platform so they're instantly saying 'Oh that's much better.' KCHS04

> 'We are not doing anything that is a risk to them or manual handling risk to us. [...] I think the ramp is a lot more invasive, isn't it.' WMS08

Carers and clients indicated that they were pleased that a physical tilt of their wheelchair was not required for the test [27]. The stability assessment process, once practiced, was reported to be quick in comparison to ramp tests and the existing load cell systems used. Prescribers were able to visualize and demonstrate changes in stability to their clients following changes/modifications to the wheelchair configuration.

Feedback suggested that further development could enhance the ease of use of the system software. It was felt that the functionality offered by the software was useful, but the usability and accessibility of the functions could be further improved. Despite the intended simplicity of the system and 'Tuning Toolkit' it was still recommended that training in use of the system would be beneficial.

## 13.5 Discussion

This chapter has outlined how design has been brought into the development of a complex system. Traditionally development in this area has been engineering driven, but here we have described some of the methods and tools that have been used to involve design, and importantly users and stakeholders in the development process.

The project team had to be multidisciplinary to enable production of such a system. This has bought complexity in terms of working together and developing a shared language and approach. The CUbe was one of the tools used to address this challenge, as well as team meetings, workshops and stakeholder consultation. We found it important to build in time to the project plan for designing together as a team to accelerate progress and to share working methods.

The use of sketch models in a variety of different materials has been critical in the development of WheelSense®. They provided a way to quickly and relatively cheaply visualise and explore the design, make modifications and get feedback from members of the project team as well as our stakeholders as various elements were integrated. The sketch models enabled consideration of aesthetics, functionality, material selection, packaging and usability at different stages of the project.

There was a concerted effort to involve stakeholders and users throughout the project to ensure their needs and wants were met. This was complex given the technical nature of the development, the range of users (primary and secondary) and stakeholders to consult. It was also challenging in the context of a research project involving the NHS that requires careful adherence to ethical approval and data regulatory processes. The time required for this can be lengthy and led to an in-depth study with a relatively small sample rather than extensive testing of the system.

The GUI design for WheelSense® was iterated several times and much improved. However, it was still identified as an area for improvement during the evaluation. With hindsight, the GUI was developed late on in the project timetable and should have been considered as a more integral part of the service design. It may have been advantageous to take a service design approach to the development of the system as a whole, as the software as well as the hardware is integral to the user experience.

As a system WheelSense® has the potential to improve the safety and performance of wheelchairs if taken up as a tool by manufacturers, suppliers and wheelchair services. The increase in life expectancy is likely to increase the need for wheelchairs for the performance of daily activities. It is important for both the occupant of the wheelchair and any attendant that the wheelchair is easy and safe to manoeuvre. The existing ramp tests provide limited information to the prescriber and the pass-fail nature of the test is intimidating and uninformative for the wheelchair client. WheelSense® was seen as an improvement and prescribers reported that they had not experienced any problems with patients feeling uncomfortable during the testing process. It was encouraging to find that the key emotional challenges associated with ramp usage from the client and prescriber perspective were in part addressed through the design. WheelSense® has a role not just in testing and prescription but in supporting and educating wheelchair clients in the characteristics of their wheelchair, how it will perform in different conditions, on different surfaces and as additional loads (e.g. shopping) are made to the chair. This educational value is something that we hope is exploited through future development of the system.

**Acknowledgements** The Wheelchair Stability Assessment System (Wheel-SAS) project was funded by the National Institute for Health Research (NIHR) Invention for Innovation (i4i) stream (Grant II-AR-0209-10099.). The user needs, and design work was approved by the Coventry University Ethics committee and access to NHS premises and staff was given by the R&D departments of each participating NHS Trust. The system evaluation study was approved by the National Research Ethics Committee, the Research and Development Department of each of the participating NHS Trusts (Birmingham Community Healthcare NHS Trust, Kings College Hospital NHS Foundation Trust, and Abertawe Bro Morgannwg, South Wales), and the Coventry University Ethics Committee. All participants gave their informed consent prior to participation. We would like to thank those participants as well as our collaborating partners Simon Fielden, Mike Heelis, Paul Dryer, Nigel Shapcott, Jill Evans.

## References

1. Martins, A., Pinheiro, J., Farias, B., Jutai, J.: Psychosocial impact of assistive technologies for mobility and their implications for active ageing. Technologies **4**(3), 28 (2016)
2. Gell, N.M., Wallace, R.B., Lacroix, A.Z., Mroz, T.M., Patel, K.V.: Mobility device use in older adults and incidence of falls and worry about falling: findings from the 2011–2012 National health and aging trends study. J. Am. Geriatr. Soc. **63**(5), 853–859 (2015)
3. Labbé, D., Mortenson, W.B., Rushton, P.W., Demers, L., Miller, W.C.: Mobility and participation among ageing powered wheelchair users: using a lifecourse approach. Ageing Soc., pp. 1–17 (2018)

4. Mortenson, W.B., Hammell, K.W., Luts, A., Soles, C., Miller, W.C.: The power of power wheelchairs: Mobility choices of community-dwelling, older adults. Scandinavian J. Occup. Ther. **22**(5), 394–401 (2015)
5. Giesbrecht, E.M., Miller, W.C., Jin, B.T., Mitchell, I.M., Eng, J.J.: Rehab on wheels: a pilot study of tablet-based wheelchair training for older adults. JMIR Rehab. Assistive Technol. **2**(1), (2015)
6. World Health Organization: Fact sheet on wheelchairs (2010). Available from https://apps.who.int/iris/bitstream/handle/10665/205041/B4616.pdf?sequence=1&isAllowed=y. Accessed 15 Mar 2019
7. Gartz, R., Goldberg, M., Miles, A., Cooper, R., Pearlman, J., Schmeler, M., Jonassen Bittman, S., Hale, J.: Development of a contextually appropriate, reliable and valid basic wheelchair service provision test. Disability Rehab. Assistive Technol **12**(4), 333–340 (2017)
8. Kirby, R.L.: Wheelchair Skills Assessment and Training. CRC Press (2016)
9. Rice, L.A., Ousley, C., Sosnoff, J.J.: A systematic review of risk factors associated with accidental falls, outcome measures and interventions to manage fall risk in non-ambulatory adults. Disabil. Rehabil. **37**(19), 1697–1705 (2015)
10. Chen, W.Y., Jang, Y., Wang, J.D., Huang, W.N., Chang, C.C., Mao, H.F., Wang, Y.H.: Wheelchair-related accidents: relationship with wheelchair-using behavior in active community wheelchair users. Arch. Phys. Med. Rehabil. **92**(6), 892–898 (2011)
11. ISO 7176-1:2014: Wheelchairs—part 1: determination of static stability 60.60 ISO/TC 173/SC 1 4 (2014)
12. ISO 7176-2: Wheelchairs—part 2: determination of dynamic stability of electrically powered wheelchairs (2017). https://standards.globalspec.com/std/10188578/ISO%207176-2
13. Caldicott, S.J., Shapcott, N.: Validation of a software-based stability assessment system for wheelchairs and their occupants. J. Med. Eng. Technol. **32**(6), 440–447 (2008)
14. Stefanov, D., Fielden, S., Avtanski, A.: Mobility assistance vehicle stability assessment. U.S. Patent Application 14/903,863, Coventry University (2016)
15. Giacomin, J.: What is human centred design? Des. J. **17**(4), 606–623 (2014)
16. Moody, L.: User-centred health design: reflections on D4D's experiences and challenges. J. Med. Eng. Technol. **39**(7), 395–403 (2015)
17. Donetto, S., Pierri, P., Tsianakas, V., Robert, G.: Experience-based co-design and healthcare improvement: realizing participatory design in the public sector. Des. J. **18**(2), 227–248 (2015)
18. Tsekleves, E., Cooper, R.: Emerging trends and the way forward in design in healthcare: An expert's perspective. Des. J. **20**(sup1), S2258–S2272 (2017)
19. Robert, G., Cornwell, J., Locock, L., Purushotham, A., Sturmey, G., Gager, M.: Patients and staff as codesigners of healthcare services. BMJ **350**, g7714 (2015)
20. Zuber, C.D., Moody, L.: Creativity and Innovation in Health Care. Nurs. Adm. Quart. **42**(1), 62–75 (2018)
21. Norman, D.A., Verganti, R.: Incremental and radical innovation: design research versus technology and meaning change. Des. Issues **30**(1), 78–96 (2014)
22. Moody, L., Evans, J., Fielden, S., Heelis, M., Dryer, P., Shapcott, N., Magee, P., Stefanov, D.: Establishing user needs for a stability assessment tool to guide wheelchair prescription. Disabil. Rehabil. Assistive Technol. **12**(1), 47–55 (2017)
23. Moody, L., Woodcock, A., Heelis, M., Chichi, C., Fielden, S., Stefanov, D.: Improving wheelchair prescription: an analysis of user needs and existing tools. Work **41**(Supplement 1), 1980–1984 (2012)
24. Magee, P., Fielden, S., Moody, L.: CUbe: an ideas generation tool demonstrated through multidisciplinary development of a wheelchair stability assessment system. In: Recent Advances in Assistive Technology and Engineering (RAaTE) Conference. Coventry, UK (Poster) (2014)
25. Hall, D.: Fail fast, fail cheap Bloomsberg Businessweek (2007). Available from http://www.businessweek.com/stories/2007-06-24/fail-fast-fail-cheap. Accessed 22 Mar 2019
26. Stefanov, D., Avtanski, A., Shapcott, N., Magee, P., Dryer, P., Fielden, S., Heelis, M., Evans, J., Moody, L.: The development and testing of a system for wheelchair stability measurement. Med. Eng. Phys. **37**(11), 1061–1069 (2015)

27. Moody, L., Evans, J., Fielden, S., Heelis, M., Dryer, P., Shapcott, N., Magee, P., Stefanov, D.: (In preparation) The usability and acceptability of a novel wheelchair stability assessment system

**Louise Moody** is Professor of health design and human factors in the Centre for Arts, Memory and Communities at Coventry University. She is also Human Factors Theme Lead in the NIHR Devices for Dignity MedTech Co-operative. She has a BSc in psychology and sports science, and a PhD in electronic and electrical engineering from the University of Birmingham. After working as Human Factors Researcher and Consultant, she joined Coventry University in 2006, initially as Senior Lecturer in intelligent product design in the Department of Industrial Design. Since then, she has held a number of management roles within the School of Art and Design as well as leading a research theme related to arts and well-being. Her research is focused on the design and development of products, interventions and services to benefit health and well-being. She employs a range of research methods as well as art-based approaches to ensure that new solutions are functional as well as being desirable and acceptable to end-users and stakeholders. Her focus is on ensuring the involvement and representation of diverse user needs and requirements from the early stages of user need assessment, through to the design, development and evaluation of solutions. Her current projects are particularly focused on assistive and self-management technologies.

**Paul Magee** is an Industrial designer trained at BIAD, UK. His design experience includes automotive styling, exhibitions, furniture and products, digital user-interfaces, infant care and medical diagnostic-technology. Paul has combined design industry experience with academic design research developing solutions that meet real user need. Collaboration is fundamental to Paul's professional approach and equally that of his personal perspective. An experienced facilitator for groups working with co-creation methods; Paul has delivered Health and Assistive Technology (AT) conference presentations on the subject. Both Paul's design work and research are developed collaboratively with users, with particular application to vulnerable user groups such as people with learning disability. His research interests include user experience, sustainable production, democratization of design, experiential insight and co-design with end users to find ways that improve lives in real terms.

**Dimitar Stefanov** is an Associate Professor in Biomedical Engineering at Middlesex University in London. After he gained his Ph.D. in biomechanics and biomedical engineering from the Bulgarian Academy of Sciences, he worked at the Institute of Mechanics in Sofia from 1982 to 1989 as an Assistant Professor. In 1990, he achieved habilitation in biomedical engineering and biomechanics and, during the same year, he became an Associate Professor at the Institute of Mechanics and Biomechanics in Bulgaria. From January 1998 to March 1999, he was a Visiting Research Professor at the New Industrial Research Organization (NIRO), Kobe, Japan and from September 2000 to June 2003 he was a Visiting Professor at the College of Engineering of Korea Advanced Institute of Science and Technology (KAIST), Daejon, Korea. From July 2003 to May 2007, he was a Principal Clinical Scientist at the Rehabilitation Engineering Department in Cardiff, UK. From 2007 to 2014, Dr. Stefanov was a Principal Lecturer at Coventry University, UK. During that time, he was actively engaged with the management and teaching on the rehabilitation engineering programme and also, he was the Chief Investigator of a large externally-funded research project on Wheelchair stability (WheelSAS). Since April 2015, he is an Associate Professor in Biomedical Engineering at Middlesex University in London.

Dr. Stefanov's research areas include design, control and interfacing of advanced mechatronic systems to assist people with disabilities of movement and transportation. His research publications include one book, three book chapters, and 90 research papers in well-established international journals/conferences. He is an author/co-author of 17 inventions. He was also a referee for various international journals and conferences. He is a Senior Member of the IEEE Engineering in Medicine and Biology Society, Chartered Clinical Scientist and a Fellow member of the Institute of Physics and Engineering in Medicine (IPEM).

# Chapter 14
# Applying Design Methods to Promote Older Adults' Walking Activities Based on Their Hobbies and Personal Interests

**Marjolein C. den Haan, Rens G. A. Brankaert and Yuan Lu**

**Abstract** People identify with their hobbies often which stimulate physical activity and participation in social activities, which in turn relate to higher levels of subjective well-being. Hobbies are potentially a powerful tool for design; however, it is unclear how to take hobbies into the design process for example to increase engagement. By creating and using the *Leisure Time Canvas* we identified older adults' previous, current and potential future hobbies, and by closely collaborating with users we process these results into the design of a walking application called Ommetje. Thus, we demonstrate a design case to motivate physical activity. We present the design rationale of Ommetje and how we evaluated this smartphone walking application with 16 older adults for four months, and reflect on the design process of using hobbies as an ingredient for design. We show the power of hobbies in a design (process) as an intrinsic motivation to trigger people to be more socially and physically active.

**Keywords** Design · Older adults · Hobbies and personal interests · Leisure time · Personalized technology

## 14.1 Introduction

The aim of this book chapter is to address how to use hobbies and personal interests in a design case, particularly by using the Leisure Time Canvas which resulted in the creation of a research probe, the walking application Ommetje. The focus of this book chapter is not on the design process to create the application, since this description can be found in a different paper [12]. However, we want to elaborate on illustrating the design case which is built upon hobbies and personal interests. Therefore, we particularly zoom in on the results of the canvas and how these are used in the design proposal related to walking.

M. C. den Haan (✉) · R. G. A. Brankaert · Y. Lu
Department of Industrial Design, University of Technology Eindhoven, Atlas, Building 3, Floor 6, Postbus 513, Eindhoven 5600 MB, The Netherlands
e-mail: m.c.wintermans@tue.nl



A. Woodcock et al. (eds.), *Design of Assistive Technology for Ageing Populations*, Intelligent Systems Reference Library 167,
https://doi.org/10.1007/978-3-030-26292-1_14

This chapter concerns the design of a support service aimed at increasing older adults' physical activity levels. We do this by including their hobbies and personal interests in the design process because older adults indicated that next to health, keeping busy and enjoying hobbies are important contributors to successful ageing.

**Successful Ageing**

This study builds upon the concept of successful ageing, which encompasses that for individuals to age well, people should live lives which avoid disability and disease to maintain their mental and physical capacities to have a productive and social engagement in society [13]. Older adults benefit greatly from regular physical activity. Ory et al. [28] argued that multiple factors such as self-efficacy, daily life integration, social support, financial resources, and awareness could positively influence regular physical activity. Therefore, by maintaining or increasing peoples' activity levels, we may prevent further physical decline and stimulate them to keep on living the lives they are currently used to and enjoy. Generally speaking, the advantages of being active are described as physical benefits [21], while they can also include economic benefits such as lowered health care costs, and psychological benefits such as improved self-esteem and overall happiness [17]. This is in line with Chapman et al. [10] who suggest that respecting peoples "desires, wishes, priorities, and attitudes" improve their health more than conventional approaches.

**Hobbies and Personal Interests**

Leisure activities extend the years of independent living, decreases disability and enhances the overall quality of life [1]. Huang et al. [18] also claim that people who participate in leisure activities feel their quality of life is higher. Also, it is important for older adults to maintain their selfhood [9], and this is shaped through hobbies and leisure time because they are often expressed throughout a lifetime [11]. Selfhood is defined by three parts, namely (i) the self of personal identity, (ii) the self of a person's physical and mental attributes, and beliefs concerning these characteristics such as traits and skills, and (iii) the self of social roles which is developed over the lifespan in various situations [9]. It is therefore important we understand and contribute to selfhood and design for people's emotions and personality. As such design can promote and facilitate successful ageing, because the resulting solutions are more personal [10].

A variety of definitions to describe hobbies and personal interests are used in the context of older adults, but we want to make an important distinction between hobbies and interests: 'Hobbies are paradigmatic examples of interest-driven practices' [2]. To elaborate on this, the difference lies in a hobby being an activity, something you enjoy doing in your leisure time. While a personal interest can be something you are interested in as an individual but is currently not embodied in a specific activity (anymore) and is much broader than hobbies because it could pertain to things like art, architecture or nature. This means certain hobbies can become personal interests over time as they are no longer actively pursued because for example a decreasing social network to do the hobbies with. On the other hand, personal interests can grow into hobbies by becoming integrated into people's routines, for example by following piano lessons on a weekly basis or having friends engage in a specific activity.

**Successful Ageing through Hobbies and Personal Interests**

Stebbins [37, 38] identified multiple benefits of serious leisure which are comparable to the elements of successful ageing. The benefits of serious leisure are: self-actualization, self-enrichment, self-expression, feelings of accomplishment, enhancement of self-image, lasting physical products, renewal of self, self-gratification or fun, and a sense of social interaction and belongingness, while these are similar to successful ageing characteristics of growth, creativity, playfulness, happiness/contentment, health, and close personal relationships.

**Our Approach**

We were interested in understanding how hobbies and personal interests can serve as concrete pinpoints to design for and with older adults. We did not only want to use the type of activities (which hobbies) but specifically the stories behind them (context) and the people connected to the activities to understand them. Therefore, we created and used the *Leisure Time Canvas* (manuscript submitted for publication) which is a card sorting tool to facilitate storytelling between the design researcher and older adults, to identify past present and potential future hobbies, including their barriers and motivators.

Based on the results of an exploration with the canvas, we designed and evaluated a walking application for and with older adults aimed at motivating them to be more socially and physically active. Developing the concept was done through three components, namely: (i) A session to learn from first impressions on the idea of a walking application, (ii) Concept cards to discuss the idea of a walking application more in-depth specifically focusing on individual features and (iii) An early prototype of the walking application to experience the concept. Based on these three stages we created a walking application to record and share routes with your social network to motivate each other to walk more.

By explaining our design process and results in this chapter, we aim to contribute in-context practical knowledge for designers, practitioners, and caregivers, to set up a design process with older adults and also be inspired by hobbies and personal interests to create new meaningful design proposals. Through our research, we want to empower older adults through their hobbies and personal interests, to stay active and thereby maintaining the lives they enjoy having.

## 14.2 Related Work

**Using Hobbies and Personal Interests**

Hobbies and activities, sometimes referred to as leisure activities [25], are often described as 'discrete categories of pastimes' (e.g., playing poker, reading), while older adults described contextual factors such as people involved and how long they performed the activity [25]. This is something to be aware of when mentioning a particular hobby because it will most likely serve as a trigger which entails more than only the activity itself.

Furthermore, Rousseau and Vallerand [31] reviewed that participation in social activities such as volunteering and community tasks are associated with higher levels of subjective well-being [16, 34, 42, 43]. Besides, Knight and Ricciardelli [22] stressed the importance of people being able to choose their activities based on their interests and abilities. This is important to consider in the design to create an optimal match with the user because they are then more willing and able to use it. Also, the chances of participating in an activity are higher when they are related to previously activities people were engaged in [20].

**Involving Users in the Design Process**

Problems, opportunities, ideas, and concepts, can be explored and articulated by incorporating users in the early phases of a project [39]. Understanding your user is important in general, but even more so because of the diversity in needs of older adults [7, 15, 26]. However, most older adults are genuinely not trained in design, but designers can enable them through tools and methods to voice their perspective and be an active contributor to the design process which is therefore called co-design [33]. We do agree with Sanders' vision on users being able to inspire the design process [32] because they are 'experts on their experience' [41]. Empathy is thereby essential to adapt and use these tools and methods to different levels of creativity, as described by Burrows et al. [6] on four different levels: doing, adapting, making and creating. Multiple empathic methods can be used to let users reflect on their personal experiences [23. These include context mapping, to learn the users' product interaction [41], generative techniques to support users is creating artifacts [36] and probing techniques to get inspiration via maps, postcards, cameras and/or booklets [14]. With involving users in research and design activities, positive effects can be seen on (1) the quality and speed of the research and design process, (2) a better match between solution and user, and (3) an increased user satisfaction [24].

**Leisure Time Canvas** The LTC is a template consisting of three columns to sort hobbies and activities (see Fig. 14.2), the columns state from left to right: *'does not suit me'*, *'I like doing this'* and *'I would like to do more often'*. Together with this canvas a set of cards with common hobbies for older adults can be given to participants, which were derived from the Pleasant Activity List [30]. On these eleven cards the name of the activity with a corresponding icon is displayed, namely: reading, walking, visiting a museum, playing games, cooking, listening to music, drawing/painting, cycling, gardening, meeting with family/friends, making a city trip. Next to the pre-made cards, empty cards are also available for the user to add missing activities. A template of the LTC can be downloaded via this link: http://bit.ly/LeisureTimeCanvas.

## 14.3 Process

In this section, we will describe the four steps of our design process (see Fig. 14.1) by investigating people's hobbies and interests through the *Leisure Time Canvas* and how we build on these with a design proposal to address their motivations for the activities. After that, we will present the data of the process.

**Step 1: Discovering Hobbies and Activities through the Leisure Time Canvas**
In this section, we describe the tool we created and used to understand people's hobbies and activities. During a session between designer and older adult, the user can put hobby cards in the preferred column *'does not suit me'*, *'I like doing this'* and *'I would like to do more often'* (see Fig. 14.2). While putting the cards, the user may already give a small explanation, but the follow-up and deepening questions can be asked after all the cards are used, to enable the user to reflect on the entire canvas. In this way the thought process is stimulated on what activities do not fit them anymore (and why), what activities they are currently doing and/or enjoying most (and why) and what activities they would like to do more often (and why), including what they feel may block them from doing that. This conversation resulted in rich contextual stories about people's drivers, barriers, and routines regarding their hobbies, and also their personal lives and interests, and this will be portrayed in the data later in this chapter.

**Step 2 and 3: Exploring the Design Case**
In this section, we describe how we translate hobbies into a new design. The sessions in which the older adult used the canvas in collaboration with the design researcher,

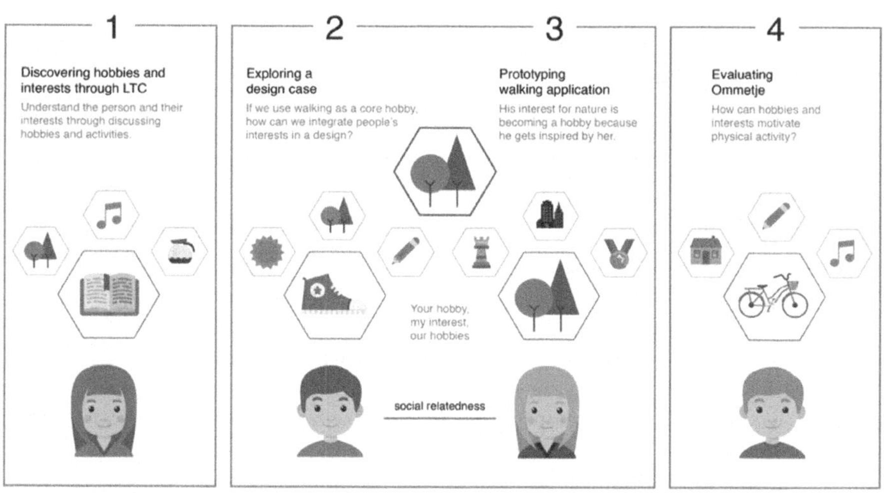


**Fig. 14.1** The four stages of our design process to go from older adults' hobbies towards the design and evaluation of a walking application

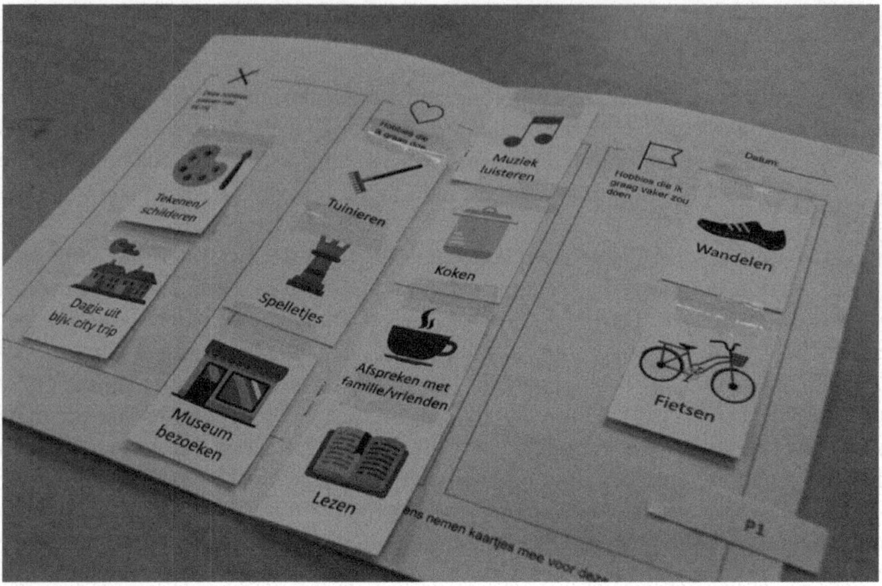


**Fig. 14.2** The Leisure Time Canvas is facilitating the user to tell stories about hobbies and activities but leads quickly to more personal stories

were audio recorded. Then a thematic analysis was performed using the approach of Braun and Clarke [5]. First, we transcribed all interviews to get familiar with the data. Second, we created initial codes across the six interviews. Third, we formed overlapping themes through connecting and comparing codes.

If you would like to know more about how we co-created the walking application in three sessions in collaboration with 42 older adults, we would like to refer to another paper focusing on the creation of Ommetje: Haan, Brankaert, and Lu [12].

**Step 4: Evaluating Ommetje**

In this section we describe how we evaluated the walking application Ommetje, to understand the result of how hobbies can be used in the design process. The purpose of the evaluation was to see (1) whether our intervention would increase the frequency of the walks by enabling them to capture and keep track of their walks and (2) whether or not people would be motivated (more) by others to go for a walk.

We installed the walking application on the older adults' smartphone and invited them to use the app for four months. We set up four focus group sessions [4] to support their learning process and to gain insights on people's experience and opinion. To conclude, we finished the study by having 1-on-1 sessions to focus on the individual's reflection on the application and if it matched their interests. In this case, other group members would not influence an individual's opinion [19]. Both the focus group sessions and the 1-on-1 sessions were subject to a thematic analysis as well.

## 14.4 Results

In this section, we will content-wise describe the four steps of our design process by investigating what are people's hobbies and interests through using the *Leisure Time Canvas* and how we build on these with a design that has been evaluated to address their motivations for the activities. In short, it is about how hobbies and personal interests were discovered, included and implemented in the design.

**Step 1: Discovering Hobbies and Activities through the Leisure Time Canvas**
The Leisure Time Canvas was used in six sessions with one older adult and one design researcher. The participants, all independently living older adults, were aged between 61 and 78 years old and consisted of 1 male and 5 females. They were recruited in a smartphone training class in Someren, a village in The Netherlands close to Eindhoven. All participants signed a consent form and the sessions lasted between 20 and 30 min.

Stories During the Leisure Time Canvas Sessions

Three participants mentioned accompanying the cards in the right column 'would like to do more often' that they did not feel like going alone and therefore struggled to do this more often. Although the barrier is in principal similar, the stories of the users show the potential to find different solutions. On the one hand, the reason why Amy did not want to have a museum card is the following: *'I once said I would love to have a museum card, but my husband is not that interested in it. And to do it by yourself… you don't do it that often. At least, not me. But it would be very nice!'*. On the other hand, Dorothy had a frightening experience: *'I chose for a less crowded road but then went this way and that way… and I completely got lost… no one came by… I did not know where I was. After a while luckily, a mountain biker came by and showed me how to get back. Since that moment I do not walk into the forest alone anymore.'* Although in both cases the barrier could potentially be solved by finding a person to join them, or facilitating going out in another way, it may be more valuable for the second lady to overcome her fear.

Furthermore, a lot of routines were found that people already performed a certain activity for many years with specific friends. For example, Patty: *'I enjoy cycling as well. Together with the elderly association once every 2 weeks. And once a month 40–50 km with my sisters (during winter walking), we have been doing that for 15–16 years already. Every time a different route, time flies! People are impressed we still keep up. And on Sundays whenever it's good weather, with my husband. I love it!'*. This shows the social network is a very relevant trigger to initiate hobbies, which could potentially be facilitated by a design concept.

In Fig. 14.3 the following information can be seen representing six users: (1) placement of hobby cards, (2) frequency of hobby cards in a specific column and (3) additional hobby cards written down separately by the participants. This figure shows some common trends within the small user group about the activities they already engage in.

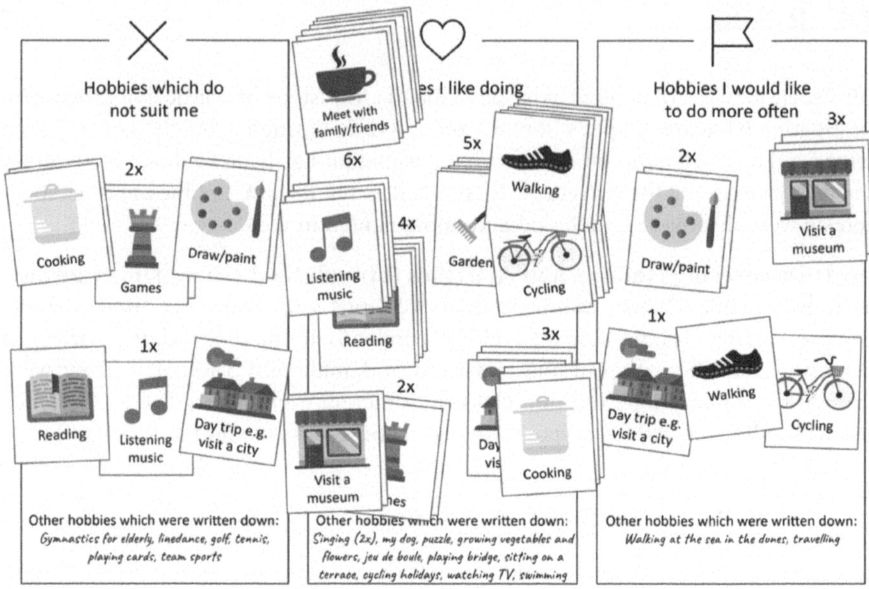


**Fig. 14.3** The results of six sessions of card sorting on the Leisure Time Canvas

Design Interpretation of the Used Cards

In the left column 'do not suit me', mostly indoor activities were mentioned such as cooking, playing games and drawing/painting. We see in the middle column 'like doing' that all participants picked meeting with friends, as social network of course is a strong motivator. Then next to this, walking, gardening and cycling were most often picked, which are all outdoor activities. In the right column 'would like to do more often', where we want to focus on to facilitate through design, we see that the cards point to a need to go out more: (1) three participants picked visiting a museum and (2) day trip, walking and cycling were mentioned once. Although we have two drawing/painting cards here as well, we decided not to continue with this for now because two participants very clearly stated this as hobby which didn't suit them.

Reflection on the Use of the Tool as Designer

As we said before, the participants were recruited via a smartphone class and for this they received a notebook to evaluate the smartphone. Therefore, they already had the Leisure Time Canvas in their possession without the cards, as we wanted to get to know the participants on a personal level. Four participants unexpectedly prepared for the meeting by writing down their hobbies already, even though we did not ask them to do so and wrote down on the canvas we would bring cards for this session. Yet this enabled us to reflect on whether or not the cards were correct.

For example, Ella did write down hobbies prior to the meeting and although most of the answers were indeed also covered by the cards, there were some new suggestions. For example, she connected walking, cycling and gardening with nature. And stated that 'gymnastics for older adults' and golf were hobbies for the grey headed. For Frederick, most written answers are similar to the cards, but he also added cards in the column 'the hobbies he likes doing': jeu de boule, bridge and sitting on a terrace. Furthermore, he did not have hobbies he would like to do more often because he can make his own decisions so he can prioritize himself any time what he likes to do more. Tamara wrote down playing cards and doing team sports as hobbies which don't suit her, next to the cards cooking and playing games. Furthermore, the written answers were quite similar in the box 'hobbies I like doing' except for she added watching television. Furthermore, we see in the right column that the user wrote down just one hobby (travelling), but put three other cards next to it.

Furthermore, we could observe that all participants easily understood the task and did not feel hesitant to share their hobbies and even anxious experiences related to these hobbies. What was somewhat surprising is that people elaborated on the left column 'do not suit me' mostly as hobbies they used to do previously but did not feel as fitting anymore, for example due to physical restraints. Furthermore, two participants mainly talked about certain hobbies being a stereotype for older adults, which was both accepted *'I have old people's hobbies now'* and rejected *'I don't feel at home with gymnastics especially for elderly. So, I went to gymnastics before, but it was not for elderly. Golf is also a bit for elderly I think, for people who are retired. The gray headed.'*. We also observed that people were proudly sharing the longitudinal aspect of some of their hobbies *'I go swimming for 35 years already every Monday afternoon, I really enjoy this!'* and surgery which they have overcome by having a strong drive to stay active: *'The recovery [knee surgery] went really well, because I thought what if I cannot go cycling and walking anymore! Home all day! What would I have to do then?'*.

**Step 2 and 3: Exploring the Design Case**

The reason why we created a walking application ('Ommetje') lies in the results of the *Leisure Time Canvas* described in the previous step. With our walking application, we expected to be able to easily include personal drivers for walking based on the results from the LTC sessions, such as other people's recommendations (social) and popular places to go to: *'I learned to play bridge when I came out of the hospital. At the start it was quite minimal, it didn't work. But my wife told me she read about a bridge course, so this replaced soccer.'*

Combined with the fact that an increasing number of older adults use internet outdoors on their smartphones and messaging services like WhatsApp [8], we saw potential in creating a walking application to connect people. While benchmarking other walking applications such as Strava and MapMyHike, we found that for novel smartphone users, there is an opportunity to simplify by decreasing the number of interaction possibilities and the amount of data presented per screen specifically for older adults. In the next section we will elaborate upon the functionalities of the walking application that we designed and implemented.

Design Case: Ommetje

Ommetje is a walking application in which people can record their walking activities and share these in a walking community. It was developed during three concept discussions with 42 users in total. The meaning in Dutch is 'small' walk as the intention of the application was to record everyday walks, even to the supermarket for example. We aimed at stimulating physical activity through social triggers to go for a walk. While designing the application, a low threshold to initiate use was important and therefore the application was free and making an account was not neccesary. If desired only a user name could be made and their location was automatically tracked by GPS. The main menu consists of four main functions (see Fig. 14.4, left):

(1) Recording routes, to keep track of walks. From the concept card discussion, our users suggested among other things to be able to add the surface and width of the path, which they can through adding a description after their recording.
(2) Reviewing the personal walking history with distance and timestamp of the walk. The users can send their walk to another user and receive walks from others.

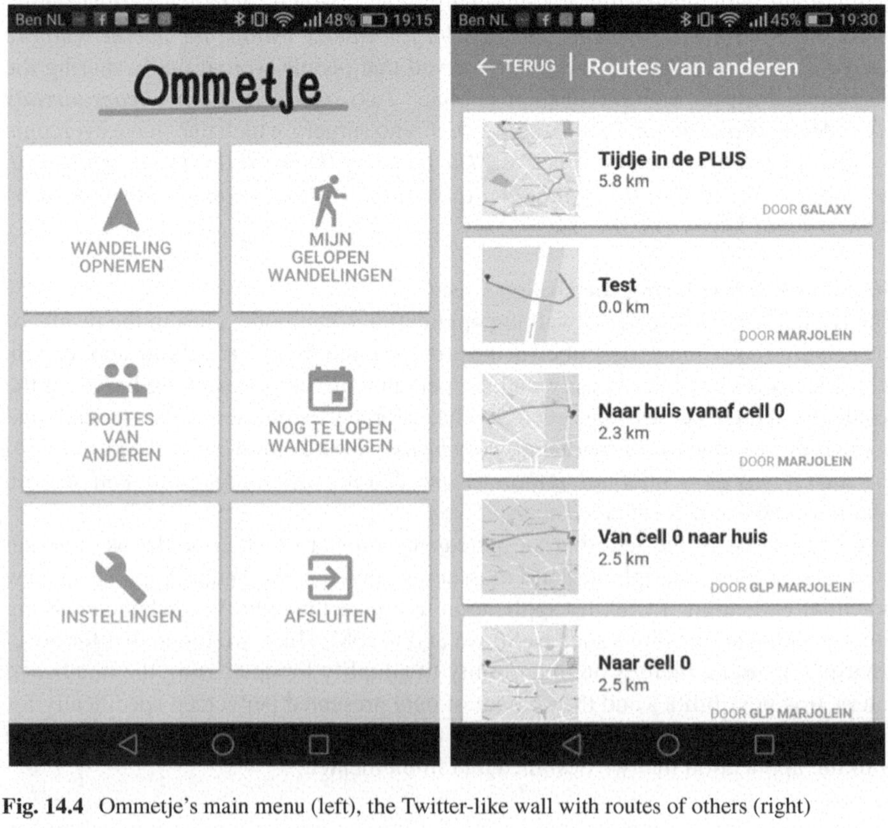


**Fig. 14.4** Ommetje's main menu (left), the Twitter-like wall with routes of others (right)

(3) Getting inspiration from other users by seeing the routes which are shared via a Twitter-like wall (Fig. 14.4, right). We aimed to provide opportunities to explore new places. Users could decide to turn the sharing feature on or off per walk, when not shared the walk would only be visible in their personal walking history. This is because we learned in the first session with users that there were concerns about privacy, particularly who could see their walk.
(4) Making a list of next walks, to downscale the list of shared and personal walks to the most interesting ones. From the concept card discussion, we learned that people were particularly interested in guidance on which places to visit in an unknown city.

The two remaining tiles are to go to settings or close the application. Furthermore, during the contextual inquiry there were some suggestions on the User Interface, for example the looks, recognizability and affordance of icons, which were changed in a follow-up version.

Applications for Older Adults Stimulating Physical Activity

We have seen walking applications exploring a group walking game and a group walking program to collect stamps while walking [40]. Furthermore motivating to increase older adults' physical activity has been done via environmental factors in the neighborhood [3, 27]. Though with our application we let users determine themselves where they want to walk. We incorporated social triggers in our application because this is shown to be one of the most important motivational factors for older adults to be more active [35]. And older adults more easily adopt technologies when their meaning is shaped together with others in communities [29].

**Step 4: Evaluating Ommetje**
Sixteen users participated in the pilot evaluation of the walking application prototype. As mentioned before the intention of Ommetje was to facilitate users to share their routes and thereby motivating them to walk (more). By making it easy to share a route and see these of others, we would lower the barrier to interact and aim at supporting them in doing their hobby. However, we have learned from our users that because they knew the other users of the walking application, they did not record their regular walks inside the village, because *'everyone knows these routes'*. This shows our users were very aware of who else was using the application and this even influenced their use. It would be interesting to find out if they would be motivated to record their village walks when strangers would use the application. Currently, instead of recording the routes in the village, they wanted to record special routes outside the village only (see Fig. 14.5), which did not occur that often as people saw these routes as special occasion weekend trips for example. This is somewhat surprising as we assumed the trigger to go for a walk would come from someone recommending the route either individually or by the list of everyone's walks. We see the trigger is already earlier in the customer journey, by the consideration to even start recording a walk. So, regarding social connectedness, it is not simply the direct contact between people, but also indirect the feeling of belonging to a particular

**Fig. 14.5** An older adult using the walking application Ommetje

community. It would be an interesting next step to see what happens if users do not know other members of the community (yet), and whether or not they do have the feeling of belonging to a community of strangers. Therefore, another track the application could go is that the start of using the application is always in existing friends or family groups, as a shared goal.

Another assumption we had was that their personal history would motivate them through seeing the walking distance and frequency. However, this was only clearly the case for one participant, who said that keeping track of all her walks motivated her. Combined with the fact that routes outside the village were more appealing to people than regular inside the village walks, we can conclude the quality of the walks is more meaningful to people than its quantity. This is in line with Kahlbaugh and Huffman [20] who discuss that for older adults the perceived quality can be of higher importance than the frequency. A design implication for this finding could be to decrease the focus on frequency, duration, and date of the walk, towards for example perceived quality and level of enjoyment. Potentially the application could develop towards a more picture-heavy interface with highlights which characterize the walk, to put the emphasis more on the quality than on the number of walks. This could even extend towards having a limited number of walks you can store, to decide which one was most meaningful to you.

To define the quality of the walk on a personal level, we should be aware of people's interests and hobbies on top of walking. Different users will have different preferences. For example, one user prefers to mainly walk through natural landscapes, and another user would select a cultural route with several highlights of a city. This could give people the feeling the list of walking suggestions is personalized for them. However, we have to consider how much input people want to give to be able for the app to respond to a personal profile. Perhaps because of privacy reasons, users may be more willing to receive recommendations from friends and family mostly, but this is something that needs to be investigated further.

## 14.5 Discussion

In this section, we will discuss the usage, analysis, and outcome of the *Leisure Time Canvas*, the design rationale of Ommetje and the impact on its users. Finally, we will discuss how our "Ommetje" design case illustrates that hobbies and personal interests could be used in a meaningful way in design for older adults.

**Step 1: Discovering Hobbies and Activities through the Leisure Time Canvas** Reflecting on whether the cards were correct from a perspective of how many hobbies were written separately on a new card, we see for the two people who did not write down their hobbies before the activities, that one only added one hobby (singing) and the other did not add new ones. The first lady felt like singing could be part of the music card, although she felt like creating music is different than listening to music. We learned from this that general cards can cover broad interests and can thus be used by more people. However, it can also be interpreted oppositely: for example, we wrote down 'games' but put a chess piece on the card together with it. This led to a person asking if it is about board games in particular, or it could also be playing a game on the iPad. On the one hand, because this canvas is used between researcher and older adult, there is the opportunity to discuss this. On the other hand, if a person does not dare to ask or does not associate the chess piece with him/her playing a computer game, the card may not be appropriately included in the canvas and this will give misleading information. We would, therefore, propose this particular category to be more specific on the diversity of playing games and therefore have several cards instead of one.

A limitation is that analyzing the results to suggest design requirements is now based on the analytical skills of the researcher and his/her empathic attitude to see value in the activity in itself. This can be addressed by having very concrete and transparent analyzing steps, and we would like to explore in future work to what extent the user can be involved and provide priorities in the analysis, design requirements and early design itself.

The Methodological Outcome of the Leisure Time Canvas

One woman specifically added hobbies for the gray-headed as she called them, such as gymnastics and golf. It may be interesting to add hobbies of other target groups/age categories, to identify if there are certain hobbies they see people do in their environment which they may be interested in doing in the future as well. So next to naming hobbies which are in line with the general interest of a specific age group, to purposefully add outliers as hobbies as well. This provocative act of adding 'strange' hobbies may lead to unexpected answers of people explaining why they would or wouldn't include particular hobbies, to reach more the activities they are passionate about to be able to distract their motivators and barriers even stronger.

It also works the other way around that a card can add a new hobby which has not been written down yet previously by the user, such as listening to music in the middle box and city trip, visit a museum and draw/paint in the right box (while she only wrote down traveling). This is interesting as it shows that you may reflect on your current situation and therefore it may be hard to imagine how to improve or adjust it. By offering several cards, users may start to imagine what else they would be motivated for in the future. Thus, we can conclude that it is interesting to let participants write down hobbies before they start using the cards, as it may invite them more to start from their current situation. However, the hobby cards do trigger new conversations which have not been thought of by themselves.

**Step 2 and 3: Exploring the Design Case**
We aimed to find out whether using hobbies would motivate people to be more physically active by using walking as a main hobby and invite for a walk using sub hobbies or interests. Discussing the results of these particular steps was done directly in the result section, to interpret and reflect on the results.

**Step 4: Evaluating Ommetje**
We hypothesized that our intervention would increase the number and distance of walks, but we instead found that this was difficult to measure through the influence of other users, because people assumed their peers from the same village would not be interested in their village routes. Most of them stated that they did not want to record a walk to the supermarket because everyone would already know this walk. This finding influenced our measured data as we could not identify a trend of increased walking behavior. Although it did point to an interesting variety to personalize the application to two different usage patterns. On the one hand, some participants were quantitively-driven to record their walks and see their walking history. They wanted to see their overview and therefore recorded (almost) every walk. On the other hand, most participants were qualitatively-driven by more unique walks outside of the village. However, if we would extend the user base outside an existing community, they may feel recording a 'normal village'-walk is more rewarding because they can inspire a person who doesn't know the environment with it. In the end, we believe that we made a shift in approaching the user to increase their physical activity: by moving from the display of the number of walks and distance, towards enabling to describe the quality of the walks what makes these interesting. This is in line with Rousseau and Vallerand [31] indicating to go beyond the 'more-is-better' approach and to identify elements which will boost the possibility for older adults to benefit from having an active lifestyle.

**Using Hobbies in Design**
In this chapter we illustrated how to use hobbies in the design process from the start to the end. We started by using a storytelling tool to discover people's activities and related stories, which provided us with contextual and motivational factors of the participants, which we could later on use as input for design. The tool guides users but also enables them to add new activities. Then we used this information to decide which hobbies would have a high potential to be translated into a design proposal, and illustrated how this could be done in the design case Ommetje. In this way, the design showed a strong match with commitment from the users to participate

and explore the application. By using hobbies in the design, we keep the threshold to use the application very low in the User Interface and increased usability of the application. Therefore, this chapter shows that researchers and designers should investigate further and pay attention to opportunities to let design be inspired by hobbies, and allow design to facilitate people to engage in their hobbies more often.

## 14.6 Conclusion

Based on the *Leisure Time Canvas* sessions with older adults and their involvement in testing the designed walking application, we discuss how to include hobbies in the design (process). Thus, we have illustrated how a hobby-driven design approach can lead to new concepts and proposals. We illustrated through the design case of Ommetje the power of hobbies in a design (process) as an intrinsic motivation to trigger people to be more physically active. It is essential to not only focus on increasing people's activity level as facts, but consider their lived experience when using the design. Thereby hobbies show a high potential to address and motivate older adults in changing their behavior and improving their independence.

## References


1. Aldrich, N.: CDC Targets sedentary older adults. Aging Today **25**(1), 7–8 (2004)
2. Azevedo, F.S.: The tailored practice of hobbies and its implication for the design of interest-driven learning environments the tailored practice of hobbies and its implication for the design of interest-driven learning environments, 8406 (2013). https://doi.org/10.1080/10508406.2012.730082
3. Barnett, David W., Barnett, Anthony, Nathan, Andrea, Van Cauwenberg, Jelle, Cerin, Ester: Built environmental correlates of older adults' total physical activity and walking: a systematic review and meta-analysis. Int. J. Behav. Nutr. Phys. Activity **14**, 103 (2017)
4. Barrett, Julia, Kirk, Stuart: Running focus groups with elderly and disabled elderly participants. Appl. Ergon. **31**(6), 621–629 (2000). https://doi.org/10.1016/s0003-6870(00)00031-4
5. Braun, Virginia, Clarke, Victoria: Using thematic analysis in psychology. Qual. Res. Psychol. **3**(2), 77–101 (2006). https://doi.org/10.1191/1478088706qp063oa
6. Burrows, A., Mitchell, V., Nicolle, C.: Cultural probes and levels of creativity. MobileHCI, pp. 920–23 (2015). https://doi.org/10.1145/2786567.2794302
7. Cabrita, M., Nassabi, M.H., Op den Akker, H., Tabak, M., Hermens, H., Vollenbroek, M.: An unobtrusive system to monitor physical functioning of the older adults: results of a pilot study. In International Workshop on Personalization and Adaptation in Technology for Health 2015 (PATH 2015) Held in Conjunction with the 23rd Conference on User Modelling, Adaptation and Personalisation (UMAP 2015) (2015)
8. CBS: Increasing number of elderly using social media (2017). https://www.cbs.nl/nl-nl/nieuws/2017/52/steeds-meer-ouderen-op-sociale-media
9. Cedervall, Y., Torres, S., Åberg, A.C.: Maintaining well-being and selfhood through physical activity : experiences of people with mild alzheimer' s disease. Aging Mental Health **19**(8), 679–688. Taylor & Francis. https://doi.org/10.1080/13607863.2014.962004

10. Chapman, B.P, Hampson, S., Clarkin, J.: Personality-informed interventions for healthy aging: conclusions from a national institute on aging workgroup. **50**(5), 1426–1441 (2014). https://doi.org/10.1037/a0034135.personality-informed
11. Cohen-Mansfield, Jiska, Parpura-Gill, Aleksandra, Golander, Hava: Utilization of self-identity roles for designing interventions for persons with dementia. J. Gerontol. **61B**(4), 202–212 (2006)
12. den Haan, M., Brankaert, R., Lu, Y.: What moves you? designing a walking app for and with older adults. In Design4Health (2018)
13. Foster, L., Walker, A.: Active and successful aging: a european policy perspective. Gerontologist **55**(1), 83–90 (2015). https://doi.org/10.1093/geront/gnu028
14. Gaver, B., Dunne, T., Pacenti, E.: Design: cultural probes. Interactions **6**(1), 21–29 (ACM) (1999). https://doi.org/10.1145/291224.291235
15. Gregor, P., Newell, A., Zajicek, M.: Designing for dynamic diversity—interfaces for older people. In Proceedings of the Fifth International ACM Conference on Assistive Technologies—Assets '02 (2002). https://doi.org/10.1145/638276.638277
16. Harlow, R.E., Cantor, N.: Still participating after all these years: a study of life task participation in later life. J. Pers. Soc. Psychol. **71**, 1235–1249 (1996)
17. Haworth, J.T., Hill, S.: Work, leisure, and psychological well-being in a sample of young adults. J. Commun. Appl. Soc. Psychol. **2**, 147–160 (1992). https://doi.org/10.1002/casp.2450020210
18. Huang, C., Lee, L., Chang, M.: The influences of personality and motivation on exercise participation and quality of life. Soc. Behav. Personality **35**, 1189–1210 (2007)
19. Ireland, C.: Qualitative methods: from boring to brilliant. In: Laurel, B. (ed.) Design Research: Methods and Perspectives, pp. 23–29 (2003)
20. Kahlbaugh, Patricia, Huffman, Loreen: Personality, emotional qualities of leisure, and subjective well-being in the elderly. Int. J. Aging Human Dev. **85**(2), 164–184 (2017)
21. Katz, S.: Busy bodies: activity, aging, and the management of everyday life. J. Aging Stud. **14**(2), 135–152 (2000). http://dukespace.lib.duke.edu/dspace/handle/10161/12388
22. Knight, T., Ricciardelli, L.A.: Successful aging: perceptions of adults aged between 70 and 101 years. Int. J. Ageing Human Dev. **56**(3), 223–245 (2003)
23. Kouprie, M., Visser, F.S.: A framework for empathy in design: stepping into and out of the user's life. **20**(5), 437–448 (2009)
24. Kujala, S.: User involvement: a review of the benefits and challenges. Behav. Inf. Technol. **22**(1), 1–16 (2003). https://doi.org/10.1080/01449290301782
25. Lazar, A., Nguyen, D.H.: Successful leisure in independent living communities: understanding older adults' motivations to engage in leisure activities (2017)
26. Lerouge, C., Ma, J., Sneha, S., Tolle, K.: User profiles and personas in the design and development of consumer health technologies. Int. J. Med. Inf. **82**(11), e251–268. Elsevier Ireland Ltd. https://doi.org/10.1016/j.ijmedinf.2011.03.006
27. Lockett, D., Willis, A., Edwards, N.: Through seniors' eyes: an exploratory qualitative study to identify environmental barriers to and facilitators of walking. Canad. J. Nurs. Res. **37**(3), 48–65 (2005)
28. Ory, M.G., Lee, S., Han, G., Towne, S.D., Quinn, C., Neher, T., Stevens, A., Smith, M.L.: Effectiveness of a lifestyle intervention on social support, self-efficacy, and physical activity among older adults : evaluation of texercise select. Int. J. Environ. Res. Pub. Health **15**(234)
29. Righi, V., Sayago, S., Blat, J.: When we talk about older people in HCI, who are we talking about? towards a 'turn to community' in the design of technologies for a growing ageing population. Int. J. Hum. Comput. Stud. **108**, 15–31 (2017). https://doi.org/10.1016/j.ijhcs.2017.06.005
30. Roozen, H., Wiersema, H., Strietman, M., Feij, J., Lewinsohn, P., Meyers, R., Koks, M., Vingerhoets, A.: Development and psychometric evaluation of the pleasant activities list. Am. J. Addict. **17**(5), 422–435 (2008). https://doi.org/10.1080/10550490802268678
31. Rousseau, F.L., Vallerand, R.J.: An examination of the relationship between passion and subjective well-being in older adults. Int. J. Aging Human Dev. **66**(3), 195–211 (2008). https://doi.org/10.2190/AG.66.3.b

32. Sanders: From user-centered to participatory design approaches. Des. Soc. Sci., vol. 7 (2002)
33. Sanders, E.B.N., Stappers, P.J.: Co-creation and the new landscapes of design. CoDesign **4**(1), 5–18 (2008). https://doi.org/10.1080/15710880701875068
34. Searle, M.S., Mahon, M.J., Iso-Ahola, S.E., Sdrolias, H.A., van Dyck, J.: Enhancing a sense of independence and psychological well-being among the elderly: a field experiment. J. Leisure Res. **27**, 107–124 (1995)
35. Smith, G.L., Banting, L., Eime, R., O'Sullivan, G., Van Uffelen, G.J.: The association between social support and physical activity in older adults : a systematic review. Int. J. Behav. Nutr. Phys. Activity **14**(56), 1–21 (2017)
36. Stappers, P.J., Sleeswijk Visser, F., Keller, I.: Mapping the experiential context of product use: generative techniques beyond questions and observations (2017). Accessed 14 Mar 2017. http://www.idemployee.id.tue.nl/g.w.m.rauterberg/conferences/CD_doNotOpen/ADC/final_paper/313.pdf
37. Stebbins, R.A.: Amateurs, professionals, and serious leisure. McGill–Queen's University Press, Montreal (1992)
38. Stebbins, R.A.: New directions in the theory and research of serious leisure. Mellen Stud. Sociol. vol. **28**. Edwin Mellen Press, Lewiston, NY (2001)
39. Steen, M., Kuijt-Evers, L., Klok, J.: Early user involvement in research and design projects–a review of methods and practices. In: Paper for the 23rd EGOS Colloquium, pp. 1–21 (2007). https://doi.org/10.1080/15710880701875068
40. Takahashi, M., Kawasaki, H., Maeda, A., Nakamura, M.: Mobile walking game and group-walking program to enhance going out for older adults. In: Proceedings of the 2016 ACM International Joint Conference on Pervasive and Ubiquitous Computing Adjunct (2016)
41. Visser, F.S., Stappers, P.J., van der Lugt, R., Sanders, E.B.N.: Contextmapping: experiences from practice. CoDesign **1**(2), 119–149 (2005). https://doi.org/10.1080/15710880500135987
42. Warr, P., Butcher, V., Robertson, I.: Activity and psychological well-being in older people. Aging Mental **8**, 172–183 (2004)
43. Zimmer, Z., Hickey, T., Searle, M.S.: Activity participation and well-being among older people with Arthritis. Gerontol. **35**, 463–471 (1995)

**Marjolein den Haan** is a Ph.D. candidate at the department of Industrial Design, at the University of Technology Eindhoven. Her research focuses on co-designing with older adults for positive ageing, with a specific interest in using leisure activities to create personalised designs.

**Rens Brankaert** is an Assistant Professor in the Department of Industrial Design at the University of Technology Eindhoven, the Netherlands as a part of the Systemic Change group. His studies revolve around design research for active and healthy ageing, with a specific interest and focus on designing for and with people living with dementia. Rens finished his Ph.D. around design for dementia, and continues this line of work. He set up and leads the Dementia and Technology centre of expertise at the TU/e, and is a Key Technology Partner Visiting Fellow at University of Technology Sydney. In his research Rens focuses on inclusive human-centred design, Living Lab field studies and multi-stakeholder innovation.

**Yuan Lu** is Associate Professor of Design for Healthy and Active Ageing in the systemic change group at Industrial Design at Eindhoven University of Technology (TU/e). She aims at creating new design research methods, tools and technological prototypes to support multi-stakeholder design and innovation to deal with societal challenges. She also holds a position as Guest Professor at the School of Software Technology at Zhejiang University in China.

# Chapter 15
# Future Older Adults and Mobile Applications for Health

**Berrak Karaca Şalgamcıoğlu**

**Abstract** Aging population is a global issue, affecting almost all areas of life. Health is one of the most important issues in this unprecedented demographic change, requiring innovative health management interventions. The growth of mobile phone usage among the elderly combined with demographic trends suggests that mobile health applications can be a promising tool to improve their quality of life. Consequently, mobile applications must be adapted to older adults' preferences and needs. There is an expanding number of health-related mobile interventions for older adults, but most of them have focused on a single disease or condition, whereas the majority of older adults have more than one. These interventions do not have a wide reach and are not being efficiently used. Today, however, almost all mobile phones have pre-installed health applications, which can reach wider audiences without excessive effort. This chapter presents the results of qualitative research that used verbal protocol analysis to observe the use of pre-installed health applications by future older adults in Turkey. The study's results show two gaps between the user and the mobile health applications. The first is between mobile applications and real-life health services and activities. The second is a cultural gap, where these applications use a global, digital, younger language. This cultural gap is even deeper in Turkey, where all written information is translated text. These two gaps prevent users from efficiently using these applications.

**Keywords** mHealth · Older adults · Aging · Mobile applications · Gerontechnology

## 15.1 Introduction

In 2045, the number of older people in the world is expected to exceed the number of children for the first time [1]. This change occurred in 1998 in more developed countries, where population ageing is advanced [1]. The term "ageing population"

B. Karaca Şalgamcıoğlu (✉)
Faculty of Architecture, Istanbul University, Istanbul, Turkey
e-mail: berrakk@istanbul.edu.tr



A. Woodcock et al. (eds.), *Design of Assistive Technology for Ageing Populations*, Intelligent Systems Reference Library 167,
https://doi.org/10.1007/978-3-030-26292-1_15

refers to the process by which older people (age 60 or more) become a proportionally larger share of the total population.

This unprecedented demographic change is now a global issue. The United Nations World Population Ageing 1950–2050 Report [2] shows that the older population is growing at a considerably faster rate in less developed countries. By 2050, these countries will be home to nearly 80% of the world's older population, and developing countries, such as Turkey, will have less time to adapt to these changes.

An ageing population affects almost all areas of life, from economic to social and political [1]. Health and health care are important areas—ageing increases the possibility of having several health problems, which requires innovative health management interventions. Long-term illnesses and disability rates from diseases increase with old age. Chronic conditions require health management in the home [3]. In addition, hospital stays are shorter and patients are likely to be discharged while using oxygen, on strong medications, or still catheterized [3].

Adults over age 65 are defined in three profiles as vigorous elderly (mobile, autonomous and without significant health problems), fragile elderly (fluctuating health status and chronic illnesses) and addicts (serious illnesses and usually treated in hospitals or clinics) [4]. As older adults' abilities change over time, their health status can pass from vigour to fragile and fragile to dependent. Therefore, health management systems can require regular monitoring, follow-up and intervention to delay this process [5].

Studies show that although elderly people need more health care, there is a weak link between the use and need for health services [6]. Especially in low-income countries with a high burden of disease, older people tend to use less health care than do younger adults [6].

Information and communication technologies have the potential to improve health service quality and provide maximum health care by lowering the elderly's health costs. Globally, mobile device use has been increasing exponentially—with more than 5 billion unique mobile subscribers in 2017, mobile phones have a greater reach than any other technology [7]. The growth of mobile phone usage among the elderly combined with demographic trends suggests that a solution can be to use them as a platform for health interventions. Mobile health (mHealth) is defined as "the use of mobile devices—such as mobile phones, patient monitoring devices, personal digital assistants (PDAs) and wireless devices—for medical and public health practice". Applications include treatment adherence, community mobilization, community and clinical health data collection, wellness and self-care, chronic disease management, and remote patient monitoring [8]. Globally, due to fast technology adoption, mHealth has a different and stronger impact than conventional health services.

Currently, there are an expanding number of health-related mobile device interventions for older adults, but most of them focus on a single condition or disease, whereas a majority have multiple chronic conditions. Moreover, these interventions do not have a wide reach and are not being efficiently used. Today, almost all mobile phones have pre-installed, ready to use health applications; therefore, these applications can reach wider audiences without excessive effort.

The purpose of this chapter is to illustrate the importance of mobile health technologies and present the results of qualitative research that used verbal protocol analysis to observe the use of pre-installed mobile health applications. The study's results show two gaps between the user and the mobile health applications; (1) a gap between mobile applications and real-life health services and activities, and (2) a cultural gap between global, digital, young and local, analogue and old.

## 15.2 Older Adults and Mobile Health

Older adults, like other age groups, accept and adopt technology when it meets their needs and expectations [9, 10]. When older adults adopt technology, determinants can be value, usability, affordability, accessibility, technical support, social support, emotion, independence, experience, and confidence [11].

Older adults' requirements and expectations of mobile services that were obtained from a different study are: feelings of safety and security, aids for memory and daily activities, communication functionalities, freedom of movement and enjoyment of a healthier independent life [12].

Kim and his colleagues reviewed various health-related devices, including mobile phones, and different ways of using them [13]. According to these researchers' study, there are various ways to integrate healthcare management with mobile phone use: telemonitoring and tele-coaching, transferring medical information, scheduling hospital visits and medication compliance, increasing drug compliance and treatment adherence, providing a life-support system for single-family elderly. Another study [14] divided health-related assistive technologies (to help the elderly become more independent) into three groups. The first group is technologies following behaviour: sensors and warning systems that alert caregivers when the older adult changed her/his previously designed position. The second group is smart home tools that contain potentially more complex environments and envision potentially dangerous behaviours before giving notifications to caregivers. And finally, the third is tele health tools, video conferencing systems that enable remote data sharing between health care professionals and patients.

Kim and his colleagues [13] also listed the difficulties that elderly people have with Information Technologies (IT): physical and cognitive changes due to ageing and using mobile phones while learning capabilities decline. To solve these problems, these researchers suggested that mobile phone applications should focus on convenience, have large icons and fonts and clearly distinguishable colours, and have a simple structure and intuitive visual representations. Moreover, the system initially should utilize simple words, take the user's learning ability into consideration and implement step-by-step services for elderly users.

Older adults see mobile phones as a more accessible technology than personal computers and the Internet [12], and a growing share of them currently own a mobile phone. Healthcare related mobile phone interventions mostly are based on health information monitoring, which focuses on capturing health activity information, symptoms, critical events, and monitoring states that are directly tied to individual health goals, such as the amount of physical activity and chronic disease management [15].

To examine the current state of elderly mobile phone use for health-related interventions, [16] used "older adults", "mobile phones" and related terms and synonyms to search PubMed and CINAHL for articles between 1965 and June 2012. According to their review, research on health-related mobile phone interventions has ten major clinical domains: daily life activities, Alzheimer's/dementia care, chemotherapy symptom management, palliative care symptom management, congestive heart failure, chronic obstructive pulmonary disease, diabetes, falls and fall risk, osteoarthritis and dermatology. These researchers' review found that most were pilot or feasibility studies, resulting in a lack of generalizability. Moreover, these studies focused on a single condition or disease, but the majority of the older adults have multiple chronic conditions. These adults often have at least one chronic condition, such as hypertension, arthritis, or diabetes, and many have more than one [3]. Joe and Demiris [16] stated that due to their universality, mobile phones could be leveraged for health care use without stigma in a manner that other devices may not allow. These researchers wrote that more investigation is needed to confidently establish whether mobile phone technologies can meaningfully improve an older adult's health and well-being.

Contrary to the literature, older adults are not a homogeneous group but a diverse one with diverse needs and requirements [12]. A study that exposes the typology of older adults using technology defines the elderly in three categories: the "Excluded", who are reluctant to use technology and concentrate on the areas restricted by technology; the "Entertained", who frequently use the technology, are excited and interested in learning it, and who focus on entertainment opportunities; and the "Networker", who attempts to use the technology at the highest level, participates frequently on social network sites and increases life satisfaction [17].

## 15.3 Mobile Technologies and Pre-installed Health Applications

Various health related mobile applications are available: in 2015, an initial search identified 32,614 and 4632 mHealth apps in the iOS and Android app stores, respectively [18]. In 2016, more than 97,000 health-related apps were available in the health and fitness category of the Apple app and Google play stores, with about 1000 more being created every month [19].

Three of these applications have special importance: "Apple Health", "Samsung Health", and "LG Health". All are pre-installed in smart phones, so users automatically have access to them when they buy an iPhone, Samsung or LG smart phone.

"Apple Health" combines health data from iPhone, Apple Watch, or third-party apps that the user already has with the aim of letting them view all of their progress in one place (www.apple.com). The application has four categories: Activity, Sleep, Mindfulness and Nutrition. One of the benefits of the iPhone Health application is allowing users to create an emergency Medical ID card that lets first responders to

**Table 15.1** Health app by Apple

| | | | | |
|---|---|---|---|---|
|  | 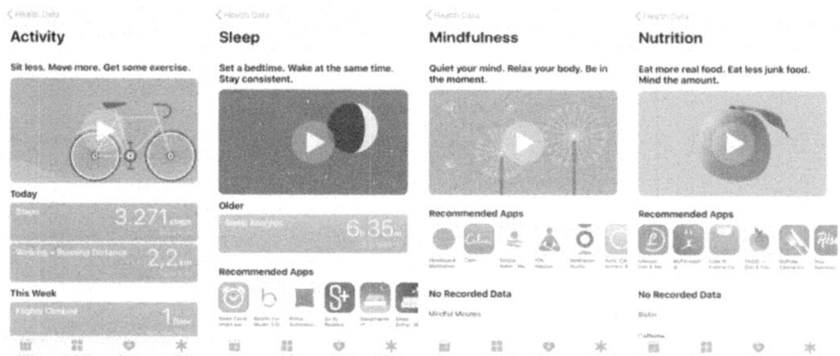 | | | |
| Health app by Apple | Activity | Sleep | Mindfulness | Nutrition |

access users' critical medical information from the lock screen without needing the passcode (www.apple.com) (Table 15.1).

"Samsung Health" aims to help users maintain healthy lifestyle and a successful diet while recording and analysing their daily activities and habits. The built-in trackers can be used to add and track various physical exercises and activities. Samsung Health allows the user to record a variety of information, such as food, caffeine and water intake. The application has a sleep and stress tracker. As with "Apple "Health" users, "Samsung Health" users can track their heart rate, blood pressure, blood glucose levels, stress and weight using third-party devices. The intuitive charts, helpful tips and physical exercise programs help users achieve fitness and diet goals. The application supports all Samsung smartphones from Galaxy S3 on and includes non-Samsung Android smart phones as well (play.google.com). The application allows users to compete with their friends and check their ranking. Users also can check daily health news with customized news feed and receive tips from health professionals (Table 15.2).

"LG Health" also tracks users' activity levels. It provides an interface and related content; therefore, all types of users can customize the app to fit their own activity levels. Different from other two applications, "LG Health" application offers two levels: beginner and advanced. At the beginner level, user can see beginner exercise information when exercise is not part of her/his daily routine. The exercise circle shows the user's progress to her/his goal over 24 h. At the advanced level, the user can see advanced exercise information when exercise is part of her/his daily routine, and the exercise circle shows progress to her/his goal by type and intensity over 24 h. Users also can manually add logs. As with the "Samsung Health" application, users can see personalized tips above the exercise circle, which are shown according to their lifestyle and exercise patterns. The application lets the user manage fitness activities, track her/his weight and monitor diet using LG Smart Watch and Tone Active. The app supports all LG smartphones and also includes non-LG Android smart phones (Table 15.3).

**Table 15.2** S Health app by Samsung

| | |
|---|---|
| <br>S Health app by Samsung | 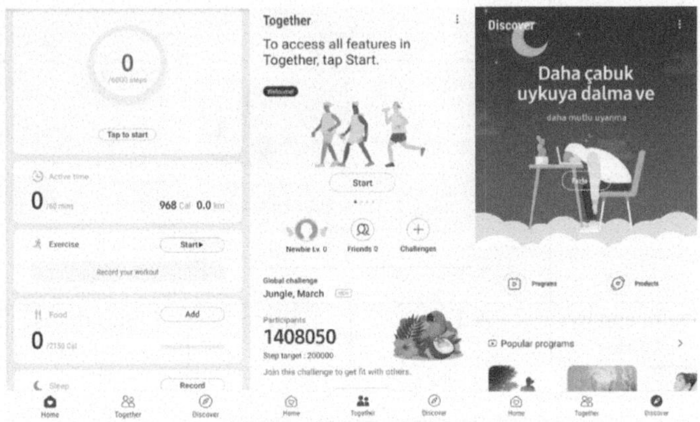 |

**Table 15.3** Health app by LG

| | |
|---|---|
| <br>Health app by LG | 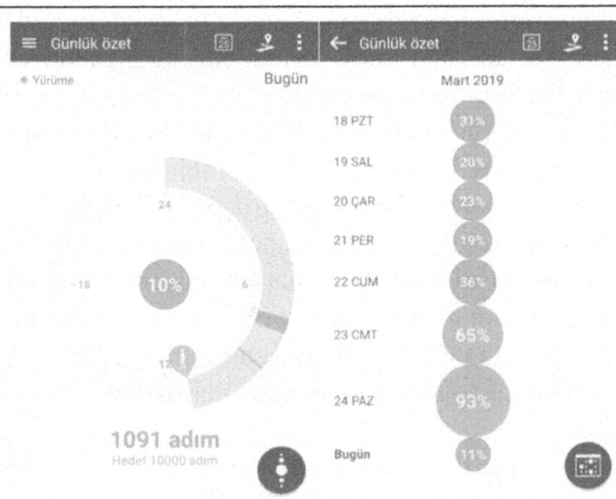 |

## 15.4 Future Older Adults in Turkey and Mobile Health Applications: Using Verbal Protocol Analysis to Understand Their Needs

The United Nations World Population Ageing 1950–2050 Report [2] estimates that in 2025, 9.5% of Turkey's population will be over 65 and this number will increase to 17.9% in 2050. According to Turkey's 1st Gerontology Atlas (conducted by the Akdeniz University Gerontology Department), 88% of the older adults that participated the study do not have regular medical check-ups, and 33% of them need special care [20].

According to Turkish Statistical Institute (TÜİK) data, 47% of Turkish people age 65 and older have chronic diseases [21], and according to the Ministry of Health's Chronic Diseases Report, approximately 22 million Turkish adults are living with one or more chronic diseases. People with chronic conditions must be closely monitored, and economic analyses indicates that every 10% increase in chronic diseases caused a five per 1000 decrease in the annual economic biomass rate and that the chronic disease burden poses a significant risk to the health system's financial sustainability [22]. Six out of 10 individuals in Turkey age 60 and older do not have any social security [21].

Mobile phone usage in Turkey is growing enormously. According to the Turkish Statistical Institute, in April 2016, 96.9% of households had a mobile phone (incl. smart phones). The "Deloitte Global Mobile User Survey 2017" was held on six continents and attended by 53,150 people from 33 countries including Turkey, and it offers significant mobile consumer data (www.deloitte.com). Of the 1005 people surveyed from Turkey, 92% said they have smart phone access and users looked at their smart phones an average of 78 times a day, which exceeds the European average of 48 by 1.5-fold. According to the report, 64% of the respondents from Turkey changed their phone in the last 18 months. Since usage is notably high in the country, there is great potential in integrating smart phones to health care management for the growing number of elderly people.

Today's middle age adults, between age 45 and 65, will be the older adults of the future. It is clear that these older adults will be more demanding than today's older adults, especially in terms of technological advances. Throughout their lifetime, they spend more time with technological devices, such as TVs and telephones, than today's older adults. Understanding these adults' mobile health service needs may guide future development in the area. However, these needs are not easy to understand, since the use of technological devices is very complex and personal.

Verbal protocol analysis, developed by Newell and Simon in [23], is based on the evaluation of protocols by verbally recording behaviours at a specific time interval in order to solve problems. It opens a window to the user's mental model where the user verbalizes all thoughts that occur when performing an activity [24]. Hence, verbal protocol analysis was used in this research study to observe health application use in order to understand future older adults' needs. First, pre-installed mobile health applications were reviewed and then a research protocol was designed. Each participant has to follow a certain protocol: opening the health application, testing different application properties, setting a medical ID for Apple users and viewing active steps for the day.

Undergraduate students from an Industrial Design Department were trained to do the research. Each student selected two middle age participants that are active smart phone users. All of the participants were close relatives of the students, allowing them to behave naturally during the analysis.

Nineteen individuals participated the study: Their ages range from 44 to 63 and nine of them were men. Eight were iPhone users, eight were Samsung users and the rest two were LG users. Table 15.4 shows participant demographics, health status and frequently used applications on their mobile phones. Table 15.4 also presents participants' type of technology use as determined by the researchers.

**Table 15.4** Participant demographics, health status, type of technology use and app usage

| | Gender | Birth date | Smart phone | Health status | Type of technology use: excluded/entertained/networker | Frequently used apps |
|---|---|---|---|---|---|---|
| 1 | Male | 1968 | iPhone | Vigorous | Networker | WhatsApp, Twitter |
| 2 | Male | 1957 | LG | Vigorous | Entertained | WhatsApp |
| 3 | Female | 1967 | LG | Vigorous | Entertained | WhatsApp, pinterest |
| 4 | Female | 1969 | Samsung | Vigorous | Entertained | WhatsApp, pinterest, Instagram, Facebook |
| 5 | Male | 1966 | iPhone | Vigorous | Entertained | WhatsApp, Facebook, Instagram |
| 6 | Female | 1962 | iPhone | Fragile | Entertained | WhatsApp |
| 7 | Female | 1972 | iPhone | Vigorous | Entertained | WhatsApp, Instagram |
| 8 | Male | 1955 | iPhone | Vigorous | Entertained | WhatsApp, Instagram, Facebook |
| 9 | Female | 1963 | LG | Vigorous | Entertained | WhatsApp, Youtube |
| 10 | Female | 1967 | Samsung | Vigorous | Entertained | WhatsApp, Instagram, Pinterest, twitter |
| 11 | Male | 1960 | iPhone | Vigorous | Networker | WhatsApp, Twitter, Facebook |

(continued)

**Table 15.4** (continued)

| | Gender | Birth date | Smart phone | Health status | Type of technology use: excluded/entertained/networker | Frequently used apps |
|---|---|---|---|---|---|---|
| 12 | Female | 1971 | Samsung | Vigorous | Between excluded and entertained | None |
| 13 | Male | 1962 | Samsung | Vigorous | Entertained | Various games |
| 14 | Female | 1967 | iPhone | Vigorous | Entertained | Facebook, Instagram, WhatsApp |
| 15 | Male | 1973 | iPhone | Vigorous | Excluded | WhatsApp |
| 16 | Male | 1973 | Samsung | Vigorous | Entertained | Facebook, Instagram |
| 17 | Female | 1974 | Samsung | Vigorous | Entertained | Facebook, Instagram |
| 18 | Male | 1960 | Samsung | Vigorous | Entertained | WhatsApp |
| 19 | Female | 1964 | Samsung | Vigorous | Entertained | WhatsApp, pinterest, Instagram |

All audio files were transcribed semi-verbatim. Verbal protocol data analysis began with open coding of these data transcripts, it continued with selective coding, and categories were then integrated and refined to themes.

### *15.4.1 Gap Between Real Life Health Services and Mobile Health Applications*

Results of the qualitative study revealed that there is a gap between mobile health applications and real-life health services/devices. Real-life health services, such as hospital visits/doctor appointments, laboratory tests, various scans and health reports/records have their own channels for communicating with the patient. Similarly, health devices such as blood sugar or blood pressure monitors have their own ways to show and keep health records. Eventually, these services either need third-party applications or manual entry of health data to the mobile health applications in order to be used by them (Table 15.5).

**Table 15.5** The gap between real life health services and mobile health applications

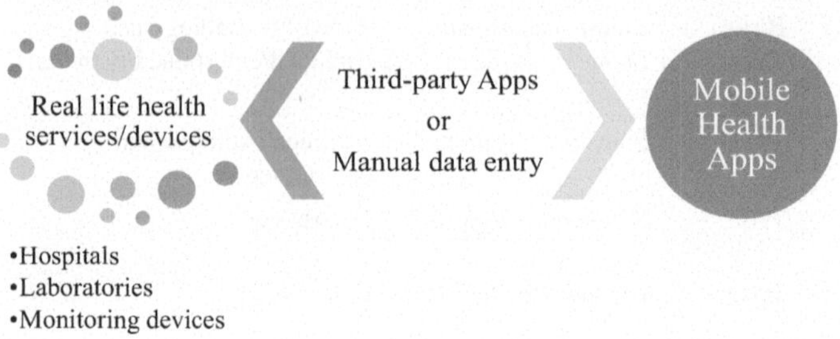


Hospitals and laboratories in Turkey have three main ways of sharing health information: 1. emailing the report/results to the patient and/or the health specialist/doctor, 2. giving the patient printed reports by hand, 3. sharing the information through websites or mobile applications by giving the patient a user ID and a password. In some cases, especially in public hospitals, test or report results are not shared with the patient, but with the health specialist/doctor. Appointments are given online or mostly by telephone. None of these health-related services are effectively linked with patients' mobile health applications.

Only 2 participants specified keeping health records with their smart phones where 13 participants specified keeping them as hardcopies (Table 15.6). Even though it is difficult and strenuous to take hardcopies of their heath records to hospital or health specialist, they still prefer to keep them as hardcopies (Fig. 15.1). Health data needs to be entered manually to these health applications, which users find extremely difficult and unnecessary.

Before the study, a participant was asked why a health application would be necessary, he expressed: "*Why it is necessary … For easy access to health services. For example, when I went to the hospital, I could see my health reports and the results of my tests by myself using the application*". It's no surprise that a user expects a link to health services from an application with the name "health".

The link between various types of health devices and mobile health applications is either non-existent or weak. Some of the health devices, such as blood sugar monitoring devices, have their own applications. Fortunately, health applications often read their data by letting the user to define these applications as a source. The participants of this study were not using these types of devices; unfortunately, it was impossible to observe their experiences. However, some of them shared their ideas on the subject while using the health applications. For example, a participant asked how the phone is going to measure her blood sugar and after learning that she had to enter it by herself, she responds: *"Oh, forget it then"*.

**Table 15.6** Participant preferences on keeping health records and usage of health applications

| | Keeping health records | Use of health apps |
|---|---|---|
| 1 | With smart phone | Yes |
| 2 | At home as hardcopies | Yes |
| 3 | At home as hardcopies | No |
| 4 | At home as hardcopies | No |
| 5 | At home as hardcopies | No |
| 6 | At home as hardcopies | No |
| 7 | At home as hardcopies | Yes (only for step counting) |
| 8 | At home as hardcopies | No |
| 9 | At home as hardcopies | No |
| 10 | None | Yes (only for step counting) |
| 11 | None | No |
| 12 | As hospital records, at home as hardcopies | No |
| 13 | As hospital records, at home as hardcopies | No |
| 14 | As photos with smart phone | No |
| 15 | At home as hardcopies | No |
| 16 | Only as hospital records | No |
| 17 | Only as hospital records | No |
| 18 | At home as hardcopies | No |
| 19 | At home as hardcopies | No |

#### 15.4.1.1 Step Counter

The step counting feature turned out to be one of the favourable aspects of health applications for these participants. Only four participants were using health applications in their phones before the study and they were using it mostly for the step counting. A participant mentioned that she likes the way health application warns her if she didn't take a walk for a couple of days. Another mentioned how another step counter application in his old phone was motivating him to compete with friends about daily activities: *"You were adding your friends and you could see how much they were running, walking, and going downhill. This was obviously motivating me. We were sitting with friends saying, 'you walked a lot yesterday'. That's really encouraging me"*.

Other participants were excited about learning how many steps they had in a day. As the step counter automatically counts the steps of the user, the link between real life and the application is strong. Thus, users have the opportunity to experience this feature efficiently.

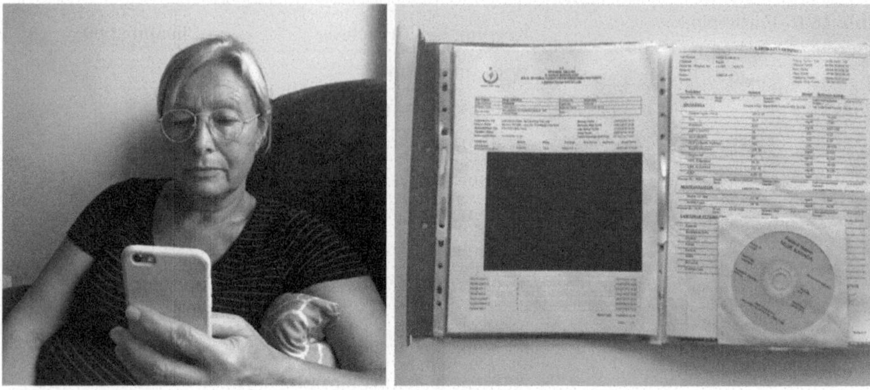

**Fig. 15.1** Author's mother (age:63), using Apple's iHealth app (on the left) and her hardcopy health records' file (on the right)

#### 15.4.1.2 Medical ID

Setting up a Medical ID in Apple phones, even after participants were told about their uses and benefits, seemed unnecessary to study participants. They stated that using a Medical ID is not worth the efforts to create it. A few participants stated that they didn't use it because they couldn't understand it. One mentioned that he would create it if he would find an appropriate time to enter the data in the future. Another participant stated mobile phones are special devices and did not want her phone to be looked at by foreigners in an emergency state. Moreover, she shared her doubts about using a Medical ID:

> Why would I create a Medical ID? My phone may be broken in case of an accident, or other things may happen, who will use the Medical ID? All kinds of information are available in the hospital through various test procedures. They can learn my blood type immediately by doing tests, it's nothing. Who will think of using Medical ID in an emergency situation?

Most importantly, "Medical ID" and it's Turkish translation is a combination of two important and serious words: "Medical" and "Identification". These two words make this feature feels more critical than it really is. An identification is what you carry with you to prove your identity and that is why a participant asked: "*Is this the Medical ID? Why didn't we put the photo in?*" Medical is a serious word as well, relating to medicine or to the treatment of illness or injury. Thus, the creation of a Medical ID seemed to be a formal and serious job that mistakes that cannot be reversed.

#### 15.4.1.3 How Smart Mobile Phones Are?

Participants were sceptical about how smart mobile phones were. They were not sure if the devices were intelligent enough to track health data without any additional effort or if they needed additional tools. For these people, the technology behind these mobile phones is still a mystery, which brings up trust issues.

For example, one participant asked if phones can use the camera to track daily food intake or if it needed to be entered manually: "*Can it see that too? I mean, how's it going to figure out how much I eat? The phone is usually away from me when I eat. I don't carry it on me*".

Other two participants were curious about how the application measures or knows the steps they took. Another, who was discovering the step counting function was surprised when he was asked to look at how far he walked on other days and he asked: "*did it follow them all, even when not being open?*" Later, he opened the application and tried to figure out how the phone counts the step as he walks through the room.

A technological system's credibility is based on whether it is believable and if users should rely on it, and system users must determine whether the information they provide is credible [3]. Although participants were unsure, they still could guess how the step counter works, so it seemed as a credible service but for other health related issues, mobile health applications were not credible. They thought that if they use some of the health application features, the smart phone could track their daily lives. It is important to note that, all of the participants were using social networking or messaging applications without these credibility issues—all of them except four stated using WhatsApp and have active WhatsApp groups with friends and family, and most of them are very active in social networking sites.

### *15.4.2 Gap Between Cultures: Young, Global, Digital Versus Old, Local, Analogue*

Another gap between future older adults in Turkey and mobile health applications is the cultural one. These applications' visuals and words seem to belong to a different a young, global, digital culture. Visual clues seem to fail, words have different meanings. Although participants tried to understand words and visuals, they found their selves misinterpreting the application and doing errors most of the time using these applications.

First of all, visual clues were not functioning properly in these applications for these participants. Before using the application, when asked to guess what the application is about, a participant confused about the icons and stated, "*One guy running, one pointing to the bottom of his feet: I didn't understand*". While using the application, he tried to guess the meanings of the icons: "*This one is sports, someone is running. The other one is cutlery, it teaches how to eat, I guess. And the other one is someone reading a book?*". Considering that the meaning of a simple gesture differs from culture to culture, it is normal for health-related applications require local visual cues to make users not to feel alien.

Creating information about a product means creating a cognitive representation of items related to this product within a symbolic language frame [25]. Although words used in the application is mostly Turkish, the meanings of the expressions are not easily understood by these participants. Most of the written information seems like directly translated text and there are many sentences that are grammatically incorrect or might cause misunderstandings. Therefore, the cultural gap is even bigger for Turkish users. For example, translation of the word "track" is used as "izle" in the Turkish version of one of these health applications, which means "watch" as well. In many ways it feels like the application is trying to talk in Turkish, but unfortunately, it fails.

Furthermore, these applications behave their users as they are kind of experts in sports. A participant was exploring the application when she saw the titles: "*Do you have fitness goals?*", "*Balance and stability improvement*", *Get in shape*". Such questions stressed users and make them feel like a noob. Another participant was reading the expressions in the application: *"Light walking, jogging, fat burning, starter. Walk at a constant speed to allow you to sing lightly for 30 min"*. It was confusing for this participant what it means to "walk at a constant speed to allow her to sing lightly". She was confused if she would really have to sing during the training. Even an expression aiming the user to understand something easily could be more confusing if it is disconnected from the cultural context.

One of the most important issues is that the warnings and explanations used in these health applications are also barely comprehended by these participants. Sounds that are being used as notifications could be disturbing as well. A participant stated why she was not using the "water intake" feature anymore: *"Of course, you live in a certain social environment. It makes the sound of glass, sound of water with a message. I feel uncomfortable. I don't forget to drink water anyway. I closed that feature"*.

Norman [26] stated that human beings are analogue beings who are trapped in a digital world. Accuracy is the most important thing in the digital world, and it is against human nature. Unlike digital world, people care about meanings, not the details of the signals. When using health applications, participants try to ascribe a meaning to different parts of the application, unfortunately they fail. This situation leads to numerous mistakes and misunderstandings when attempting to use these applications. The results of the study revealed that these users do not have control over the cause-effect relationships in their phones and the processes related to what they aim to do. After adding her daughter as an emergency contact during the Medical ID creation, one participant asks; "*Where does this number come from?*" looking at her daughter's phone number.

While trying to find the health application, same participant clicked the "Settings" first, then "Reminders" and finally the "Notes". Another was asking "*Where do you hide this application?*" to her daughter when it had asked her to enter the health application in her phone. "*I clicked and a lot of things happened*" said another. A participant spent some time looking for the health application:

> Music, settings … not in the store, right? Messages … Can it be in the settings then? It is not here, not here, not inside camera, not inside map, not photos, apple store, weather, hours, messenger … It can be inside the settings, where will it be? Well, I've never looked at the folder called lifestyle. Hah, here it is, stock market, health.

In contrast to learning knowledge, learning to use a product means acquiring regular practices with the items related to the product, or rather, dominating their functionality [25]. The study revealed that the participants neither have the motivation nor the time for acquiring regular practices about these health applications, since the cultural gap is making this harder than they thought.

The study's participants do not want to be labelled as "too old to use technology" and when a younger relative was helping them, they wanted to cover their mistakes. As an example, a participant asked what the expression "manage items" means in a low voice to the researcher.

A few participants stated the fonts are too small in the application, they could not see. One tried to zoom into the application but failed to do it.

These applications do not have local clues to help users trust, understand, and use them easily. But above all, these people belong to a culture in which spending too much time looking at mobile phones is a shame. A participant mentioned: "*I don't like those people who carry their phone in their hands all the time*". Another stated: "*It's something new that the phones are always in people's hands. I don't get used to it. I only use it when I need it*". Being addicted to health-related issues, tracking every activity is seen as being self-centred: "*Health is sometimes about not seeing some of the things. Knowing about ignoring. If you turn this into a kind of obsession, you're going to have to take care of it all the time and we don't have time for that*". Additionally, recording every step, trying to control life was also an awkwardness for these participants. When it was asked why people use the step counter function a participant expressed: "*People now want to control everything*". Another identified his ideas about why people use a Medical ID: "*I don't know, everybody started recording everything on their phones. The phones know us better than we do*".

## 15.5 Conclusion

This chapter presents a literature review of the current state of mobile health applications for older adults, and the results of qualitative research to identify what determines the use of pre-installed mobile health applications by future older adults in Turkey. A range of issues was identified using verbal protocol analysis.

A significant amount of research recently has been dedicated to integrating health management with mobile phone use for older adults. However, none of it focused on pre-installed applications on our smart phones. These applications are ready to use on almost all smart phones without the need to download from the online application stores. Considering the number of potential users, these applications can help us manage health with our mobile phones.

There was a clear consensus among the participants of the study that mobile health applications are beneficial, especially for step counting. However, the gap between real life health issues and these health applications seems to prevent users from benefiting from them. Integrating mobile phones into the health management system requires the use of various applications, sensors or products that allow health status to be checked regularly. Moreover, the visual and verbal language difference between these users and the applications creates another gap. These applications use a global, digital and younger language that seem to belong to a different culture.

Besides, health is a critical area for human beings. Fisk and colleagues [3] denoted that a typical user does not know the health care field well, and it is filled with unfamiliar terms and phrases. The consequence of making an error in health-related technology may mean illness, an additional hospital visit, or worse. For example, a healthcare technology user who was recently diagnosed with an illness may be in a more stressful and emotional situation than any other user [3]. Thus, during their use of health-related applications, people are more stressful and more excited than they are when they use other applications, so they tend to make more mistakes. It is clear that these applications need to comfort their users. As one participant expressed during the study, all they want is simplicity from mobile technologies: *"The phone will say hello to me, I will talk, it will say hello to me, I will hang up"*.

Technology undoubtedly will play an increasing role in future health care and management systems. However, social interactions and relationships continue to be important drivers of health, behaviour, and health care utilization worldwide. [6] Besides, culture and real-life is and will be always a central part of the health-related issues. The results of the study showed that there is great value in studying the needs and behaviours of elderly mobile health service users, especially with a focus on cultural and social needs.

## References

1. United Nations, Department of Economic and Social Affairs, Population Division. World Population Ageing 2009 New York, US (2010)
2. United Nations, Department of Economic and Social Affairs, Population Division.: World Population Ageing: 1950–2050. Retrieved from http://www.un.org/esa/population/publications/worldageing19502050/ (2002)
3. Fisk, A.D., Rogers, W.A., Charness, N., Czaja, S.J., Sharit, J.: Designing for older adults: principles and creative human factors approaches. CRC Press, Boca Raton (2009)
4. Le Deist, F., Latouille, M.: Acceptability conditions for telemonitoring gerontechnology in the elderly: optimising the development and use of this new technology. IRBM **37**(5–6), 284–288 (2016)
5. Öney Doğanyiğit, S.: Yaşlı Sağlığında Geronteknolojik Ürünlerin Kullanımı (Use of Gerontechnological Products in Elderly Health). In: Duben, A. (ed.) Yaşlanma ve Yaşlılık (Aging and old age), pp. 135–152. Istanbul Bilgi University Publications, Istanbul (2018)
6. He, W., Goodkind, D., Kowal, P.R.: An aging world: 2015. United States Census Bureau, Washington, DC (2016)
7. GSMA Intelligence.: The Mobile Economy 2018. https://www.gsma.com/mobileeconomy/ (2018). Accessed 2 Jan 2019
8. World Health Organization.: Global diffusion of eHealth: making universal health coverage achievable: report of the third global survey on eHealth. http://apps.who.int/iris/bitstream/handle/10665/252529/9789241511780-eng.pdf?sequence=1 (2016). Accessed 2 Jan 2019
9. Conci, M., Pianesi, F., Zancanaro, M.: Useful social and enjoyable: mobile phone adoption by older people. Lect. Notes Comput. Sci. **5726**, 63–76 (2009). https://doi.org/10.1007,7978-3-642-03655-2-7

10. Venkatesh, V., Morris, M.G., Davis, G.B., Davis, F.D.: User acceptance of information technology: toward a unified view. MIS Q. **27**(3), 425–478 (2003)
11. Lee, C., Coughlin, J.F.: Perspective: Older adults' adoption of technology: an integrated approach to identifying determinants and barriers. J. Prod. Innovation Manage. **32**(5), 747–759 (2015)
12. Plaza, I., Martín, L., Martin, S., Medrano, C.: Mobile applications in an aging society: status and trends. J. Syst. Softw. **84**(11), 1977–1988 (2011)
13. Hun-Sung, K., Lee, K.-H., Kim, H., Kim, J.H.: Using mobile phones in healthcare management for the elderly. Maturitas **79**(4), 381–388 (2014)
14. Blaschke, C.M., Freddolino, P.P., Mullen, E.E.: Ageing and technology: a review of the research literature. Br. J. Soc. Work **39**(4), 641–656 (2009)
15. Klasnja, P., Pratt, W.: Healthcare in the pocket: mapping the space of mobile-phone health interventions. J. Biomed. Inf. **45**(1), 184–198 (2012)
16. Joe, J., Demiris, G.: Older adults and mobile phones for health: a review. J. Biomed. Inf. **46**(5), 947–954 (2013)
17. Gjevjon, E.R., Øderud, T., Wensaas, G.H., Moen, A.: Toward a typology of technology users: how older people experience technology's potential for active aging. In: Nursing Informatics, pp. 25–31 (2014)
18. Sunyaev, A., Dehling, T., Taylor, P.L., Mandl, K.D.: Availability and quality of mobile health app privacy policies. J. Am. Med. Inf. Assoc. **22**(e1), e28–e33 (2014)
19. Peng, W., Kanthawala, S., Yuan, S., Hussain, S.A.: A qualitative study of user perceptions of mobile health apps. BMC Public Health **16**(1), 1158 (2016)
20. Ayman, O. (Ed.).: Türkiye Yaşlılık Atlası (Aging Atlas of Turkey). National Geographic Türkiye, 133, Appendix. (2012, May)
21. Sungur, N.: Türkiye'de ömür 20 yılda 10 yıl uzadı (In Turkey, life expectancy extended 10 years in the last 20 years). National Geographic Türkiye, **133**, 78–109 (2012, May)
22. Ünal, B., Horasan, G.D., Kalaça, S., Sözmen, K.: Türkiye Kronik Hastalıklar ve Risk Faktörleri Sıklığı Çalışması (Turkey (TR), Study of chronic diseases and risk factors) Ministry of Health Publications No: 909 (2013)
23. Newell, A., Simon, H.A.: Human problem solving. Prentice Hall, Upper Saddle River, NJ (1972)
24. Nemeth, C.P.: Human factors methods for design: making systems human-centered. CRC Press, Boca Raton, FL (2004)
25. Nohl, A.M.: Eşya ve İnsan: Bir Pratik İlişkinin Felsefesi, Pedagojisi ve Sosyolojisi (Goods and Human-beings: Philosophy, Pedagogy and Sociology of a Practical Relationship). Ayrıntı Yayınları, İstanbul (2018)
26. Norman, D.: Affordance, conventions, and design. Interactions **6**, 38–42 (1999)
27. Deloitte Global Mobile Consumer Survey 2017. Deloitte Turkey. Last modified 11 Jan 2018. https://www2.deloitte.com/tr/tr/pages/technology-media-and-telecommunications/articles/global-mobile-consumer-survey-2017.html
28. IOS—Health.: Apple. https://www.apple.com/ios/health/. Accessed 23 July 2018
29. Samsung Health—Apps on Google Play.: Google Play. https://play.google.com/store/apps/details?id=com.sec.android.app.shealth. Accessed 23 July 2018
30. Coughlin, J.F.: Understanding the Janus face of technology and ageing: implications for older consumers, business innovation and society. Int. J. Emerg. Technol. Soc. **8**(2), 62–67 (2010). Available at: http://www.swinburne.edu.au/hosting/ijets/journal/V8N2/vol8num2-GuestEditorial.html
31. Zickuhr, K., Madden, M.: Older adults and Internet use. http://www.pewinternet.org/2012/06/06/older-adults-and-internet-use/ (2012). Accessed 2 Jan 2019

**Berrak Karaca Şalgamcıoğlu** is currently an Assistant Professor at Istanbul University, Department of Industrial Design. She got Mechanical Engineering BSc, Industrial Product Design BSc, and Mechanical Engineering MSc Degrees from Istanbul Technical University (ITU). Karaca Şalgamcıoğlu received her PhD degree from ITU, Department of Industrial Product Design. Prior to taking up her current role in Istanbul University, Karaca Şalgamcıoğlu was employed as a R&D Engineer in various companies including Arçelik and as a researcher in University of Michigan and ITU. Between 2015 and 2019 she was the founding Chairperson of the Department of Industrial Design at Istanbul Medipol University. Her research interests include qualitative methods in design research, ethnography, inclusive design, and ageing.

# Chapter 16
# Mobility Information Services for Older People

**Cindy Mayas and Heidi Krömker**

**Abstract** Mobility is an essential prerequisite for the meeting of social needs of senior people. Nevertheless, mobility reduces in the advanced age. Next to changing destinations and the health situation of senior people, one reason for that reduction might be the lack of orientation of mobility services towards the needs of senior people. In particular, the fulfillment of information needs is a necessary condition for the use of public mobility services by senior people. Based on the example of two senior personas this chapter clarifies typical characteristics of the information needs of senior people in public transport. In order to consider efficiently these information needs in the development of existing or innovative mobility information systems, this chapter presents guidelines to analyze, integrate and evaluate the mobility and information needs of senior people. Finally, a case study for interactive public displays at stations shows the characteristics according to the usage context and the satisfaction of senior people with interactive public mobility information.

## 16.1 Introduction

Mobility is one of people's basic needs. For senior persons, mobility becomes even more important than in youth. Especially in the later stages of life, mobility ensures social participation and preserves health.

The definition of 'senior persons' varies depending on the context and geographical factors. In context of employment, people are already considered as 'older' from an age of 50 years. According to a Eurobarometer survey from 2011, people are described as old from the age of 64 years on average [7]. New research shows the trend that perception of "old age" postpones over the years. For example, women

C. Mayas (✉) · H. Krömker
Technische Universität Ilmenau, Ilmenau, Germany
e-mail: cindy.mayas@tu-ilmenau.de

H. Krömker
e-mail: heidi.kroemker@tu-ilmenau.de

 
A. Woodcock et al. (eds.), *Design of Assistive Technology for Ageing Populations*, Intelligent Systems Reference Library 167,
https://doi.org/10.1007/978-3-030-26292-1_16

in France are considered "old" at an age of 74 years [18]. In addition to the age in years, ageing process is also accompanied by typical changes in the way of life [4]:

- Change of participation: The most serious change in participation is the retirement, which starts in most European countries between 60 and 67 years [20]. In addition to this complete disappearance of everyday working life, there are other typical changes for senior people that can start earlier or concern part-time jobs for old-age pensions — these include, above all, shortening working hours or changing work tasks. Therefore, new activities can appear, for example voluntary work or free time activities.
- Change of connectedness: Social connections to other people often change continuously throughout the life. Typical changes for senior people are the move-out of the children from the family home or even the death of a life partner. These incidents often also take effects on the housing situation.
- Change of independence: Ageing is often accompanied by the deterioration of the health status. The major impairments are related to the three parts senses, body, and mind [1]. The senses concern seeing, hearing, tasting, smelling and feeling, limitations of the body concern the flexibility, power, and dexterity, while the mind have to come along with slowed coordination, memory, and reaction. Because of these impairments, senior people need help with their daily life activities, resulting in a reduction of independence in their lifestyle.

These life changes are also typical indicators for changes of the mobility behavior, for instance the choice of the mobility service [25], and mobility needs, which are described in Chap. 2.

To ensure that mobility services can be used, their accessibility and interconnection must be as simple as possible. The mobility services and information must be methodically developed in such a way that the users' needs are concerned, at all touchpoints of a journey [24]. As a result of the authors' analytical and empirical studies, a user-centered guideline set has been derived that can be used to integrate the needs of both younger and senior people in the development of mobility services. These guidelines are described in Chap. 3.

Within a case study with a German transport company, senior people represented an important target group for the development of mobility technologies, showing the practical impact of the guideline set in Chap. 4.

Final insights and recommendations are summed up in Chap. 5.

## 16.2 Mobility Information Needs

In order to analyze the mobility information needs of senior people, it is essential to understand their mobility behavior and reasons. The results need to be communicated intelligibly to integrate them into the development of mobility systems. Therefore, the method of personas is adopted for the construction of mobility personas and information needs for different mobility categories are derived.

### 16.2.1 Mobility Behavior of Senior Mobility Users

The MID-study collects data of mobility diaries of 316 361 persons from varying age groups and varying regions in Germany [11]. The analysis of the data according to their age group shows the following three trends in the change of mobility behavior beginning in the age group of persons who are older than 50 years.

**Change of Reasons and Destinations** The most serious change in reasons of mobility is the reduction of business trips and trips to get to work from 42% to two per cent between 50 and 80 years. Accordingly, the share of trips for shopping, errands, and free time activities rises, each per 13%. According to the changing reasons, the daytime of mobility and the length of the trips of senior people are changing. The majority of trips are rearranged from the rush hours to the late morning or the early afternoon. In addition, the majority of destinations are chosen in the local area. Therefore, the trips shifts from long-term trips (more than 30 km) to short-term trips (less than 10 km) by the majority.

**Reduction of Number of Trips** The age group "40–49 years" reaches with 3.7 trips every day the highest average value. Up to the age group with persons, who are more than 80 years old, the number of trips reduces continuously to 1.9 trips a day on average. The main group with two trips per day remains constant over all age groups at a share of nearly 30%. However, the number of persons without any trips per day rises from 12 to 35% in the group with persons, who are more than 80 years old.

The reasons for that are diverse. On the one hand, the number of trips reduces according to the change of mobility reasons. When regularly tasks, such as going to work or picking up the children, disappear, the number of daily trips reduces. On the other hand, other daily tasks change or appear and can increase the ways for shopping, errands, or free time activities.

**Change of Mobility Service** Among all persons investigated, who conduct at least one trip a day, the share of trips by motorized individual mobility decreases from 64 to 48% of all trips of the group between 50 and 80 years. Though, the relation from trips as driver to trips as passenger changes from 8:1 to 2:1. Accordingly, the share of trips on foot doubles up to 34% with 80 years and the trips with public transport slightly increase from 8 to 11% of the group between 50 and 80 years.

Altogether, the data shows a reduction in mobility of senior people, although the basic social needs keep a high need for mobility for senior age groups. Therefore, the reduction in mobility could be rather caused by a lack of service of these basic needs in the current mobility offers and information. The challenge is to improve mobility services in such a way that they better meet the basic needs of older people, in order to enable more mobility to senior people.

### *16.2.2 Personas for Senior Mobility Users*

In order to improve the potential of public transport for senior people, the needs of senior people have to be integrated in the development and advancement of public mobility and information systems. To illustrate these distinctions of special user groups, the method of persona descriptions [5] became an established method of user-oriented development [6]. Personas are stereotypes derived from characteristics and behavior of real people. The described personas are fictitious and clearly distinguishable characters with a typical behavior, motivation and goals of users in order to personify a group of users. Typical usage situations of personas can be described in scenarios [17].

The methods of personas and scenarios of senior people were successfully applied for the development of ambient assisted living systems [16, 21]. The research project IP-KOM-ÖV[1] developed personas for different mobility behavior [13]. The persona set also includes the following two senior people and their mobility behavior.

**Senior Casual User Hildegard Krause** Hildegard (see Fig. 16.1) is retired and does not have a driver's license. Until the death of her husband, she went with him by car or by bicycle. Nowadays, she relies on public transport, because cycling and walking becomes more and more difficult for her due to her knee replacement. She stays in contact with her friends, children, and grandchildren via smartphone. Hildegard uses the public transport independently for her daily errands and visits of

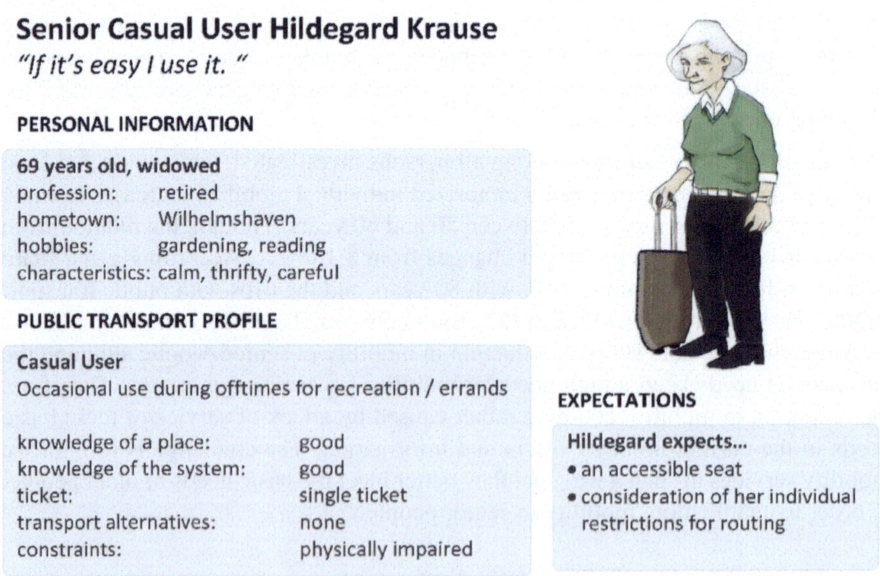


**Fig. 16.1** Persona senior casual user Hildegard Krause [13]

[1]Source: https://www.vdv.de/ip-kom-oev.aspx, retrieved 01.02.2019

**Senior Ad-hoc User Bernd Lorenz**

*"By car I'm quicker and more flexible!"*

**PERSONAL INFORMATION**

**58 years old, two children**

| | |
|---|---|
| profession: | senior project manager |
| hometown: | Düsseldorf |
| hobbies: | sailing, classic cars |
| characteristics: | impatient, pragmatical |

**PUBLIC TRANSPORT PROFILE**

**Ad-hoc User**

Infrequent use of public transport for business trips

| | |
|---|---|
| knowledge of a place: | moderate |
| knowledge of the system: | none |
| ticket: | single ticket |
| transport alternatives: | car |
| constraints: | none |

**EXPECTATIONS**

**Bernd expects...**

- individual navigation
- easy use of public transport
- safety and security, punctuality, and comfort

**Fig. 16.2** Persona senior ad hoc user Bernd Lorenz [13]

her garden. She knows the local area and the network connections of her trips well, but she requires detailed information about the accessibility of the vehicles and the availability of seats. In cases of disturbance, she feels very unconfident and prefers personal assistance.

**Senior Ad hoc User Bernd Lorenz** Bernd (see Fig. 16.2) is working in a leading position and lives together with his life partner, who cares for the daily errands. Bernd loves cars. He drove each trip, private and business trips both, by car. Since his children moved out few years ago, he dedicates himself to his passion for classic cars. Therefore, he uses a smaller car for private trips and public transport for the business trips. Bernd's knowledge of the network connections and the local station areas is very low. He expects navigation information for public transport similar to his individual car navigation.

### *16.2.3 Information Needs of Senior Mobility Users*

Each persona is involved in two scenarios describing both a routine and an extraordinary trip. Next to the extraction of use cases, information needs [9] are derived from the scenarios. The resulting 81 information needs are systematized into seven mobility information categories. Each normalized category value consists of 5–20 single information needs.

Figure 16.3 shows the comparison between the senior personas Hildegard, Bernd, and a commuter persona. Whereas for a commuter mobility serves as a means to an end and is often restricted to few routes, mobility of senior people is more multi-faceted and requires more information. Six out of the seven categories of mobility information show a higher information need of the personas of senior mobility users. Due to the lower knowledge of the public transport system, persona Bernd has the highest information need in the categories ticket information, location information, time information, and network plan information. Persona Hildegard's information needs are highest in the categories vehicle information, especially regarding the accessibility and seat availability, and connection information, to ease the choice of connection alternatives.

Mobility information supports and ensures the use of according mobility services, which preserves a wider range mobility for senior people. Therefore, mobility is an important impact for the basic social needs for senior people resulting from the life changes described in Chap. 1: participating in social activities or volunteer work, feeling of connectedness to others, and feeling of independence [4].

The challenge for the development of mobility information systems for senior people is to connect these information needs with other social needs, for instance communications with others, in order to facilitate the access to the systems.

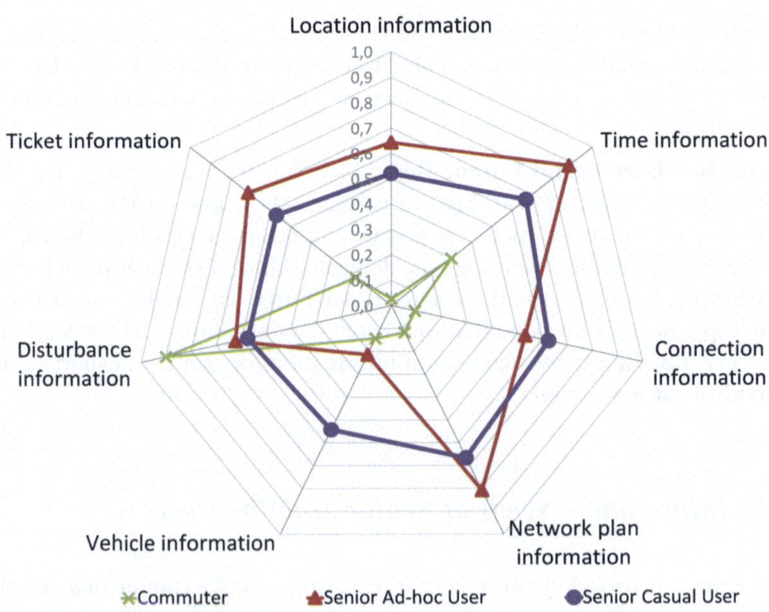


**Fig. 16.3** Information needs in public transport (according to [9])

## 16.3 Guidelines for the User-Oriented Development of Mobility Information

The authors conducted a range of research projects for the development of mobility information standards,[2] mobility information systems,[3] and innovative mobility services.[4] The key challenge of the development is to ensure that the requirements of younger and older target groups remained methodically present for all actors during the entire development process. The precise understanding of the different people's habits and their conditions formed the basis for all design decisions in the development process.

The developed set of 12 guidelines consists of recommendations for the analysis of the usage context, the conceptual design, and the evaluation of mobility information systems. In the analysis phase, the persona method is combined with scenarios and suitable data gathering methods. Workflows and utility are derived from user journeys for the conceptual design of mobility systems. Finally, the method of field-testing was specified for the use in the mobility context for the evaluation of mobility information systems. The application of these methods in several research project detected, that transparency, consistency, and intelligibility are essential indicators of quality of mobility information [8], which can be applied in particular for senior people.

Next to the focus on senior people, the following guidelines are applicable for development processes for all age groups. While other user groups might be more adaptive to diverse test situations, the nonobservance of the guidelines for senior people can result in falsified data and decisions, and as a consequence, a lower acceptance of the developed mobility information system.

### *16.3.1 Analysis of Senior Usage Context*

The goal of the analysis of usage context is the understanding of the user, their tasks and motivations, as well as the environment of the usage according to ISO 9241-210 [12]. In addition, the analysis should focus on relations between information needs and social needs of senior people, in order to improve the mobility of senior people, in general.

---

[2]The research and standardization projects IP-KOM-ÖV, DIMO-OMP, DIMO-FuH developed interface standards for passenger information in public transport with focus in the connection between personal mobile devices, vehicles systems, passenger information systems and control centers, from 2010 to 2018.

[3]The research project DynAPSys developed a planning system for individual task and day planning, aiming to make the use of public passenger transport easier and more flexible, from 2013 to 2016.

[4]The research project Move@ÖV developed a productive service system in public transport by means of service engineering, from 2014 to 2017. The project focuses on the individualization of mobility services in public transport in order to make them more flexible and to integrate them efficiently into existing mobility systems.

1. **Observe Senior Users** Senior people often don't express critical or negative opinions, due to social desirability and politeness. In order to detect typical workflows in user behavior, workarounds, or incidents, observations are necessary in the real usage environment. Furthermore, it is recommended to observe the integration of the tested system into other tasks and the day structure, in order to detect relationships between the mobility and other participating and communicating activities.
2. **Interview Senior Users** Senior people are less likely to take part in online surveys. Therefore, the results are often not representative for the whole target group. In order to cover the range of the target groups, interviews with a direct personal contact are necessary. In interviews, an open discussion atmosphere can also provide further information on motivations and expectations on mobility. For instance, the questions "How do you like to travel?", "Who would you like to go to?", "Who would you like to talk to while traveling?" detect further information on the expectations on the independence of mobility and the impact of mobility for social and participatory activities.
3. **Interview Experts from Senior Interest Representations** Many interest groups and associations, for example passenger advisory boards, represent the needs of senior people in a targeted manner. The integration of their holistic experience into the analysis can provide valuable indications of development potential regarding the improvement of social needs for senior people.
4. **Describe Senior Users in Personas** The method of personas is well suited to document the results of the analysis in a sustainable and understandable way. Examples of senior mobility personas are shown above in Chap. 2. The description should also cover the social needs of senior people regarding independency and participation. Personas support the user-oriented determination of product characteristics based on goals and tasks. Personas can be used as universal communication tool for different stakeholders within the development process, especially to discuss challenges and solutions, based on a common understanding of the users.
5. **Describe Senior User Behavior in Scenarios** Scenarios describe the interaction of a certain user with the system in a certain environment with a certain goal [17]. Scenarios are stories, which orientate at users and their needs. Therefore, the user in a scenario can be directly connected to the persona method. Regarding senior people scenarios should not only focus on interactions but also context factors which might influence the social needs of senior people.

### *16.3.2 Conceptual Design*

According to the usability engineering lifecycle of Mayhew [15] the conceptual design is the first step the iterative design and testing phase. The guidelines refer in particular to the conceptual design, in order to guarantee the integration of the analyzed requirements into the design process. Further design activities should orientate on the actual state of the art of user interface design.

6. **Define Utility** Based on the analyzed scenarios and information needs, key utilities and according functions of a mobility information system for senior users should be derived. The utility is a major factor for the suitability of a task and the according user acceptance of a mobility information system. Development potential can be detected especially in the area of security, infrastructure, service frequency, and new service combinations, which support the fulfillment of social needs.
7. **Create User Journeys** In order to integrate the utilities in a holistic context, user journeys [26] enrich the task-oriented perspective with other activities and events related to the users' mobility. Mobility experience maps [24] should consider all touchpoints of a user trip to detect relations between functions to fulfill the mobility needs and further social needs, for example communication with others while traveling or a personal assistance after a consultation. In this way, the functions should be related to each other along user-oriented workflows.
8. **Define Use Cases** Use cases specify all possible interaction sequences of different users in different situations for a certain functionality [3]. Use cases can be extracted from scenarios and user journeys, where each description is a specific instance of different use cases. For the further development, it is important to maintain the background information of functionalities for user-oriented communication and decision processes.

### *16.3.3 User-Centered Evaluation*

Personas, scenarios, user journeys, and use cases support the integration of the user requirements in the design and development process. However, the developed concepts and prototypes have to be evaluated with users throughout the development process according to established methods of user-oriented testing. In particular, the evaluation criteria for evaluations with senior people should refer to the analyzed social needs.

9. **Create Confidence** In order to obtain as unbiased as possible user opinions of senior test persons, it is important to gain the trust of the test person. Therefore, it is recommended to provide sufficient information about the test background and the planned test procedure. A familiar test environment also strengthens the sense of security of senior test persons. Allow sufficient time for the test and the interviews even if user opinions or questions are digressive. Some personal anecdotes might detect important indicators of the expectations of independency and participation of the test persons. In contrast, time pressure and rigid guidelines makes senior test persons uncertain and suppress user opinions.

10. **Let Seniors Think Aloud** Standardized assessment questionnaires are answered much more positively by some senior people than the test procedure would suggest. For reasons of social expectation, senior people tend to attribute operating errors to their own mistakes and therefore express little criticism of the system. Therefore, it is recommended to use a kind of thinking aloud method [2], within users are asked to express their opinions and emotions aloud, for example direct thinking aloud during the test or retrospective thinking aloud after the test.
11. **Inquire Reasons of Senior Behavior** Out of courtesy during the thinking aloud moments, some senior people only express little and cautions their critism concerning the tested system. A constant asking for the causes of operating errors, e.g. "What did you expect?", "How did you understand that?", "Why did you do that?" can provide important hints for the design of workflows and the designation of interaction areas.
12. **Test in a Field Environment** Usability field tests analyze the use of systems in their real context of use. On the one hand, the common environment might decrease uncertainty of senior users. Concurrently on the other hand, there are many influencing factors, which might irritate especially senior test persons, for instance regarding the social environment. Nevertheless, in particular, these variable actions and influences provide a holistic perspective of the behavior of the senior test person, which is necessary to evaluate the fulfillment of mobility needs in relation to further social needs.

## 16.4 Case Study: Interactive Public Mobility Information

The case study "Interactive Public Mobility Information" shows how the requirements of the Convention on the Rights of Persons with Disabilities [22] can be systematically implemented. Interactive public displays enable a digitalized paperless mobility information at the stations of public transport. The case study shows how the mobility information needs of senior people are met with the new information system.

**General Objective** The interactive mobility information should provide the same collective information as the hardcopies at the same place before. The interactive application should be controlled via a 46" touchscreen. Information on timetables, route maps, ticket information, station map, and information on disturbances are defined as the most relevant mobility information for all target group according to an external analysis.

**Design Concept** The design concept bases on an analysis of the ergonomic characteristics of touch displays, which defined comfortable areas for interaction and content in standard and special accessible modes. Each main content could be reached via a 2-level-navigation with a maximum of two interactions.

**Fig. 16.4** Example of an interactive public display at the paperless test station in Stuttgart. Reproduced from ST-VITRINEN Trautmann GmbH & Co. KG

Some supportive functions, as zooming and searching, are implemented next to the navigation functions (Fig. 16.4).

**Method** The entire user-oriented design process included several expert interviews, user interviews and observations in interactive usability studies [10, 23]. The results of the usability study were included in the subsequent final development of the interactive public displays. Therefore, next to the detection of further usability issues, the first field evaluation focuses on the aspects of context of use and satisfaction in different age groups.

According to these objectives, the field evaluation [14] consists of a method mix considering data on different levels. In a first step, system logfiles are analyzed from six months in order to find out general information about the extent of use. Within the logfile analysis no age information about the users is available. In a second step, an online survey is conducted in order to reveal the different aspects of acceptance and some factors of the context of use. This method is applied, in order to integrate all age groups. The participation of senior people, who reached at least the age of 50 years, reached only a share of 16% of participants. Finally, users of the interactive displays are observed and interviewed in the field, in order to gather specific data about the usability and to increase the information from senior people. The share of observation participants, who are older than 50 years, reached finally 58%.

**Results** As a part of the usage context, the intention and information need of the usage is analyzed, next to usage times and environment. Figure 16.5 shows, that senior users show a greater need for ticket information, local information, disturbance information, and connection information than younger users. Younger users are more likely attracted by time relevant information and even use the information system

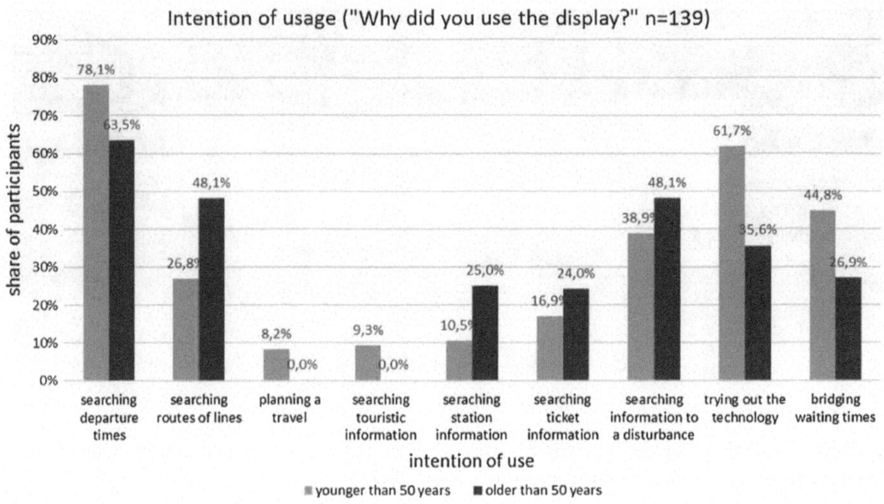


**Fig. 16.5** Usage context of interactive public displays

for fun or out of curiosity. These results confirm the analysis of information needs of senior users in Chap. 2.

In contrast, the analysis of the satisfaction regarding the usability of the system shows hardly any differences between younger and senior users (see Fig. 16.6). These results are further indicators of the main relevance of the consideration of the mobility information needs as an indicator for utility of the whole system. A suitable utility also transfers to a positive usability and acceptance of the system.

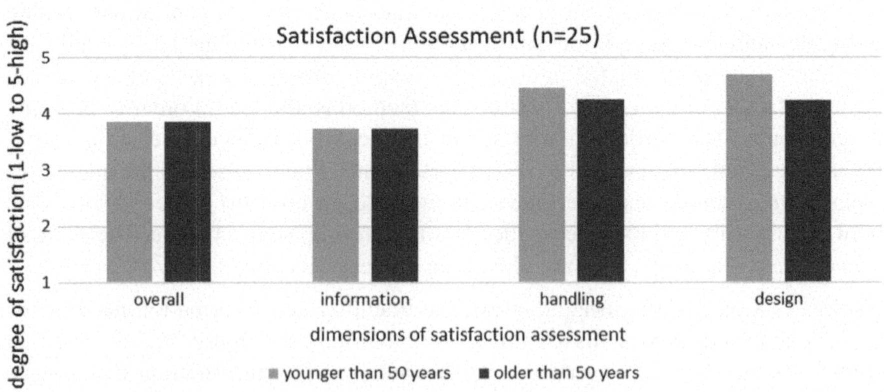


**Fig. 16.6** Assessment of satisfaction with interactive public displays

## 16.5 Recommendations

"How might the growing Internet of Things help older adults improve quality of life?" is one of the questions for the grand HCI challenge "Support successful aging Strategies" [19]. The authors' studies show that the mobility information needs of senior people can be successfully addressed with mobility information systems for all target groups. The derived development guidelines include flexible methods which can be applied for diverse mobility information systems and even further information systems for senior people. In order to address Shneidermans challenge, it is necessary to combine information and communications systems for several social needs of senior people, in order to strategically improve the quality of social life. The case study with public interactive displays shows that differences in utility hardly influence the usability for senior people. Transparency, consistency, and intelligibility [8] of information are important indicators for successful utility of mobility information systems in particular for senior people.

However, the mobility information at the vehicle, at the ticket vending machine, at the station, at loudspeaker announcements and also on the smartphone must be methodically developed in such a way that all age groups are taken into account. Based on identical sources of information, the information technologies should be closely linked to each other. This is the only way to make mobility information and technology an integral element that is used continuously throughout all phases of life.

**Acknowledgements** Part of this work was funded by the German Federal Ministry of Education and Research (BMBF) grant number 02K12A175 within the project Move@ÖV, the German Federal Ministry of Transport and Digital Infrastructure (BMVI) grant number 19E16007F and 19E16009F within the projects DIMO-OMP and DIMO-FuH, and the German Federal Ministry of Economy and Technology (BMWi) grant number 19P10003L and 19P12013B within the IP-KOM-ÖV und Dynapsys project.

## References

1. Biermann, H., Weißmantel, H.: Der SENSI-Katalog. Benutzerfreundliches und seniorengerechtes Design. http://www.emk.tu-darmstadt.de/~weissmantel/sensi/sensi.html (2003). Accessed 20 Jan 2019
2. Boren, T., Ramey, J.: Thinking aloud: reconciling theory and practice. Prof. Commun. IEEE Trans. **43–3**, 261–278 (2000). https://doi.org/10.1109/47.867942
3. Bruegge, B., Dutoit, A.H.: Object-Oriented Software Engineering—Using UML, Patterns and Java. Pearson, Boston (2010)
4. Bruggencate, T., et al.: Social needs of older people: a systematic literature review. Ageing Soc. **38**(9), 1745–1770 (2018). https://doi.org/10.1017/S0144686X17000150
5. Cooper, A.: The Inmates are Running the Asylum. Why High-Tech Products Drive Us Crazy and How to Restore the Sanity. Macmillan, New York (1999)
6. Goodwin, K.: Designing for the Digital Age. How to Create Human-Centered Products and Services. Wiley, Indianapolis (2009)

7. Davies R et al.: Older people in Europe—EU policies and programmes, 140811REV1. http://www.eprs.ep.parl.union.eu (2014). Accessed 18 Jan 2019
8. Hörold, S.: Instrumentarium zur Qualitätsevaluation von Mobilitätsinformation. Springer, Wiesbaden (2016)
9. Hörold, S., et al.: Identifying the information needs of users in public transport. In: Stanton, N.A. (ed.) Advances in Human Aspect of Road and Rail Transportation, pp. 331–340. CRC Press, Boca Raton (2012)
10. Hörold, S., et al.: Interactive displays in public transport—challenges and expectations. Procedia Manufact. **3**, 2808–2815 (2015). https://doi.org/10.1016/j.promfg.2015.07.932
11. Infas Institut für angewandte Sozialwissenschaft GmbH.: Mobilität in Deutschland. Kurzreport. http://www.mobilitaet-in-deutschland.de/pdf/infas_Mobilitaet_in_Deutschland_2017_Kurzreport.pdf (2018). Accessed 30 Jan 2019
12. ISO International Organization for Standardization: DIN EN ISO 9241-210:2010: Ergonomics of human-system interaction—Part 210: Human-centred design for interactive systems, German version. Beuth Verlag GmbH, Berlin (2010)
13. Mayas, C. et al.: Das Begleitheft für den Entwicklungsprozess. Personas, Szenarios und Anwendungsfälle aus AK2 und AK3 des Projektes IP-KOM-ÖV. urn:nbn:de:gbv:ilm1-2012200028 (2012)
14. Mayas, C., et al.: Interactive public displays for paperless mobility stations. In: Kurosu, M. (ed.) Human-Computer Interaction. Interaction in Context. HCI 2018. Lecture Notes in Computer Science, vol. 10902, pp. 542–551. Springer, Cham (2018)
15. Mayhew, D.J.: The Usability Engineering Lifecycle: A Practitioner's Handbook for User Interface Design. Morgan Kaufmann Publishers Inc., San Francisco (1999)
16. Meyer, S., et al.: Das intelligente Haus—selbständige Lebensführung im Alter. Möglichkeiten und Grenzen vernetzter Technik im Haushalt alter Mensche. Campus Verlag, Frankfurt (1997)
17. Rosson, M.B., Carroll, J.M.: Usability Engineering: Scenario-based Development of Human-Computer Interaction. Morgan Kaufmann Publishers Inc., Burlington (2002)
18. Sanderson, W.C., Scherbov, S.: Faster increases in human life expectancy could lead to slower population. Aging. PLoS ONE **10**(4), e0121922 (2015). https://doi.org/10.1371/journal.pone.0121922
19. Shneiderman, B., et al.: Grand Challenges for HCI Researchers. Interactions **23**(5), 24–25 (2016). https://doi.org/10.1145/2977645
20. Statista—Das Statistik-Portal.: Durchschnittliches reales und gesetzliches Renteneintrittsalter sowie durchschnittliche Lebenserwartung in ausgewählten europäischen Ländern (in Jahren). https://de.statista.com/statistik/daten/studie/173132/umfrage/renteneintrittsalter-in-europa/ (2011). Accessed 30 Jan 2019
21. Sträter, O., et al.: Qualitätskriterien im Umfeld von AAL—Produkte, Dienstleistungen, Systeme. VDE VERLAG, Berlin (2011)
22. United Nations UN.: Convention on the rights of persons with disabilities. https://www.behindertenrechtskonvention.info/vertragstexte (2006). Accessed 30 Jan 2019
23. Verband Deutscher Verkehrsunternehmen VDV: VDV-Report 7036: User Interface Design für die elektronische Aushanginformation. Beka Verlag, Cologne (2014)
24. Wienken, T., Krömker, H.: Experience maps for mobility. In: Kurosu, M. (ed.) Human-Computer Interaction in Context. HCI 2018. Lecture Notes in Computer Science, vol. 10902, pp. 615–627. Springer, Cham (2018)
25. Zemlin, B.: Das Entscheidungsverhalten bei der Verkehrsmittelwahl. EUL VERLAG, Köln (2005)
26. Zomerdijk, L.G., Voss, C.A.: Service design for experience-centric services. J. Serv. Res. **13**(1), 67–82 (2010)

**Cindy Mayas** works as a scientific assistant at the Media Production Group of the Technische Universität Ilmenau since 2009. After her university degree in Applied Media Science, she focusses on research projects for the development of user-centered mobility information services and digital knowledge spaces for various user groups. Her special research interest is the improvement of the integration of usability engineering methods for development processes and project management.

**Heidi Krömker** is Director of the Institute for Media Technology at the University of Ilmenau and holds a professorship for "Human-Technology Interaction" with the research focus "User-centered Mobility Information". Prof. Krömker investigates the user-centered design of mobility. This includes the collection of requirements, prototyping, evaluation and analysis of quality and acceptance factors of intermodal mobility. She headed the international User Interface Design Center of Siemens Corporate Technology in Princeton NY (USA), Beijing (China) and Munich (Germany) from 1995 to 2001. After studying sociology, she led various projects in Siemens Corporate Technology, such as the "Introduction of the usability concept in the Siemens Group". She received her Ph.D. in the field of user interface design from the University of Bamberg, Germany.

# Part VI
# Assistive Technology and Dementia

# Chapter 17
# Improving the Quality of Life of Individuals with Dementia Using Personal Digital Media

**Kanvar Nayer and Selby Coxon**

**Abstract** Older populations are prone to chronic degenerative diseases, amongst which dementia, in its various forms, is on a steep rise [1–3]. Dementia deprives individuals of many cognitive faculties; loss of short-term memory and agitated behaviours are two common manifestations of the disease. Persons with dementia may find themselves depressed, bored, isolated and lonely, leading to behaviours that are challenging for carers to manage. Medication is used as the primary method of treating the symptoms of dementia. More than 40% of people with dementia at aged-care facilities are being prescribed unnecessary medication [4], which leads to reduced well-being and quality of life [5] and, may even accelerate cognitive decline [6]. There is broadening evidence of the efficacy of non-medicinal interventions to alleviate the stressful effects of dementia [7], in particular the positive stimulus of interacting with music and digitally mediated imagery. The challenge presented by these types of intervention is how to make them easily accessible to, and operable by those with cognitive impairment. This chapter describes the research and development of a touchscreen-based interface providing a variety of media experiences from music and photographs to family messages, all capable of being autonomously interacted with by elderly individuals.

**Keywords** Improving quality of life · People with dementia · Multimedia system · Personalised media · Digital media · Non-medicinal intervention · Independence · Self-reliance · Reduce staff burden · Family members · Dedicated form factor · Touchscreen · Infra-red technology · Information architecture · Interface design · Tactile buttons · Alleviating symptoms of dementia · Depression · Boredom · Loneliness · Isolation · Aged-care facility

---

K. Nayer (✉) · S. Coxon
Faculty of Art, Design and Architecture, Monash University, Melbourne, VIC 3145, Australia
e-mail: kanvar.nayer@gmail.com



A. Woodcock et al. (eds.), *Design of Assistive Technology for Ageing Populations*, Intelligent Systems Reference Library 167,
https://doi.org/10.1007/978-3-030-26292-1_17

## 17.1 Introduction

The term 'dementia' describes a number of conditions caused by neuro-degeneration in excess of that resulting from natural ageing [2]. It can occur at any stage of adulthood but is far more common in the geriatric population [8]. Approximately 10% of adults over the age of 65 and 50% of those over 90 are affected [9–11]. Most of the conditions responsible for late-life dementia are progressive, degenerative and irreversible [12]. In 2010, the population of people with dementia was 35.6 million; this number is estimated to nearly double every 20 years to approximately 65.7 million in 2030 and to 115.4 million by 2050 [13]. The number of people diagnosed with dementia increases by one new person every 15 s [1–3].

Dementia has many symptoms. The most common initial symptom is a decline in the ability to remember new information. The International Psychogeriatric Association (IPA) has grouped the symptoms of dementia under the umbrella term 'behavioural and psychological symptoms of dementia', also referred to as BPSD [14]. These may vary depending on the level of impairment. An individual with mild dementia may experience one or two symptoms such as misplacing items or repetitive questioning that have a relatively minor effect on day-to-day living. Those with moderate or severe dementia may experience a greater number of symptoms such as agitation or aggression that require constant oversight. Agitation and wandering have been reported as the most enduring symptoms of dementia [15]. The behavioural symptoms of dementia include wandering, screaming and cursing, restlessness, culturally inappropriate behaviours and sexual disinhibition. The psychological symptoms of dementia include depressive mood, anxiety, delusions and hallucinations. Other common symptoms (neither behavioural nor psychological) include loss of memory, confusion, loss of language and communication skills, loss of ability to perform everyday tasks, disorientation in relation to time and space, impaired judgement and physical coordination due to muscle rigidity, learning and concentration difficulties, altered sleeping patterns and eating disturbances [12, 16, 17]. BPSD makes life very difficult for affected individuals, family members and caregivers [18, 14]. Untreated BPSD may contribute to premature institutionalisation [19], increased financial cost [20], reduced quality of life for both caregivers and individuals with dementia [18], and significant caregiver stress at aged-care facilities [21, 22].

Moving from the comfort of family homes to residential care can be very challenging. The sudden change of routine, loss of self-reliance and lack of familiarity in a new environment often result in depression, which can affect a resident's quality of life and increase the risk of cognitive impairment [23, 24]. The situation may be compounded by family members gradually finding it more difficult to communicate with their loved ones in the facility. Shortfalls in staffing levels due to low pay and lack of adequately trained personnel may result in limited personal attention, assistance and social stimulation for residents [25], creating a sense of boredom and isolation and/or loneliness in the interned population [26, 25]. This leads to a lack of meaningful activities at aged-care facilities [27, 28]. By providing more stimuli and activities, the quality of life of people with dementia could be improved [29].

Even though there is currently no cure for dementia [1], medication is prescribed for BPSD. Despite evidence of high placebo response rates, medication continues to be the primary form of treatment for BPSD [30, 31]. In the developed world, more than 40% of people with dementia are being prescribed unnecessary medication. Prescribing these drugs without first considering other treatment options can be of particular concern [4]; they may produce adverse side effects such as sedation, falls, reduced well-being, reduced quality of life [5], and even cognitive decline [6]. Other side effects include drowsiness, shaking and unsteadiness, increased risk of infections, blood clots, stroke, worsening of other dementia symptoms and possibly an increased risk of death [5, 6]. Drugs may help reduce BPSD, but at the expense of a person's quality of life [30, 31]. According to Alzheimer's Society [32], all medication used for treating the symptoms of dementia has side effects that may worsen those symptoms. Douglas et al. [31] point to an increasing recognition for medicinal treatments to be used as a second-line approach and for non-medicinal options, in best practice, to be pursued first.

O'Connor et al. [7] reported that non-medicinal therapies, in particular those tailored to participants' backgrounds, interests and skills, were effective in reducing some symptoms of dementia. A literature review reported that reminiscence therapy was most effective when using photographs [33–40], personalised music was more effective in engaging people with dementia than randomly selected music [41–45], videotaped simulated presence was far more effective in reducing agitation than audiotaped simulated presence [46], one-on-one interaction was more effective in reducing verbal agitation than personalised music or simulated presence, and a combination of therapies was more effective in reducing some symptoms of dementia than any one therapy by itself [47]. The literature also reported that touchscreen technology has potential in providing a degree of autonomy to people with dementia [33, 48].

Based on this evidence, research carried out at Monash University by the primary author resulted in a non-medicinal intervention to address the challenges posed by the common symptoms of dementia in residents at aged-care facilities, in particular loss of self-reliance, depression, boredom and isolation and/or loneliness.

Nowadays, computer vision techniques play a significant role in many applications. In spite of pioneer investigations in remote sensing since the 1970s, the development of theories and algorithms for this purpose remains the crucial issue yet.

## 17.2 A Personalised Multimedia System for Persons with Dementia

There is a growing acceptance for the use of technology in activities that provide reminiscence. However, an ageing cohort has a different attitude towards technology than the technologically experienced younger generation. This is due mainly to the impact of age-related physical and perceptual problems which include loss of hearing, loss of sight, cognitive impairment, loss of physical strength, and loss of manual dexterity [49]. Upton et al. [50] suggest the importance of designers anticipating such difficulties experienced by older people. These age-related challenges encourage design responses that provide sufficient auditory amplification, screen illumination, large visual detail and the need for users to be seated if a screen-based non-medicinal intervention is to be used by older people.

Early field visits revealed that aged-care facilities had their own sets of restrictions. These included restrictions on any devices being suspended from the ceiling, screwed onto the wall or bolted into the floor, suggesting the need for a non-obtrusive, free-standing physical device. Usability experiments with residents with mild or moderate dementia highlighted significant challenges with a touchscreen tablet's hardware and software. The results of these experiments implied that people with dementia would benefit from tablets with larger screens that were securely mounted and adjusted for height. One resident with Parkinson's disease had difficulty in activating and thus using the tablet's touchscreen. The introduction of tactile buttons that corresponded to the digital selections on the screen quickly addressed this problem. A study by Favilla and Pedel [51] reported that some people with dementia had difficulty using an iPad due to a lack of moisture on the surface of their fingertips or because they used their fingernails. 'Tactile agnosia' is the inability to recognise objects by touch. It can affect interaction with standard touchscreen tablets—a majority of which employ capacitive touchscreen technology—because of too much or too little pressure exerted on the screen. This presented the opportunity to explore alternative technologies that may better cater to people with dementia, infra-red being one such technology that addressed the issues of tactile agnosia and accommodated fingernail contact.

Given the residents' varying levels of impairment, the usability tests demonstrated the need for specially considered information architecture of varied complexity to provide autonomy to people with mild, moderate or severe dementia while maintaining an immersive calming experience or 'flow'. Flow is the deep cognitive involvement in an activity that makes a person lose track of time [52]. In order to maintain flow, the level of challenge of an activity should match a person's ability. If the challenge is greater than the ability, the person becomes overwhelmed and anxious, whereas if it is lower, the person gets bored. It is therefore important to have the perfect balance when designing an intended mentally immersive activity [53].

The interaction with the screen media is managed by the hierarchy of the information architecture to have the right ratio of depth and breadth. Depth refers to the number of steps in the information architecture, while breadth refers to the number of options at each level. When considering breadth, it is important to ensure that information overload on users is avoided, content is grouped and organised where possible, and designs are subjected to rigorous testing [54].

It was apparent from the aged-care facility visits that some individuals with mild dementia were able to navigate through information architecture of greater depth, while others with mild or moderate dementia were only able to get as deep as one level. While the information architecture was not tested on people with severe dementia, it was presumed that they, too, would require a maximum depth of one. People with dementia could therefore benefit from a multimedia system that would provide both 'narrow and deep' and 'broad and shallow' hierarchies. As the name suggests, a narrow and deep hierarchy provides users with a minimal number of selections on a homepage (as few as one) and requires multiple steps to access its deepest content; while a broad and shallow hierarchy provides users with a greater number of selections on a homepage and requires fewer steps (as few as one) to access its deepest content [54].

The narrow and deep hierarchy may better cater to people with mild dementia who would have more control and options available to them, and a broad and shallow hierarchy may better cater to people with moderate and severe dementia where the deepest content would be just one click away. For purposes of the research, the narrow and deep hierarchy was called 'level 1 architecture' and the broad and shallow 'level 2 architecture'.

## 17.3 The Information Architecture

The level 1 architecture catered to people with mild dementia. It provided a narrow and deep hierarchy and a top-down approach where users could discover content by navigating through the depth of hierarchy at their convenience. A review of literature suggested that people with dementia did not cope well with multiple sources of information [33]. Given that the multimedia system aimed at presenting four media types (thus, a breadth of four), the level 1 architecture limited its breadth to a maximum of four selections at any depth. In this way, users would be able to navigate through greater depth without a reduction in flow. However, it was important to monitor depth for each user to ensure that flow was not lost before the media was accessed. Figure 17.1 displays an example of the architecture from the four media types (homepage) to media output/format (media operation). The levels between the homepage and media operation were referred to as 'folder', 'sub-folder' and 'file', these determining the architecture depth (four levels in this example).

The level 2 architecture used a broad and shallow hierarchy with the aim of assigning a low level of complexity to people with moderate or severe dementia (and to people with mild dementia who were unable to use the level 1 architecture). It used a bottom-up approach where users could access media instantly, without going through the folder, sub-folder and file levels between the homepage and media operation. This is shown in Fig. 17.2.

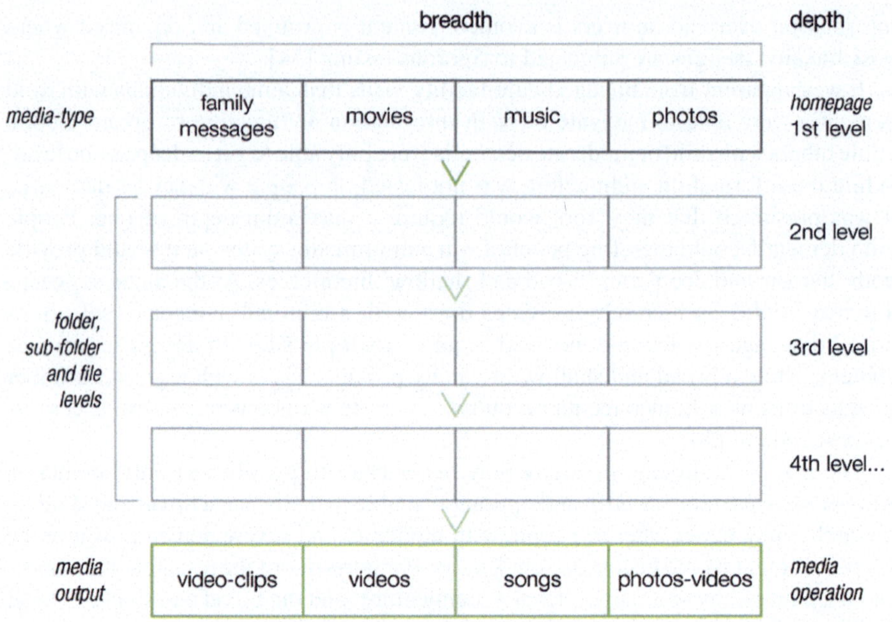


**Fig. 17.1** The level 1 architecture

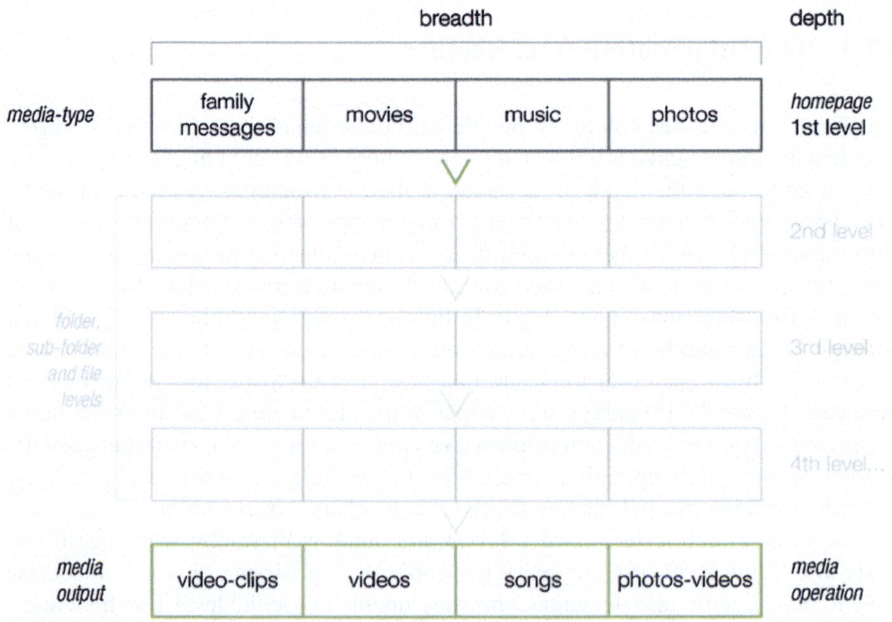


**Fig. 17.2** The level 2 architecture

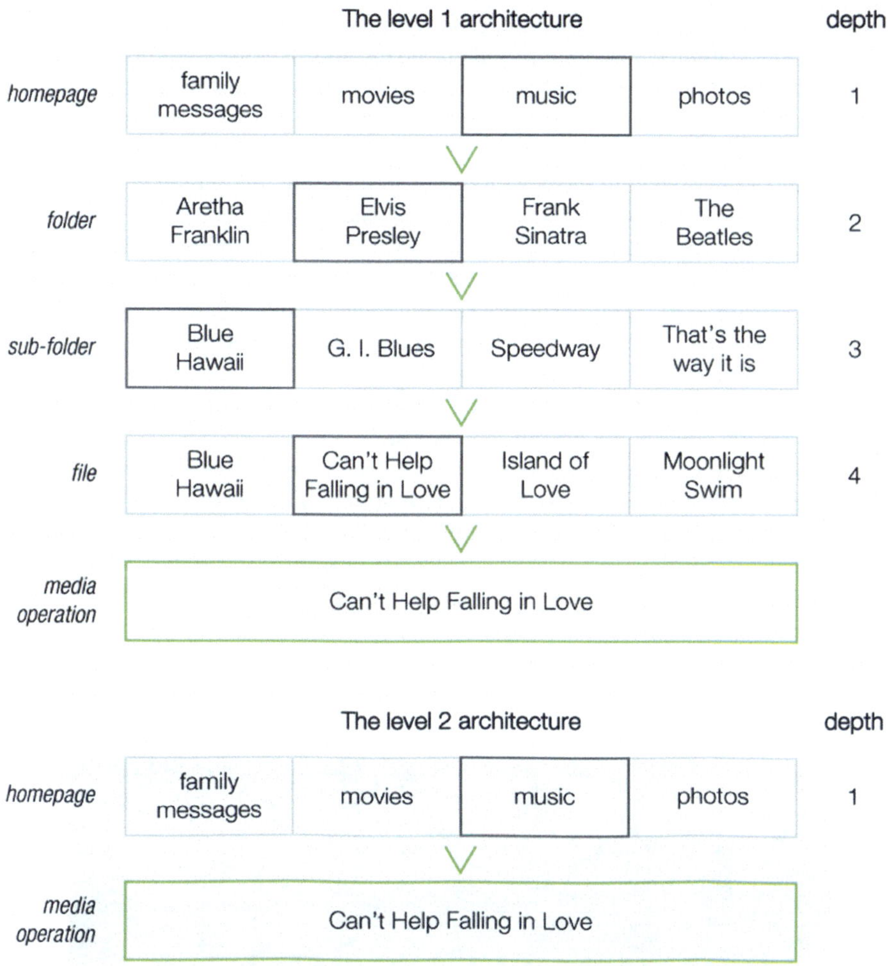


**Fig. 17.3** Example of the level 1 versus level 2 architectures when music is selected

As Fig. 17.2 shows, users would be presented with the four media types on the homepage, similar to the homepage in level 1. However, on media selection, users were instantly rewarded with media operation. Figure 17.3 shows the difference between the level 1 and level 2 architectures when 'music' is selected.

## 17.4 The Interface Design

During usability tests an infra-red touchscreen (23-in.) addressed the problems associated with tactile agnosia. This technology was far less 'slippery' or 'fast' compared with more common tablet hardware. Due to age-related physical impairment, all selections were aligned horizontally and placed towards the lower half of the screen's base for easy and comfortable access. Figure 17.4 shows the infra-red touchscreen and the defined area within which all media selections and controls (both digital and tactile) were positioned.

To ensure consistency in the design of the architecture, the screen provided designated areas for all media selection, controls and displays presented to users at any level. These consisted of the four media selections, an optional 'More' selection, 'Stop' and 'Pause' controls, time and date, media operation and tactile buttons. According to Jin et al. [55], designers should consider spacing adjacent selections between 6.35 and 12.7 mm apart for enhanced accuracy when designing interfaces for older people; zero spacing reported the lowest accuracy and preference among participants in a study [55]. A spacing of 9.5 mm between media selections and an aspect ratio of 4:3 for media operation were used.

Media selections were the dominant elements of the information architecture and were therefore positioned at the bottom-centre of the screen. The 'Pause' and 'Stop' controls (sub-dominant elements) sat at either end of the media selections, and the time and date (subordinate element) were placed towards the top of the screen as they did not require to be accessed. Figures 17.5 and 17.6 display the multimedia system's visual hierarchy and designated areas where these elements were positioned.

The tactile buttons were also positioned towards the base of the screen for enhanced comfort. To distinguish between media selections and controls, buttons of

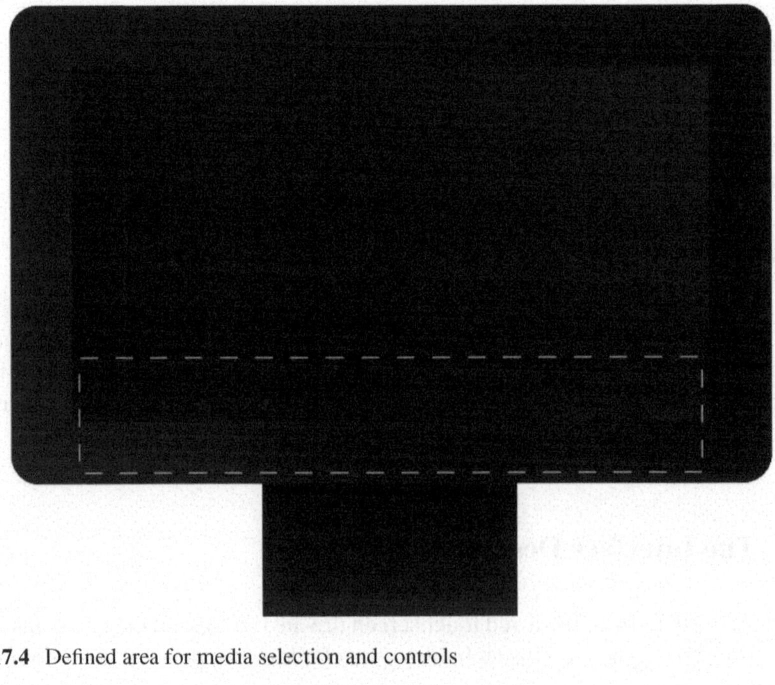

**Fig. 17.4** Defined area for media selection and controls

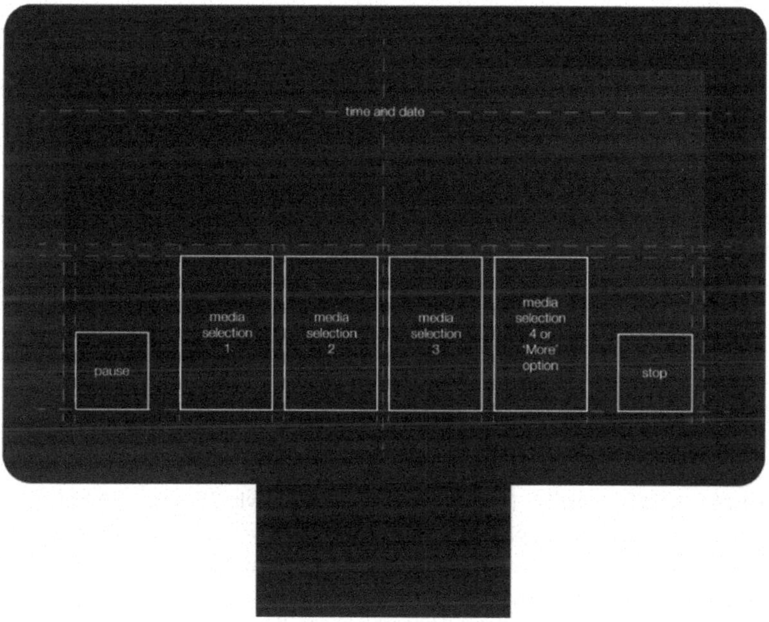


**Fig. 17.5** Visual hierarchy

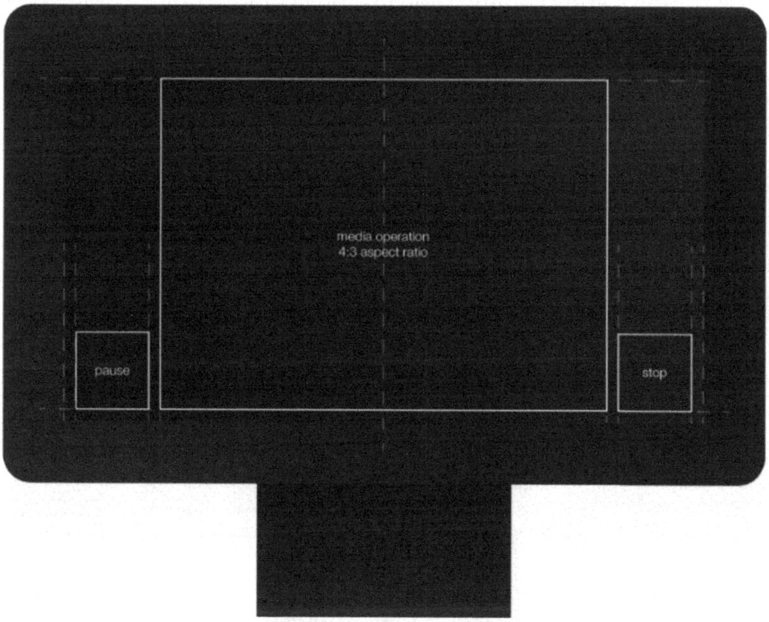


**Fig. 17.6** Designated areas for media selection, controls and operation

different shape were used; rounded-square buttons corresponded to media selections and circular buttons to media controls. The buttons had a force range of 0.30–0.48 N and a travel range of 0.2–3.9 mm based on results from a force-travel review on push buttons for older people [56, 57]. A volume knob with diameter 38 mm and a torque lower than 0.037 N m was positioned below the media selection and control tactile buttons [58]. This is shown in Fig. 17.7.

The media selection and control tactile buttons were manufactured using a translucent material that glowed due to a hard-wired back-lit assembly. This allowed for a button to be blanked out if there were fewer than four media selections on the screen (for example, when a user did not have enough media uploaded in a particular media category), thereby not distracting from valid, illuminated options.

The multimedia system's homepage consisted of the four selected media types (family messages, movies, music and photographs) and the time and date. As the system aimed to provide personalised media to people with dementia, to enhance personalisation and provide a sense of ownership to users, the media names included the prefix 'My': 'My Family' (for family messages), 'My Movies', 'My Music' and 'My Photographs'. The homepage, being the first level of the information architecture depth, did not require any media controls. Therefore, the tactile control buttons were not illuminated. This is presented in Fig. 17.8.

Astell et al. [33] reported that people with dementia were able to recognise vector-based representations of a music player, encouraging the use of this styling in the multimedia system. During the initial usability tests, when asked to select media icons

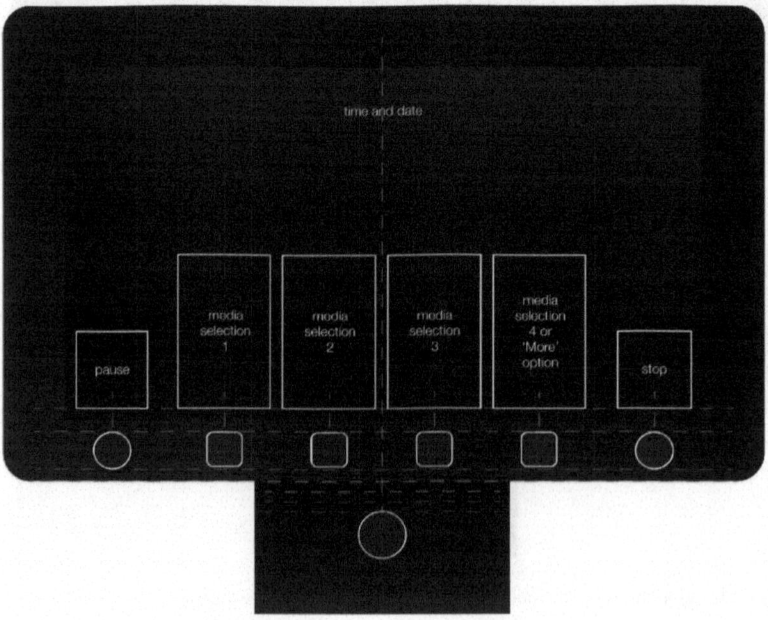


**Fig. 17.7** Placement of tactile buttons

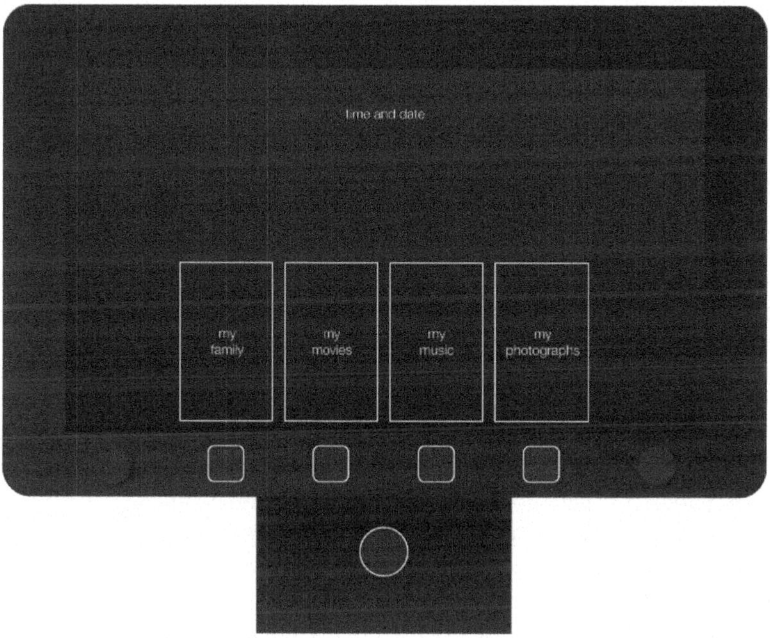


**Fig. 17.8** Information on the homepage

that best represented media types, all participants selected familiar and traditional representations that dated back approximately fifty years. Although a small sample size, the results supported the theory that long-term memories were better preserved than short-term ones [33, 59]. Therefore, the following traditional representations were designed for the four types of media (Fig. 17.9).

The literature review reported that a combination of titles/text and associated photographs may have provided more options for a participant to better recognise a photo-video given the challenges of text legibility [60]. Therefore, each media

**Fig. 17.9** Vector-based media icons

selection consisted of a photograph/icon and its associated text. According to Bernard et al. [61], older people perceive sans serif fonts more easily than serif fonts. When presenting text to the elderly on a screen, a minimum font size of 14 points should be considered, as individual characters at this size are easier to perceive [62]. Based on this knowledge, the multimedia system employed a 30-point sans serif font (Helvetica Neue). Figure 17.10 shows the combination of the media icons and their associated text.

Due to macular degeneration, it is important that media selections on the screen are clearly perceivable by users. Figure 17.11 presents the media selections in achromatic colours to contrast them against the black background.

While the media selections in Fig. 17.11 provide more contrast than those in Fig. 17.10, there remains a lack in contrast between each media selection. Figure 17.12 shows a digital colour-wheel where tetrad colours (four colours spaced evenly on a colour wheel) are explored and a desired range integrated in the media selections. Complementary colours are placed next to one another (blue–orange, green–pink), dividing analogous ones (blue–green, orange–pink).

Figure 17.13 presents the homepage interface design based on the visual hierarchy, icon development, review of typography and colour selections in the previous sections.

The time and date were presented using the format: time (12-h), day, date, month and year. This layout was adjustable via a secondary interface to suit a user's preference. In addition, the homepage included animated voice prompts that guided a user through the information architecture of the multimedia system, with the intention of enhancing the overall user experience (for users without hearing impairment). The voice prompt issued the following message in the user's preferred language: 'Please make a selection from one of the options below', after-which each media selection flashed one after the other, acting as a visual reminder [63]. According to Nass and Yen [64], the human brain prefers a female voice over a male voice and it is easier to find a female voice that people prefer than a male voice. Therefore, the multimedia system used a female voice for its voice prompts. The voice prompt repeated itself every half hour if no selections were made on the homepage. The homepage also included media selection feedback which, according to a study by Koskinen [65], reported that despite the type of technology used, it is important to receive feedback

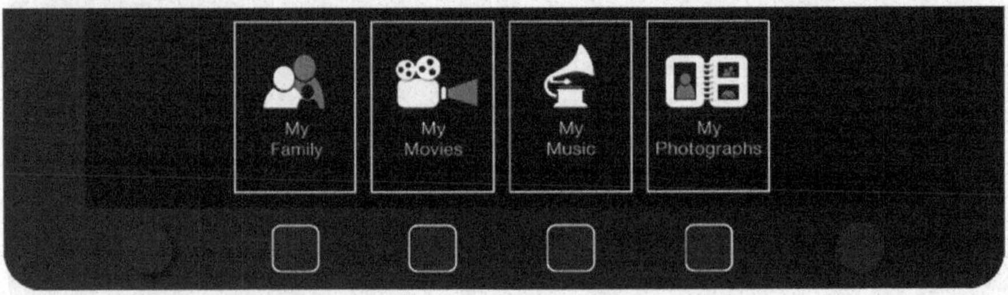


**Fig. 17.10** Combination of media icons and associated text

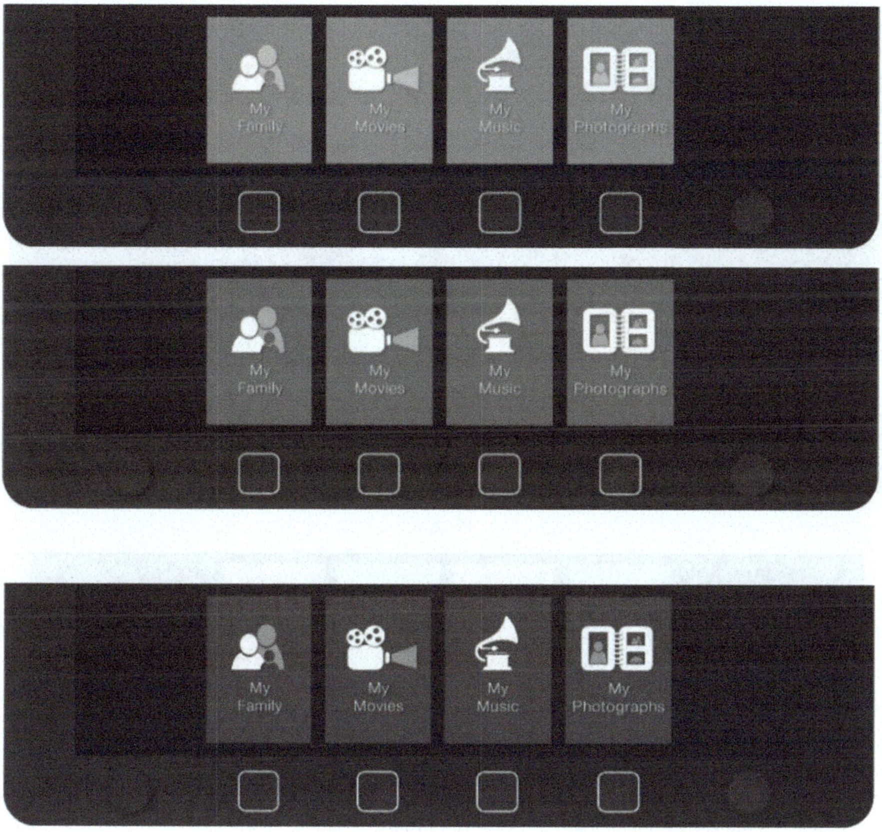


**Fig. 17.11** Media selections colour test

on successful registration of tactile action. Therefore, a 'click' sound was generated every time a selection was made.

The multimedia system's folder, sub-folder and file levels consisted of:

- Four media selections—the media selections used the same layout as that of the homepage; the icons, however, were replaced by photographs relating directly to their folder, sub-folder or file levels. To provide users with optimal engagement, a media file was not accepted on the secondary interface of any media type unless an associated image was uploaded alongside it.
- Page title—the title of the page appeared in place of the time and date to remind users where they were in the information architecture.
- 'More' option—where there were more than four media options for a particular media type, the 'More' option presented itself as the fourth media selection.
- 'Stop' option—the 'Stop' option was illuminated, taking users back to the homepage when selected.
- Animated voice prompts—these were played once, upon landing on a page.

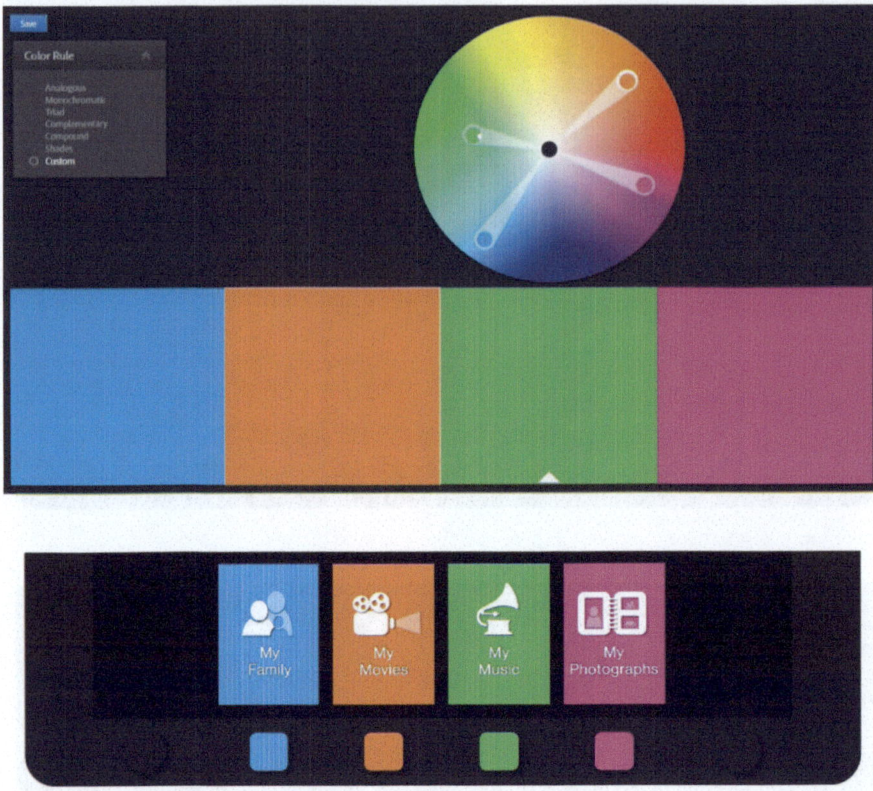


**Fig. 17.12** Media selections in a tetrad range of colours

- Media selection feedback—this was the same as on the homepage.
- 30-s selection period—if no media selection was made within 30 s, the system reverted to the homepage.

Figure 17.14 presents the interface design using an example of the 'My Movies' option being selected. The movies displayed were selected from the public domain database. Rather than colour-coding each media-type, for example blue for 'My Family', each depth used the four colours within its breadth to provide contrast and to allow any page to act as a homepage in a scenario where users forgot where they were.

On the first screen in Fig. 17.14, three movie files are presented, suggesting that only three movie files have been uploaded on the system. Therefore, the area where a fourth media selection would normally appear is blanked out and its corresponding tactile button not illuminated. On the centre screen, four movie files are available, suggesting only four movie files have been uploaded on the system. Finally, on the third screen, the appearance of the 'More' option suggests that a user has more than four movie files available on the system. Figure 17.15 shows the information

**Fig. 17.13** The homepage interface

**Fig. 17.14** The folder, sub-folder and file level interfaces

architecture behaviour when the 'More' option is selected repeatedly, assuming five movies are available on the system, while Fig. 17.16 shows the interface during movie operation on the level 1 and level 2 information architectures.

Figure 17.17 shows the interface during music operation on the level 1 and level 2 information architectures.

During music operation, the system presented a combination of audio and visual components to users. If a file was purely aural (no accompanying video-clip), the visual component was a spinning record with the artist's/album's image cropped onto it to create a sense of where the music was coming from [33]. The level 1 interface also presented the other media selections, while the level 2 interface did not.

The photo-video interface had the same layout as that of family messages and movies. Each photograph in a photo-video was displayed for 15 s before gently

Fig. 17.15 The 'More' option in operation

Fig. 17.16 Movie operation interface

fading into the next one, as positively reported in a study by Yasuda et al. [66]. It also had a pan and zoom effect as shown in Fig. 17.18.

## 17.5 Testing the Efficacy of Multimedia Interaction

The purpose of creating and developing a digital media interface was to determine the feasibility of providing a high degree of autonomy to people with a range of levels of dementia, and to measure if it improved quality of life for residents and caregivers at care homes. Usability testing was carried out with a group of eleven

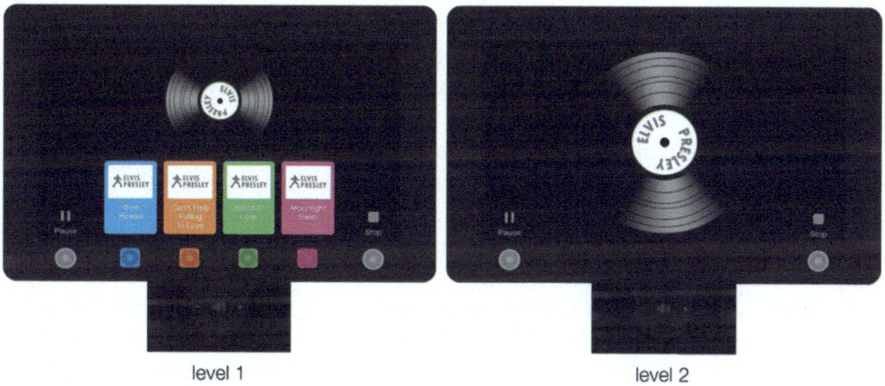


**Fig. 17.17** Music operation interface

**Fig. 17.18** Photo-video interface

residents (degree of dementia: 1 mild, 9 moderate and 1 severe) who had consented to participate in the tests, some having their families consent for them. All residents were first introduced to the level 1 interface a few hours prior to the tests. For the residents for whom flow was not achieved, the level 2 interface was engaged instead.

In addition to the multimedia system intervention, a social control intervention was conducted for half an hour each week during the testing period. During these sessions, a psychologist read from a newspaper and discussed current affairs with each resident. The aim of this intervention was to provide non-personalised social engagement, this being contradictory to the aim of the multimedia system intervention.

Results of the multimedia system intervention showed that all residents were able to use the system with a high degree of autonomy via its adaptable interfaces; the 1 resident with mild dementia and the 6 with moderate dementia were able to use the level 1 interface. In addition: symptoms of depression decreased significantly during the multimedia system intervention; symptoms of anxiety also decreased significantly; quality of life rated by the residents, caregivers and family members all increased slightly during the multimedia system intervention; and scores on staff burden decreased during the multimedia system intervention [67]. These are shown in Table 17.1.

With the social control intervention, there were no statistically significant differences before and after the intervention, with the exception of QOL–AD as rated by caregivers which decreased during the social control intervention. There was a small decrease in scores on the CMAI, RAID, and staff-rated Burden Interview, but these changes were insignificant (Table 17.2).

Residents with dementia found the multimedia system to be enjoyable and pleasurable, and engaged well with it during training sessions, thereby indicating high levels of satisfaction with the experience. When the multimedia systems were placed in their bedrooms for the testing period, the residents with dementia were reported to use them on most days of the trial, according to automatic logging of their run-times by the systems. Units recorded a daily run time of between 25 min and 5 h, with a mean of 2.6 h. Family members and caregivers reported high levels of enjoyment when observing the residents using the systems. The ability to access favourite music and photographs reportedly caused great delight in many residents. Some content of the multimedia systems was not found interesting by two of the eleven participants, which may have resulted from the devices not being set up optimally for them. These reports, however, were inconsistent with their family members' positive observations.

Many family members of the residents were optimistic about the multimedia system, expressing a desire to purchase the device if and when it became available. Residents commented mostly on the functionality of the multimedia system, while their family members talked about its usability and its impact on the residents. They

**Table 17.1** Multimedia system intervention scores [67]

| Outcome measure | Pre-memory box mean (SD) | Post-memory box mean (SD) | Post-memory box minus pre-memory box (SD) | $t$ |
|---|---|---|---|---|
| CMAI | 45.82 (19.39) | 37.27 (9.86) | −8.55 (13.70) | −2.07 |
| QOL-AD resident rated | 42.36 (5.54) | 43.09 (4.76) | 0.73 (3.74) | 0.64 |
| QOL-AD staff rated | 34.82 (6.37) | 35.55 (8.14) | 0.73 (5.04) | 0.48 |
| QOL-AD family rated | 38.90 (6.49) | 41.10 (7.25) | 2.20 (4.44) | 1.57 |
| Cornell scale | 5.73 (3.10) | 3.45 (2.54) | −2.28 (2.24) | $-3.36^1$ |
| RAID | 6.64 (4.18) | 3.18 (2.93) | −3.46 (4.03) | $-2.84^2$ |
| Staff Burden Interview | 8.00 (5.93) | 7.09 (3.96) | −0.91 (4.83) | −0.62 |

*Note* $^1p < 0.01$, $^2p < 0.05$

**Table 17.2** Social control intervention scores [67]

| Outcome measure | Pre-control mean (SD) | Post-control mean (SD) | Post-control minus control score (SD) | *t* |
|---|---|---|---|---|
| CMAI | 42.82 (9.51) | 40.09 (9.19) | −2.73 (6.78) | −1.33 |
| QOL-AD resident rated | 43.09 (4.01) | 42.09 (6.41) | −1.00 (4.84) | −0.69 |
| QOL-AD staff rated | 36.18 (8.23) | 32.27 (7.93) | −3.91 (4.74) | −2.73[1] |
| QOL-AD family rated | 39.09 (5.97) | 39.45 (6.61) | 0.36 (4.72) | 0.26 |
| Cornell scale | 5.18 (5.95) | 5.36 (2.20) | 0.18 (5.33) | 0.11 |
| RAID | 7.55 (7.54) | 4.64 (3.38) | −2.91 (6.33) | −1.52 |
| Staff Burden Interview | 8.09 (6.67) | 5.91 (4.42) | −2.18 (5.84) | −1.24 |

*Note* [1]$p < 0.05$

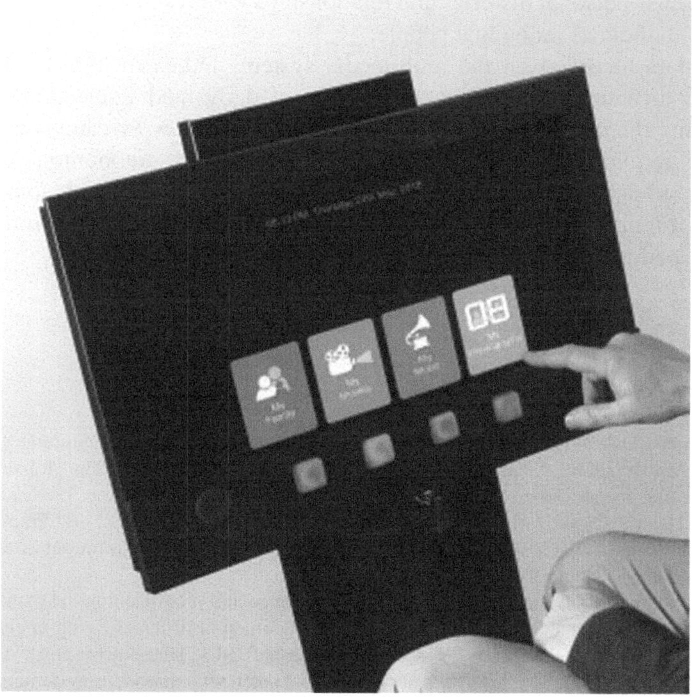

**Fig. 17.19** Multimedia system in use

described it as normalising, motivating, engaging, providing companionship, and facilitating adjustment to the new environment. They reported that access to media, especially photographs and music, triggered pleasant memories, and the ease of operating the system afforded the residents a sense of control. After the multimedia system trial, some family members perceived positive changes in the residents' well-being and motivation. This perception could not be confirmed, as only data on cognitive impairment at baseline was collected. Residents and caregivers believed the multimedia system to be beneficial to residents who had tended to stay in their rooms and may have been socially isolated. Those who were active outdoors during the day were believed to find it less beneficial.

In addition to its primary aim of facilitating easy access to personalised media for persons with dementia, the multimedia system provided some unexpected fringe benefits of a social nature. The residents felt that the multimedia system enhanced their engagement with family members by providing new conversation topics and opportunities to reminisce. Several residents appeared to enjoy taking visitors through the media on their devices. Caregivers found that the multimedia system allowed them to discuss personal histories and relationships with residents and become more familiar with them as individual people.

While data recorded on the multimedia systems indicated high usage, it was difficult to differentiate between the number of daily media selections made by the residents themselves and those made by their families or caregivers. Further research is required in identifying residents' capacity for autonomous use of the multimedia system and facilitating regular use by those who require some level of assistance. Figure 17.19 presents an image of the multimedia system prototype that was developed for the usability tests.

## References


1. Alzheimer's Disease International: World Alzheimer's Report Executive Summary (2009)
2. Prince, M., Bryce, R., Albanese, E., Wimo, A., Ribeiro, W., Ferri, C.P.: The global prevalence of dementia: a systematic review and meta-analysis. 68–69 (2013)
3. World Health Organization (WHO): Women, Ageing and Health: A Framework for Action: Focus on Gender. Available from www.whqlibdoc.who.int/publications/2007/9789241563529_eng.pdf (2007)
4. Margallo-Lana, M., Swann, A., O'Brien, J.: Prevalence and pharmacological management of behavioural and psychological symptoms amongst dementia sufferers living in care environments. Int. J. Geriatr. Psychiatry 39–44 (2001). Marshall, M.J., Hutchinson, S.A.: A critique of research on the use of activities with persons with Alzheimer's disease: a systematic literature review. JAN: Inform. Pract. Policy Worldwide Through Res. Sch. 488–496 (2001)
5. Ballard, C., McKeith, L., Harrison, R., O'Brien, J., Thompson, P., Lowery, K., Perry, R., Ince, P.: Dementia with Lewy bodies and Alzheimer's disease. Int. Psychogeriatr. 381–388 (1997)
6. McShane, R., Keene, J., Gedling, K.: Do neuroleptic drugs hasten cognitive decline in dementia? Prospective study with necropsy follow-up. BMJ 211–212 (1997)
7. O'Connor, D.W., Ames, D., Gardner, B., King, M.: Psychosocial treatments of behavior symptoms in dementia: a systematic review of reports meeting quality standards. Int. Psychogeriatr. 225–240 (2009). https://doi.org/10.1017/s1041610208007588

8. Graham, J.E., Rockwood, K., Beattie, L.B., Eastwood, R., Gauthier, S., Tuokko, H., McDowell, I.: Prevalence and severity of cognitive impairment with and without dementia in an elderly population. Lancet 1793–1796 (1997)
9. Bäckman, L., Small, B.J., Wahlin, Å., Larsson, M.: Cognitive functioning. Very old age. In: Craik, F.I.M., Salthouse, T.A. (eds.) The Handbook of Aging and Cognition. Lawrence Erlbaum Associates, Mahwah, New Jersey, USA (2000)
10. Jick, H., Zornberg, G.L., Jick, S.S., Seshadri, S., Drachman, D.A.: Statins and the risk of dementia. Lancet 1627–1631 (2000)
11. Jorm, A.F., Korten, A.E., Henderson, A.S.: The prevalence of dementia: a quantitative integration of the literature. Acta Psychiatr. Scand. **76**, 465–479 (1987)
12. Alzheimer's Australia: Keeping Dementia Front of Mind: Incidence and Prevalence 2009–2050. Report by Access Economics Pty Limited (2009)
13. World Population Ageing: Population Division, 2013 World Population Ageing 2013ST/ESA/SER.A/348, p. 17. United Nations, Department of Economic and Social Affairs (2013)
14. Finkel, S.I., Costa e Silva, J., Cohen, G., Miller, S., Sartorius, N.: Behavioural and psychological signs and symptoms of dementia: a consensus statement on current knowledge and implications for research and treatment. Int. Psychogeriatr. **8**(Supplement S3), 497–500 (1996)
15. Devanand, D.P., Jacobs, D.M., Tang, M., Castillo-Castaneda, C.D., Sano, M., Marder, K., Bell, K., Bylsma, F.W., Brandt, J., Albert, M., Stern, Y.: The course of psychopathologic features in mild to moderate Alzheimer disease. Arch. Gen. Psychiatry 257–263 (1997)
16. Alzheimer's Australia: What is Dementia? Help Sheet. https://www.dementia.org.au/sites/default/files/Help_Sheet__About_Dementia_1_fin.pdf. Accessed 21 Apr 2012
17. International Psychogeriatric Association (IPA): Behavioral and Psychological Symptoms of Dementia, Specialists Guide (2012). ISBN 978-0-9829509-0-6
18. Burgio, L., Scilley, K., Hardin, J.M., Hsu, C., Yancey, J.: Environmental "white noise": an intervention for verbally agitated nursing home residents. J. Gerontol. 364–373 (1996)
19. O'Donnell, B.F., Drachman, D.A., Barnes, H.J., Peterson, K.E., Swearer, J.M., Lew, R.A.: Incontinence and troublesome behaviors predict institutionalization in dementia. J. Geriatr. Psychiatry Neurol. (1992)
20. Cohen-Mansfield, J.: Stress in nursing home staff: a review and a theoretical model. J. Appl. Gerontol. 444–466 (1995)
21. Draper, B., Snowdon, J., Meares, S., Turner, J., Gonski, P., McMinn, B., McIntosh, H., Latham, L., Draper D., Luscombe, G.: Case-controlled study of nursing home residents referred for treatment of vocally disruptive behavior. Int. Psychogeriatr. 333–344 (2000)
22. Rodney, V.: Nurse stress associated with aggression in people with dementia: its relationship to hardiness, cognitive appraisal and coping. JAN: Inform. Pract. Policy Worldwide Through Res. Sch. 172–180 (2000)
23. Baldwin, R.: Mood disorders: depressive disorders. In: Jacoby, R., Oppenheimer, C., Dening, T., Thomas, A. (eds.) Oxford Textbook of Old Age Psychiatry, pp. 529–556. Oxford University Press, Oxford (2008)
24. Boyle, P., Wilson, R.: Risk factors for Alzheimer's disease. In: Ames, D., Burns, A., O'Brien, J. (eds.) Dementia, 4th edn, pp. 398–404. Hodder Arnold, London (2010)
25. Cohen-Mansfield, J.: Nonpharmacologic interventions for inappropriate behaviors in dementia: a review, summary and critique. Am. J. Geriatr. Psychiatry 361–381 (2001)
26. Alm, N., Astell, A., Gowans, G., Ellis, M., Dye, R., Campbell, J.: Designing an interface usable by people with dementia. In: CUU '03 Proceedings of the 2003 Conference on Universal Usability, pp. 156–157 (2003)
27. Brooker, D., Foster, N., Banner, A., Payne, M., Jackson, L.: The efficacy of dementia care mapping as an audit tool: report of a 3-year British NHS evaluation. Aging Ment. Health 60–70 (1998)
28. Chung, J.C.C.: Activity participation and well-being of people with dementia in long-term—care settings. OTJR: Occup. Particip. Health 22–31 (2004)

29. Robinson, L.J., Thompson, J.M., Gallagher, P., Goswami, U., Young, A.H., Ferrier, N.I., Moore, B.P.: A meta-analysis of cognitive deficits in euthymic patients with bipolar disorder. J. Affect. Disord. 105–115 (2006)
30. Ballard, C.G., O'Brien, J.T.: Pharmacological treatment of behavioural and psychological signs in Alzheimer's disease: how good is the evidence for current pharmacological treatments? BMJ 138–139 (1999)
31. Douglas, S., James, I., Ballard C.: Non-pharmacological interventions in dementia. Adv. Psychiatr. Treat. J. Contin. Prof. Dev. APT 171–177 (2004)
32. Alzheimer's Society: Drugs Used to Relieve Behavioural and Psychological Symptoms in Dementia. Dementia Australia, fact sheet, p. 3 (2006)
33. Astell, A.J., Alm, N., Gowans, G., Ellis, M.P., Dye, R., Campbell, J.: CIRCA: a communication prosthesis for dementia. In: Mihailidas, Normie, L., Kautz, H., Boger, J. (eds.) Technology and Aging. IOS Press (2008)
34. Bohlmeijer, E., Roemer, M., Cuijpers, P., Smit, F.: The effects of reminiscence on psychological well-being in older adults: a meta-analysis. Aging Ment. Health 291–300 (2007)
35. Cappeliez, P., O'Rourke, N., Chadhury, H.: Functions in reminiscence and mental health in later life. Ageing Ment. Health **9**, 295–301 (2005)
36. Cohen-Mansfield, J., Werner, P.: The effects of an enhanced environment on nursing home residents who pace. Gerontologist 199–208 (1998)
37. Dochterman, J.M., Bulechek, G.M. (eds.): Nursing Interventions Classification: Iowa Intervention Project, 4th edn. Mosby, St. Louis (2003)
38. Kuwahara, N., Abe, S., Yasuda, K., Kuwabara, K.: Networked reminiscence therapy for individuals with dementia by using photo and video sharing. In: Assets '06: Proceedings of the 8th International ACM SIGACCESS Conference on Computers and Accessibility, pp. 125–132. ACM Press, New York, NY, USA (2006)
39. Sarne-Fleischmann, V., Tractinsky, N.: Development and evaluation of a personalised multimedia system for reminiscence therapy in Alzheimer's patients. Int. J. Soc. Hum. Comput. 81–96 (2008)
40. Woods, B., Spector A., Jones C., Orrell, M., Davies, S.: Reminiscence therapy for dementia. Cochrane Database Syst. Rev. (2005)
41. Clark, M.E., Lipe, A.W., Bilbrey, M.: Use of music to decrease aggressive behaviors in people with dementia. J. Gerontol. Nurs. 10–17 (1998)
42. Gerdner, L.A.: Effects of individualized versus classical "relaxation" music on the frequency of agitation in elderly persons with Alzheimer's disease and related disorders. Int. Psychogeriatr. 49–65 (2000)
43. Music and Memory: www.musicandmemory.org. Accessed 05 Mar 2012 (2010)
44. Sherratt, K., Thornton, A., Hatton, C.: Music interventions for people with dementia: a review of the literature. Aging Ment. Health 3–12 (2004)
45. Thomas, D.W., Heitman, R.J., Alexander, T.: The effects of music on bathing cooperation for residents with dementia. J. Music Ther. 246–259 (1997)
46. Cohen-Mansfield, J., Werner, P.: Management of verbally disruptive behaviours in nursing home residents. J. Gerontol. 369–377 (1997)
47. Remington, R.: Calming music and hand massage with agitated elderly, Nurs. Res. 317–323 (2002)
48. Cahill, S., Begley, E., Hagen, I.: Health technologies for people with early dementia: the ENABLE project. In: Moniz-Cook, E., Manthorpe, J. (eds.) Psychosocial Interventions in Early Dementia: Evidence-Based Practice. Jessica Kingsley, London, United Kingdom (2008)
49. Mulvenna, M.D., Doyle, L., Wright, T., Zheng, H., Topping, P., Boyle, K., Martin, S.: Evaluation of card-based versus device-based reminiscing using photographic images. J. CyberTher. Rehabil. **4**(1) (Virtual Reality Medical Institute) (2011)
50. Upton, D., Upton, P., Jones, T., Jutlla, K., Brooker, D.: Evaluation of the Impact of Touch Screen Technology on People With Dementia and Their Carers Within Care Home Settings, p. 5. University of Worcester (2011)

51. Favilla, S., Pedell, S.: Touch screen collaborative music: designing NIME for older people with dementia. In: Proceedings of the International Conference on New Interfaces for Musical Expression (2014)
52. Gee, J.P.: Learning by design: good video games as learning machines. SAGE J. E-Learn. Digit. Media. https://journals.sagepub.com/doi/abs/10.2304/elea.2005.2.1.5 (2005)
53. Csikszentmihalyi, M.: The Creative Personality, Psychology Today. http://www.cre8ng.com/wp-content/uploads/2011/05/Mihaly-Csikszentmihalyi-10-variables.pdf (1996)
54. Rosenfeld, L., Morville, P.: Information Architecture for the World Wide Web. O'Reilly Media Inc., California, USA (2002)
55. Jin, Z.X., Plocher, T., Kiff, L.: Touch screen user interfaces for older adults: button size and spacing. In: Stephanidis, C. (ed.) Universal Access in HCI, Part I, HCII 2007, LNCS 4554, pp. 933–941. © Springer, Berlin, Heidelberg (2007)
56. Cushman, W.H., Rosenburg, D.J.: Human factors in product design. In: Advances in Human Factors/Ergonomics. Elsevier (1991)
57. Rahman, M.M., Sprigle, S., Sharit, J.: Guidelines for force-travel combinations of push button switches for older populations. Appl. Ergon. **29**(2), 85–155 (1998)
58. Tilley, A.R.: The Measure of Man and Woman. Human Factors in Design. Wiley, New York (2002)
59. Rentz, C.A.: Reminiscence a supportive intervention for the person with Alzheimer's disease. J. Psychosoc. Nurs. Ment. Health Serv. 15–20 (1995)
60. Topo, P., Mäki, O., Saarikalle, K., Clarke, N., Begley, E., Cahill, S., Arenlind, J., Holthe, T., Morbey, H., Hayes, K., Gilliard, J.: Assessment of a music-based multimedia program for people with dementia. 331–350 (2004)
61. Bernard, M., Liao, C., Mills, M.: Determining the Best Online Font for Older Adults. http://usabilitynews.org/determining-the-best-online-font-for-older-adults/. Accessed 13 Apr 2012 (2001)
62. Tullis, T.S., Boynton, J.L., Hersh, H.: Readability of fonts in the windows environment. In: CHI '95 Proceedings (1995). http://dx.doi.org/10.1145/223355.223463
63. Alm, N., Riley, P., Newell, A.: An interactive tool to promote musical creativity in people with dementia. Comput. Hum. Behav. **25**(3), 599–608 (2009). https://doi.org/10.1016/j.chb.2008.08.014
64. Nass, C., Yen, C.: The Man Who Lied to His Laptop: What We Can Learn About Ourselves from Our Machines. Penguin Group, USA (2012)
65. Koskinen, E.: Optimizing Tactile Feedback for Virtual Buttons in Mobile Devices. Helsinki University of Technology, Faculty of Electronics, Communications and Automation (2008)
66. Yasuda, K., Beckman, B., Yoneda, M., Yoneda, H., Iwamoto, A., Nakamura, T.: Successful guidance by automatic output of music and verbal messages for daily behavioural disturbances of three individuals with dementia. Neuropsychol. Rehabil. 66–82 (2006)
67. Davison, T.E., Nayer, K., Coxon, S., de Bono, A., Eppingstall, B., Jeon, Y., van der Ploeg, E.S., O'Connor, D.: A personalised multimedia device to treat agitated behaviour and improve mood in people with dementia: a pilot study. J. Geriatr. Nurs. **37**(1), 25–29 (2016). (Elsevier)

**Dr. Kanvar Nayer** is a lecturer and coordinator of the industrial design program at Monash College in Melbourne. He recently completed his Ph.D. which resulted in the development of 'MemoryBox'—a personalised multimedia system for individuals with dementia. Kanvar is also Director (Collaboration Design & Research) at CollabCare, currently working on the commercialisation of MemoryBox and other related platforms. His design and research interests lie in improving the quality of life of the geriatric population, their families and caregivers. Kanvar was the Monash University finalist in the 3-Minute Thesis competition in 2013 and was shortlisted in the James Dyson Awards in 2010 for the design of a torque-reducing fire-fighting nozzle.

**Selby Coxon** is an Associate Professor and founding director of Monash University Mobility Design Lab. His career spans thirty years in professional industrial design, research and higher education, most of which have been in leadership positions. During his time in the commercial sector, Selby worked for leading consultancies and brands with a wide range of products and services in Europe, including Philips Electronics and Schlumberger. Selby has published widely on a range of design responses to challenging problems, but in recent years the focus of his research has been in the pursuit of improved public and private transport, particularly in new and emerging modalities.

# Chapter 18
# Promoting Personhood for People with Dementia Through Shared Social Touchscreen Interactions

**Sonja Pedell, Stu Favilla, Andrew Murphy, Jeanie Beh and Tanya Petrovich**

**Abstract** In this chapter we describe an approach aimed to promote personhood in technology via touchscreen interactions. The approach we propose employs co-design methods aimed at giving people living with dementia a strong voice to overcome usage barriers. Secondly, we propose interest-based design to engage and create meaningful social interactions. We illustrate the approach through two case studies demonstrating its successful application within playing games (two player games) and music (small collaborative ensembles). The chapter concludes with six recommendations for the design of creative and shared interactions for touchscreen technologies.

**Keywords** Dementia · Personhood · Social interactions · Co-design · Interest-based design · Well-being · Assistive technologies · Touchscreen interactions · iPad games · Music application

## 18.1 Introduction

This chapter presents a continuation on Kitwood's [1] theory of personhood for dementia and how this may be applied in the co-design of interactions and technology.

---

S. Pedell (✉) · S. Favilla · A. Murphy · J. Beh
Future Self and Design Living Lab, Swinburne University of Technology, Hawthorn, VIC 3122, Australia
e-mail: spedell@swin.edu.au

S. Favilla
e-mail: sfavilla@swin.edu.au

A. Murphy
e-mail: amurphy@swin.edu.au

J. Beh
e-mail: jebeh@swin.edu.au

T. Petrovich
Centre for Dementia Learning, Dementia Australia, Parkville, VIC 3052, Australia
e-mail: Dr.Tanya.Petrovich@dementia.org.au



A. Woodcock et al. (eds.), *Design of Assistive Technology for Ageing Populations*, Intelligent Systems Reference Library 167,
https://doi.org/10.1007/978-3-030-26292-1_18

We propose that understanding personhood is essential in creating meaningful and engaging interactions to increase the quality of life for people living with dementia. This understanding encourages individual expression, social interactions and content creation right throughout the design process embracing support networks (family members and carers) and harnessing the interests of the dementia participants. Hence our projects apply co-design processes with people with dementia and their support network focusing on interest-based design. Our approach also promotes shared social interactions (within dyads and small groups) through touch technologies typically designed for a single user such as iPads. Here we report on two case studies employing co-design with people in the context of active ageing groups for one year in average per technology application. The settings for the case studies included both day-care and full-time residential care homes. The process in each case study involved an understanding of the environment, the participant's interests, the social dynamics and impact.

Kitwood [1] proposed that actions of care staff have a critical role determining the wellbeing of people living with dementia. It has been shown that social isolation and depression, as a consequence of dementia, has a severe impact adding to the physical decline caused by the condition. Therefore, there is a clear necessity for positive, meaningful and engaging social interactions and experiences in the dementia space.

We present two case studies that focus on the social potential and engagement of technologies.

One case study presents touchscreen apps (games and activities) specifically for the interaction dyad of visitors (life partner/carer) and resident (person with dementia). Our aim was to create a positive pleasurable shared experience between the resident and visitor(s) hence promoting a better visit. The two player apps developed were found to facilitate equal engagement and mutual social interaction with the visitors and residential staff. They provided new social opportunities with volunteers, or other people unknown to the older adults.

A second case study formed small collaborative music interactions and ensembles (three or more participants) for touchscreen tablets in day care settings. This study demonstrates the power of music to bring rich engaging and positive experiences for people with dementia, including novel interactions and dementia participants performing alongside professional musicians. Interaction studies explored the development of gestural touches (gestalts) over time signifying a shift from exploratory to performatory (practised) interactions.

We conclude with general recommendations for technology development with older adults living with dementia across the case studies and propose these are useful in guiding future touchscreen technology development to support personhood.

## 18.2 Current Research and Literature

### 18.2.1 Personhood and Challenges in Care Environments

Kitwood [1] stressed that every person should be treated with deep respect: "It is a standing or status that is bestowed upon one human being, by others, in the context of relationship and social being. It implies recognition, respect and trust" [1, pp. 8]

Professional care-staff working with persons with dementia are under constant pressure to provide activities for stimulation "to support function to maintain physical, mental and social well-being" [2, pp. 266]. It is challenging to repeatedly come up with fresh ideas and new activities when attention spans are brief. Staff are also continually responding to the high demands of ageing well groups and increasing group sizes. At the same time, person-centred care regimes require staff to be warm and respectful to the people they care for [2]. Within the residential and day care spaces volunteer staff and family members also deliver care. Visitors who have known the person prior to dementia are often challenged to be patient and respectful in the light of all the changes that have reshaped their lives.

Persons living with dementia should be enabled to socialise, share experiences, interact and learn from each other via meaningful activities. Ryan et al. [3] suggest active communication as a means for supporting personhood. Craig [4] recommends that meaningful activities need to be informed by interests and aspirations to increase quality of life. We concur and are particularly interested in how touchscreen tablet technologies can support engaging interactions and promote social encounters for the person with dementia with people from their social and support networks.

More accessible than conventional computers, touchscreen and tablet technologies have been playing a valuable role in dementia care for over a decade [5, 6] being used to support reminiscence, aid recall, increase interpersonal interactions, promote intergenerational relationships, improve staff and resident relationships, and enhance quality of life. Easier to use than a computer and more portable, the touchscreen tablet also "encourages increased interpersonal interaction and a positive social environment." [6, pp. 27].

Furthermore, these studies demonstrate the effectiveness of touchscreen tablets in both care facility and home environments [5, 6]. In many cases technologies have demonstrated clear benefits to quality of life through supporting and maintaining a sense of personhood [1] for people living with dementia.

Implementation of technology should also consider those who help facilitate it. Despite reporting positive interactions and engagements for participants, studies have also documented that care staff find touchscreen tablets time-consuming and stressful. Staff can be worried about breaking devices and have reservations on how to go about engaging and disengaging persons with dementia using apps [7]. Upton et al. [6] suggest that touchscreen tablets require a scaffolded (one-to-one mentoring) interaction approach in dementia care. A one-carer-to-one-dementia user model is neither practical, nor possible to provide in most care facilities. However, situated groups of people with dementia, around tables or in sitting areas are commonplace.

In these scenarios, collaborative and multi-user activities would be valuable, where a single caregiver could mentor and facilitate an activity for a group of people with dementia. Hence, we must strive for ways to support *group* interactions through touch technology with the *individual* in mind. Lastly, we should also cater for different people who may be available to share these interactions without a sole focus on staff and carers.

### *18.2.2 Benefits of (Touch) Technologies for Dementia*

The use of touchscreen interactions (often games) as a therapeutic recreation tool for persons living with dementia has shown promise for delivering positive outcomes across a myriad of categories including cognitive, physical and emotional-social [6, 8, 9]. A review by Cutler et al. [10] on research exploring the effects of off-the-shelf technologies such as the Wii and iPad suggests health benefits for people with dementia living in the community and those residing in assisted living and nursing home environments. However, how do we address the individual person with such technologies and create meaning?

Like cinema, theatre, music or books, digital interactions can be persuasive and expressive with the power to change attitudes [11]. According to Bogost and Wright [11] these media influence and change us, contributing to the type of person each of us becomes. We propose that tablets afford interactions supported by film, photo-media, animation and music, each with potential of stimulating minds and stirring memories, stories and conversations. Hitch et al. [12] summarised that use of touch-screen technology in reminiscence therapy yielded even better results than memory books and structured conversion. And while more in-depth research is required the current trend indicates the use of touchscreen technology improves quality of life due to the elevated levels of interaction.

Touchscreen (tablet) music interactions have also been used to promote reminiscence activities between people with dementia and their carers [13] or have been designed specifically for *cognitive stimulation* such as memory, attention and concentration training [14]. The use of digital applications as an occupational or therapeutic tool has been shown to have a positive impact on mental health conditions including depression and anxiety [15]. Other studies demonstrate positive impact with problem dementia behaviours [7] presenting new strategies to manage challenging behaviours without pharmaceuticals.

Joddrell and Astell [16], argue that touchscreen technology's role has so far been mostly used "as a method to deliver assessments and screening tests or to provide an assistive function or cognitive rehabilitation" [16, pp. 1]. They suggest that "more use could be made to deliver independent activities for meaningful occupation, entertainment, and fun" [16, pp. 1]. We agree with this and further propose that there are opportunities for touchscreen technologies to support the individual through personal *interests* and *preferences* (continuing personhood) in social situations and less focus on the declining properties caused by dementia or undesirable behaviours when developing technologies.

### 18.2.3 Wellbeing and Social Interaction (Social Connectedness)

Existing touchscreen interactions for dementia and related studies have often focused on cognitive "brain-training" or used as assessment tools with social-emotional aspects being noted in the background [16, 17]. This is surprising as the "use of the iPad as a group activity for persons with dementia appears to be as viable as traditional group activities in promoting well-being." [2, pp. 270]. Engaging with digital gaming in a group setting can elevate social interaction organically through discussion about the technology itself, as it is used, or by the application presenting topics of conversation either implicitly or explicitly [10]. We expect that positive social experiences facilitated through meaningful touchscreen technology interactions may lead to benefits for self-esteem, identity and connectedness.

### 18.2.4 Technology Development and Ease of Use

Digital applications designed for people with dementia should avoid overly complicated mechanics that require lengthy instruction, limited time scenarios, complex vocabulary and fast-moving graphics. Where possible they should be designed for devices with adequate display size with accessible input methods and employ a graphic-based user interface [10, 18, 19]. Touchscreen applications that contain advertisements should be avoided as they can be distracting and interrupt the flow of the application, [18]. Joddrell and Astell [16] demonstrated that prior familiarity with a game, particularly a non-digital version, is not a guarantee for usability.

Despite these recommendations it is important to note that the ability of people with dementia to interact and engage with technology, regardless of being unfamiliar with it, often surprises persons with dementia and carers alike [10, 18]. Joddrell and Astell [16] have shown when the touchscreen game is suitable (simple, straight forward and casual) that only minimal instruction is required for people living with dementia to successfully play it. Additionally, they [16] suggest that people with dementia can enjoy tablet interactions independently from the level of success achieved in game, implying the core interaction has a greater impact over winning or a high score. Likewise, Cutler et al. [10] found participants could derive a great deal of enjoyment from the act of learning technologies. However, Groenewoud et al. [20] discovered touchscreen tablet games can be the cause of negative reactions when they do not fit the users' preferences or capabilities or if the game is considered too childish. This highlights the importance of finding the right balance of challenge and frustration [18]. Cutler et al. [10] also advocate that interactions should encourage users to think and not to be solely focused on the touchscreen tablet but rather promote social interactions.

To strike a fine balance between preferences, non-patronising content and layout and to promote personhood, we suggest two design approaches; interest-based design and co-creation.

### 18.2.5 Interest-Based Design

Hitch et al. [12] outline that a large amount of existing research is based around applications designed for an entire population. Interests and abilities of people with dementia are greatly varied and what might be engaging for one person might be uninteresting or even frustrating to another, leading to negative experiences. For a person with dementia to successfully participate with an application or technology supported activity it is important to find one that works for them [9]. Leng et al. [2] argue that activities should reflect people's past experiences, interests, and remaining capabilities to facilitate engagement in meaningful activities, but also to promote well-being.

Leahey and Singleton [9] suggest that playing games together or observing may provide carers with insights regarding an individual's cues, interests and abilities—potentially leading carers to discover new suitable activities. In a similar manner we started our technology development projects with this first phase of interest-based design. This approach has been informed by Beh and Pedell [21] who discovered that older adults take up and learn touchscreen technology more easily when it supports their pre-existing interests.

To facilitate interest-based design, initial field trips were spent getting to know the participants living with dementia and their loved ones. This is an essential and important step in our research process usually taking several weeks to establish trust and a good rapport. We believe that as researchers we need to follow the same premises of personhood as carers to both successfully interact and develop applications with people living with dementia. Activities undertaken during this phase of the research included interacting with the iPads together, exploring existing commercial apps and singing around a piano (played by one of the researchers). After we established a better understanding of interests we then moved to an iterative co-design process with technology to support these interests.

### 18.2.6 Co-design

Co-design is a design process which focuses on including all stakeholders in the design process to ensure that the results meet the user needs [22]. Effectively, the role of the user transcends that of "user as subject" and becomes one of partnership with the designer [23, pp. 1]. Co-design is highly beneficial for designers, as it gives a unique insight into user needs, preferences and ideas. The concept of co-design with the end-user can be challenging for both designers and end-users. However, when it is carefully organised and implemented, co-design can result in successful sustainable products that enhance quality of life [23].

To address the needs and interests of people living with dementia it is equally important to include them in the design process of new technologies. This also avoids "doing things to people with dementia rather than with them" and eroding their ability to do things themselves [24, pp. 15]. There are examples where the integration of people with dementia in a co-design process has been done very successfully in terms of quality outcomes that increase self-esteem, personhood and wellbeing (e.g. see Rodgers [25] and Treadaway et al. [26]). Although Tsekleves et al. [27] argue that for people with advanced dementia involved in design, the term co-creation is more suitable than co-design, all authors agree that the strong involvement is both possible and necessary to create products that are meaningful and useful for people living with dementia [25, 26, 28]. It is important to note that in order to follow a good development process and give people with dementia a very strong voice we engaged the whole team (including the developers) in the visits and the co-design process of the described projects.

## 18.3 Aim of Our Work

The main objectives of both projects described in the two case studies was to (i) utilise touchscreen technologies in innovative ways to create positive social interactions and benefit people with dementia and (ii) to promote active engagement (opposed to passive consumption such as TV) of the players in emphasising creativity, skills and participation while easy to use. Such shared co-located social experience is unique as most iPad games or apps emphasise single player use or playing with people remotely.

## 18.4 Case Study 1: A Better Visit Application

### *18.4.1 Context and Environment*

In this project we aimed to help families have a positive shared social experience when they go to visit their family member living with dementia in residential care. We co-designed interactive touchscreen apps (games and activities) with older adults living with moderate to advanced dementia and their visitors. The project, a collaboration with Dementia Australia and Lifeview Residential Care, aimed to enable increased and meaningful social activity through a yearlong co-design process.

We saw an opportunity to co-design touchscreen apps (games and activities) specifically for the dementia care centre setting and the *dyad* (pair) of visitor (often the partner) and resident (person with dementia). Our aim was to develop games to be played by two people using interest-based design and co-creation. We wanted to create a positive pleasurable shared experience between the resident and visitor(s) hence promoting a better visit.

We specifically wanted to learn about the interests of the residents—in many cases about the hobbies or activities they had pursued in earlier years. Our rationale was that the iPad apps should not cater primarily for people with dementia, but for people with certain interests who also live with dementia. At the same time we considered it to be relevant to also address the interests of the visitors.

Early interviews with staff and visitors of three Lifeview homes established that a good visit is defined through interaction and communication. One staff elaborated on the quality of interaction: "So having a good visit is being there in that moment and not about what has happened before or what is happening tomorrow […] so having something, a common ground, something that is not reflecting: 'this is what happened in the past, or this is what's happening now'. So, taking away all of that and just having that moment; there and then". One visitor defined it in a similar manner: "Oh, I went and saw Dad and it was really good. We talked about this".

Consequently, a bad visit was defined as spending time without interaction. Often family members do not want to visit their family member; coming in and leaving quickly, because the family member in residence no longer recognises them and they do not know what to do and underestimate what the resident is able to do. They are at loss how to connect and so they leave without much or even no engagement. One staff member expressed it this way: "When they just sit next to them and there is nothing. The family member will just sit there on their phone and do nothing else. They don't talk. Don't do anything, just look around the room and then leave and that is their visit."

Another staff member said: "A not so good visit that I've seen, and it mainly occurs in the evening, is a resident and a family member sitting there and watching TV, the resident is nodding off and the family member is just sitting there, sort of not really knowing what to do or what to say. That's one. Another one is when you've got someone living with dementia and they're in that state of confusion, and you've got a family member who is constantly trying to bring that person back into reality and that person who is living with dementia is just not having a bar of it, 'this is my reality now, not that!' ".

This application is to assist with engaging all family members and have a 'Better Visit' as a result.

### *18.4.2 Participants*

Eighteen residents living with moderate or advanced dementia and their main visitors (often partners, children and carers) were recruited from four separate residential care homes. The project consisted of three phases; (i) understanding the needs of both residents and visitors during a visit (interest-based design), (ii) systematic co-creative development and investigation of interactions and (iii) iterative evaluations.

Many of the participants had no previous technology experience and a range of conditions including loss of touch and tactile sensitivity, restricted vision, aphasia, and shortened attention spans. Hence a wide range of interaction solutions needed to be found to accommodate these challenges to achieve an enjoyable and social experience.

### *18.4.3 Iterative Co-design of Touchscreen Application*

Interests explored were related to Australian landscapes (beach, mountains and cities), typical animals and plants (parrots, kangaroos, native flowers) and habitual activities such as bowling, fishing and ball room dancing, but also family-oriented themes such as backyard cricket, BBQ parties and holidays. Besides these interests, later depicted as interaction media content, we also explored music favourites. Through our design sessions with the dyads we realised music could support our interactions in several ways. Friday afternoons at the homes would see all the home's residents come together for happy hour during which time a musician would play songs by request. The residents became so animated, sang and talked so jovially amongst themselves and some even dancing that we knew music had to play a large part in our design solutions. Music is effective at prompting memories and stories, yet also stimulates attention and engagement. Favourite songs would create positive emotions lifting mood and raise activity levels. During our initial play-testing phase we observed how all the participants enjoyed skill and competition games. Frustratingly many existing commercial games although visually stimulating and appealing, are not playable by people with moderate to advanced dementia. Games that focus on complex scoring, goals and fast competitive action are not suitable either.

Finally, we observed preferences in interaction mechanisms with technologies. Overall, there was a lack of technology experience amongst the group. However, there was no fear of technology. Problems with technologies were the following:

Amongst our participants we frequently observed loss of connectivity to the iPad's touchscreen during our preliminary trials using existing apps. Later, during the development of our first prototypes we observed how this intermittent connectivity could result in tens of fast multiple touches being registered by the iPad, creating significant lag in interaction flow and even leading to software crashes. For button pushing and simple switch interactions we found an excellent and robust solution setting our software interactions to register touch-offs rather than touch-ons.

Accidental touches through other fingers and resting palms on the screen also proved very challenging with our first prototypes. These accidental touches would often activate interactive edge zones resulting in the app window closing. Even with these zones deactivated and multi-touch gestures deactivated, accidental touches would still unexpectedly and critically interrupt play. To retain navigation but limit accidental interaction with said navigation we placed a single menu button in the top left-hand corner of the screen, away from the centre of interaction. Additionally, we restricted the orientation of the apps to landscape only; facilitating side-by-side sharing of the screen between users whilst avoiding unintentional rotation changes during use. We also decided to embrace accidental and palm touches. Our challenge then became how to build successful apps that would utilise and even reward these touches.

### 18.4.4 Final Designs

The two-player app developed, titled *A Better Visit* (available on the App Store), bundles together eight mini-games, including picture guessing, dance hall tic-tac-toe; skill-based games of bowls, fishing and gyroscopic marble maze; as well as creative activities where users trace lines of colour, clean windows to reveal destinations or colour scenes to play animations.

The bundling of multiple distinct mini-games allows the app to deliver different activities based on and catering for the explored interests and abilities, to be played at different times.

The content of the games has been shaped around familiar themes, interests and popular activities all of which are recognisable across the user demographic.

*A Better Visit* employs a wide range of easy interaction solutions including exploiting multiple and accidental palm touches, registering button presses on touch-offs, removing edge zones and supporting interactions with music.

The mini-game **Co-colouring** (Fig. 18.1 left) presents a black and white picture outline which the users magically colour in by touching anywhere on the screen. The game then rewards the participants by coming alive through animation once the screen is completely coloured over. In **Washing-windows** the users work together to wipe off a white foreground to reveal a high-resolution photo-image underneath.

**Reveal** (see Fig. 18.1 right) was resourced by a rich photo library exploring themes such as animals, food, interesting objects and famous people. The interaction is simple with users clicking to remove large tiles revealing a picture underneath. Touch-offs register reliably in case a user leaves their finger resting on the screen and

**Fig. 18.1** Co-colouring (left) and reveal (right)

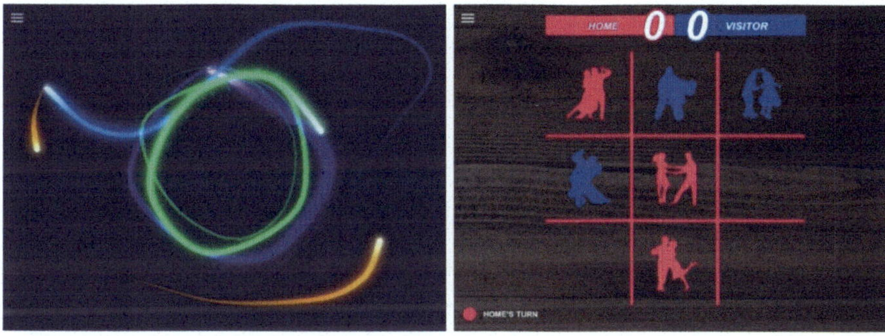


**Fig. 18.2** Traces (left) and Tic Tac Tango (right)

upon completion a prompt for a conversation is added with the answer to stimulate conversations.

**Traces** (Fig. 18.2 left) is developed for people with advanced dementia and is intended as a simple yet non patronising, stimulating audio visual experience. Once again, all touches are processed by the app as participants explore combinations of touches and full-screen gestures together. Traces then replays the users' gestures gradually fading ready for the next gesture. Background music animates the traces creating surges and pulses of brightness and colour. Although intended for advanced dementia users, Traces was popular amongst all participants and provided an excellent collaborative exercise and warm-up app prior to the other games.

Playing a simple tic-tac-toe game can be challenging as the symbols (O's and X's), the turn sequence and the rules such as the X always having the first move, are often confused by people with dementia. Trialling different mechanisms, we noted how colour became a much more dominant cue than the symbol for prompting whose turn it was. We decided to embrace this with **Tic Tac Tango** (Fig. 18.2 right), a tic-tac-toe game backed by dance hall music loops that layer up as players take turns. The music was intended to stimulate thinking and add a layer of excitement and fun to the game. It also made the app a portable musical instrument allowing participants to play the app's music alone exploring different dance and rhythm styles.

Lawn bowls (and indoor carpet bowling) remains one of the most popular activities of older adults in Australia, including our participants. We created **Bowls** (Fig. 18.3 left) as another competitive game accommodating various levels of player skill. Providing the home team (resident) with some gravitational assistance and the option for a one touch shot (to include advanced dementia participants too) we created a game that is quite challenging for a visitor to win.

**Gone Fishing** (Fig. 18.3 right) is another game that adheres to favourite Australian activities. Players can reel in fishes of different size and types bragging about a good or even better catch than their co-player. We also explored interactions where the resident would pick up and hold the iPad. **Marble Maze** makes use of the iPad's accelerometer and gyroscope sensors to move a marble through a short maze. Residents enjoyed the gameplay, however it was probably the least social of the games

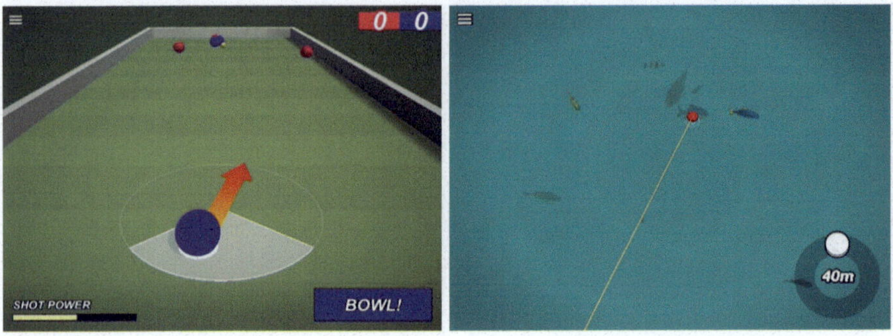


**Fig. 18.3** Bowls (left) and gone fishing (right)

due the single player physical focus of the interaction and the concentration required by the user to play it. Therefore, we added an option to play using touch controls, allowing each player to control the tilting of the maze co-operatively.

Finally, we also created an instrumental music album of favourite songs providing a soundtrack for the games. This created another channel of familiar content supporting mood, concentration and social interaction. As this music was designed to play in the background, we opted for purely instrumental melodic arrangements of songs without vocals. Many of the songs were also remixed at slower tempos to create relaxation and ambient mixes for the art activities.

### *18.4.5 Results of Final Evaluation*

We evaluated the games over a period of six months, remotely logging play sessions from multiple iPads used by twenty-four residents across four Lifeview homes to gain insight into favoured games, frequency and duration of play. To track individual play sessions, we implemented a login screen within the app to record the participant's name and who was playing with them. Evaluating game data is challenging, time on task alone is not an effective measure of engagement and other interaction measures suggested by Yasini and Marchand [14], including the number of games launched, success rate and scoring are neither suited to studying dementia users nor social interactions.

Therefore, In addition to data logging, we also interviewed staff, volunteers, family members and the dementia participants in these homes with follow up observations. We were not able to interview all family members and visitors as a significant number of dementia participants had passed away during the course of the project.

The data logging revealed that game sessions varied in frequency and duration across the range of participants throughout our evaluation. A small number of participants established a routine time of day for play whereas other participant's sessions seemed opportunistic, spontaneous and sporadic. Moments of recognition would create a flow-on cognitive 'buzz' effect, stimulating chains of memories, stories and emotions through images and music shared by both participants.

Our data showed us which of the activities were played most by individual residents over the months hence providing a good indication on preferences on game styles and level of difficulty. The suite of games offers a wide range of activities including competitive and collaborative; art and music; exploration and skill-based activities. The game play results can also be used to inform other activities. The presentation for preferences is shown for one participant in the following spider diagram below (see Fig. 18.4). Across all residents the easy to use creative games were the most popular ones.

Interviews were transcribed and analysed. Here we report on the prominent themes evolving from the data that are relevant for the discussion on personhood.

**Ease of Use.** One of the volunteers commented on the importance of ease of use in terms of being inclusive to all residents. This enables them to experience impact through even very simple interactions without a lot of effort: "… so he really liked that (Traces). A lot of the residents liked that one because firstly it was simple, and then they can see straight away without having to guess."

But the activities were also judged to be inclusive for people who did have other conditions besides dementia such as stroke often resulting in movement of only one hand. One staff commented particular on vision problems: "The ones with the vision problems like it (Tic Tac Tango), they might need some help to tap sometimes but

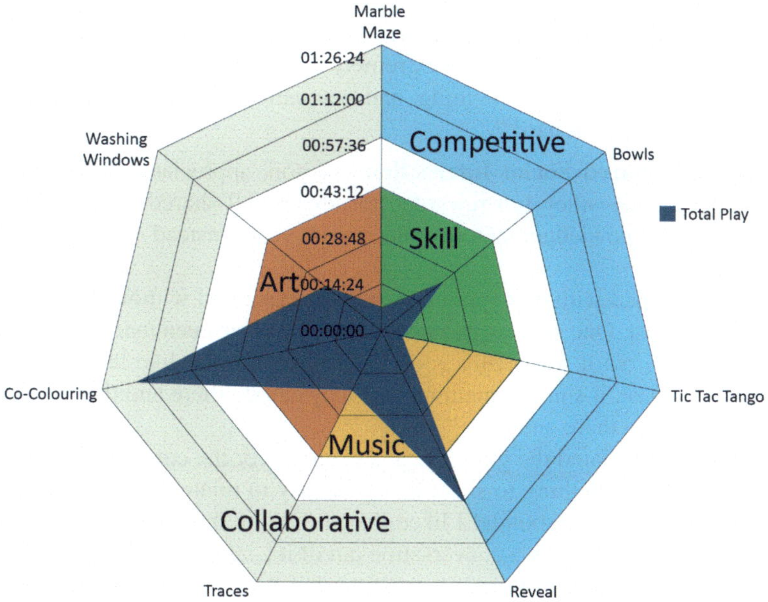


**Fig. 18.4** Spider graph of one participant's accumulated play time and his interaction preferences

with the music and the sounds it helped. I have one person and that is all she played because she had vision problems."

Visitors in particular saw the ability to use the interactive games as a display of capabilities they did not take for granted. In order for these capabilities to become apparent the interactions need to be very easy for the person with dementia to experience the impact on their actions and for the visitor to see the engagement: "well with the iPad, with what we were doing with dad, when he was capable of swiping (Traces and Co-colouring) and the window, the window washing one."

Another dimension of ease of use was brought up by a staff member in regards of the low hurdle to pick up the iPad and use the application: "it's a no brainer … you don't have to come into your visit thinking 'what am I going to talk about, what is my family member going to remember? What are they going to do?' it's there for you; you just have to move your fingers basically."

**Engagement and Experiencing Impact**. Through engagement it is possible for care staff as well to gauge capabilities: "It helped me with some of the residents to find out their cognitive abilities which helped me judge what activities I could take them to. We have a lady that we did it with and after doing the reveal with her you realised that she could remember or that she knew a lot so she wasn't coming to activities that she should be really. That helped us out."

Even though many the residents would not communicate a lot throughout the activities staff and visitors were sure they could see and identify engagement. One staff member said: "I think it's facial expressions, whether that's through their eyes or a smile, it doesn't have to be a verbal 'this is really good', but if they're there and they're looking at it and showing an interest, then you know that they're engaged."

Engagement was considered to be extremely important—one volunteer said: "… they were thinking and that's the thing, and you can see it, you know, that they're thinking and looking at it hard."

**Increased and Shared Social Interaction**. Visitors appreciated the social component of the games—they felt rewarded through small shared interactions: "just anything … an acknowledgement, a smile, a nod, yeah. Instead of me just 'bla bla bla'."

The games helped with an experience in the present and without any obligation to deal with the past. One staff explained: "because you've been married to someone for so long and … so it was really good I think for couples like that, because it's irrelevant to the past, it's just something that they can do there and then and they're having that memory there and then."

One visitor appreciated the guidance and prompts for the conversation in Reveal feeling it gave her something to share that was easy to relate to for both: "oh yeah the animals. Oh yeah, the sounds! I liked that; I did like that one myself. And then because then you could make a conversation out of it […] I remember saying to dad: 'what sound would that be?' it gave a question, for me to say something, instead of just bumbling on. More about yeah, trying to have something where we're relating to this thing and speaking about it there and now instead of just waffling about 'oh the tree down the road' or whatever."

Physical contact often supported these conversations as a means to help with the interaction on the screen as expressed by a visitor: "I had to show him, I had to put my hand over his hand and then we'd do it, we traced it. With everything apart from the colouring in one, once I showed him, then he'd go and do it himself, but everything else you'd have to do it with him."

Not only during the activities, but after the session there were positive effects on social interaction which meant that there was an effect that was going beyond the actual interactions with the technology. One staff member observed: "More talkative definitely and they discuss what they have just done and with others in the room if I walk away, especially if they win! […] they are still very talkative for about twenty minutes".

Also the importance of being social with a variety of people – in this case volunteers and staff was mentioned by two staff members: "It is also a great tool like with the volunteers when they come in, to give them something to do and it is a great way for them to socialise and get to know that resident as well." And "I really loved it in our evening program to see staff with residents using it; as a one on one, because I mean, at the end of the day we're like family members to some of them. So, seeing the staff utilise it in that evening sundown period, it really had some really good benefits and staff would be like 'where is the iPad?' "

While the activities were developed with visiting family members in mind residents sometimes interacted via the iPad amongst each other. Several staff liked the idea of residents playing with each other building a relationship explaining: "The iPad will sit on their lap while they are in a social mix or small group. […] They go into their little bubble in front of the fire. These ladies there are the ones that enjoy it a lot." And "… to have that relationship between two residents is going to build, because it's an intimate time together and they don't get a lot of that."

**Skill Development**. There were clear signs of skill development in using the apps remembering use from one session to the next. One staff member gave the examples of Tic Tac Tango: "the more you play it, the better they get. They get the hang of it and they do remember, so I would get it and sit down a couple of days later and say we're going to play tic-tac-toe and she would come back from the previous fun they had." and Bowls: "We just used the one where you pulled back and used the button to release the bowl. That way they did remember next time when we came back to it that that's the action you use."

## 18.5 Case Study 2: Touchscreen Collaborative Music

### *18.5.1 Context and Environment*

Touchscreen tablet computers have been used to create powerful and expressive controllers for digital music. Contemporary iPad Apps including *Midi-touch* and *TouchOSC* [29] are well suited to developing collaborative [30], multi-user [31] performance systems whereby a number of performer iPads transmit each to a single host computer. The host computer interprets the control gestures and produces the resulting sound, which is amplified for the musicians (and audience).

The gathering of controllers to a single host computer creates opportunities where many players can actively contribute to the formation of just one sound, create a texture or part in the musical fabric, or participate in complex interactions. Additionally, the computer can provide agency in the creative process where algorithms compose melody, harmonization, audio-mixing or engage random and generative processes. The success of a system depends on the balance of complexity and expressivity [30] and on how well the player is engaged and encouraged [31].

Within dementia care, music therapy is used to control mood, problem behaviours and even reduce the need for some pharmacological and physical treatments [13]. Music therapy's aim to "promote communication, relationships, learning, mobilisation expression, organisation and other therapeutic needs" provides meaningful engagement for people with dementia continuing right to the final stages of the disease [32, pp. 30]. Both *receptive* (music listening) and *active* types of music therapy (where participants are actively involved playing instruments) are tailored to suit the individual's particular needs [6]. Music therapy sessions are often carried out in groups embracing improvisation and supporting reminiscence activities.

### 18.5.2 Participants

We set out to co-design technology to collaboratively engage a group of older adults with dementia using a range of music touchscreen applications. 14 participants and the care staff were recruited from an activity group of elderly people with dementia organised by Wyndham City Council as part of their ageing-well program. The council had purchased 12 iPads with the objective to engage older people with dementia in technology use. The management realised that the older people had no iPad experience and had difficulty finding suitable applications to engage this group over an extended period. Our aims were to explore the potential of iPads and music while focusing on the needs and capabilities amongst the participants in this group. We wanted to discover the potential and limits of such an activity and co-design with the dementia participants controllers and music interactions that were enjoyable and engaging.

The group met at a community centre facility during their routine Friday morning activity session. The 12 iPads were managed via a recharge station and a central computer and another laptop was used to develop music interaction software using Max object orientated programming language. Interactions and performances were recorded for study comparisons.

As in the previous project many of the group participants were affected by other age-related conditions posing some design constraints. Given these design constraints we opted for a simple controller approach and planned to maximise the expressive capabilities of the interactions.

### 18.5.3 Iterative Co-design of Touchscreen Application

The initial exploration of commercial music apps proved very insightful. These music apps consisted of the *Tiny Piano* (http://www.tinypiano.com/) and *Rhythm Pad* apps (http://rhythmpad.com/). Neither of these applications kept any of the participants engaged for very long (approx. 5 min). Through observations we concluded the main reasons for this were:

- Screen controls and objects were way too small (both to see and locate with the fingertip)
- iPad speaker sound was too quiet
- The sounds were not engaging (i.e. same attack and dynamic over extended periods)
- Interaction was limited to simple one-shot triggering of sounds

In addition, participants commented on the toy-like quality of these apps' sounds while most of the participants struggled to operate the apps successfully. This posed an interesting design challenge requiring a high-quality audio design solution.

Our first prototype explored abstract sounds with no tuning scales and no rhythmic beat or bar structures. The synthesis instrument consisting of nine touch buttons arranged in a grid and a 2D (x/y) touchpad controller situated right (see Fig. 18.5). Each button once touched would stay on until pressed again and would select a specific synthesis software engine. The x/y touch controller was set-up to produce many possible variations on each synthesis encouraging exploration and rewarding interaction. The sounds for each iPad were then output separately with each participant having their own portable external speaker for monitoring. The system was capable of producing a large array of abstract yet expressive sounds including tones, whines, chortles, groans, roars, whistles and pops. Moving the x/y control pad afforded a large range of transformations and modifications to be made to the sound in real-time.

**Fig. 18.5** Researcher and participant playing (left) on the first prototype (right)

"Sound is very good—very interesting. I like this!" said one of the participants who loved the controller. The participant played with the app for well over an hour and had never played a musical instrument nor heard of electronic music before. The care-staff were all surprised to see his enjoyment and engagement in the activity.

Other participants explored the controller in depth sometimes listening to each other and then joining together in impromptu jam sessions. The simple layout required a minimum in scaffolding and mentoring, which consisted of a minimal demonstration. The facilitation of a separate loudspeaker greatly aided the enjoyment of music making. "There are many different sounds and things you can do" commented one participant.

We discovered right-handed players would sometimes accidently activate the buttons with their palm or other fingers while playing the x/y pad. Other participants who had previously played music commented that the controller bore no resemblance to any musical instrument. A few participants refused to play saying that the sounds (although interesting) were just noise or just too strange. Some participants revelled in the discovery of new sounds while others were thrilled to be doing something, anything on an iPad themselves.

### 18.5.4 Final Design

Responding to ongoing feedback, several controller iterations were trialled with the participants. The buttons were arranged to resemble a piano keyboard. The x/y control pad was re-located in the lower right. A circular panning controller was added. The controllers were also rendered in four different colours for ensemble settings, green, orange, red and purple (see Fig. 18.6).

During this trial phase the iPads' music activity shifted to a more conventional stimulating musical setting, that of Bach. The interaction centred on controlling the expression of a pre-recorded keyboard performance of Bach's *Goldberg Variations*. Unlike the commercial piano roll music apps this interaction used a pre-recorded

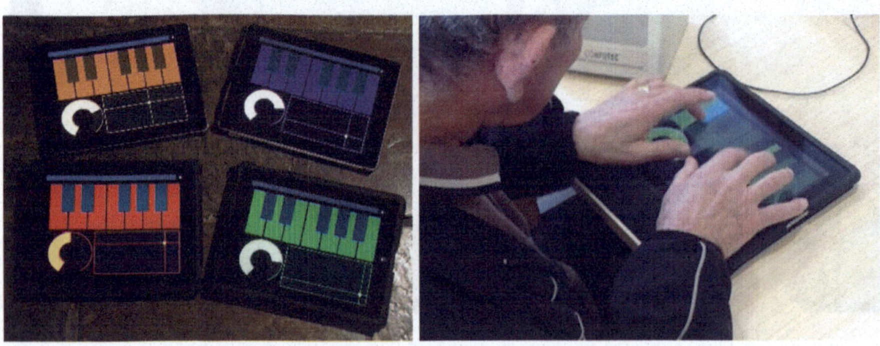

**Fig. 18.6** Final design of iPad controller for ensemble music (left)—group setting (right)

performance played by a professional musician and included beautiful Baroque ornamentation and interpretive expression throughout. The players were immersed into the performance through mapping the x/y joystick controller pad to control the attack dynamic (loudness and brightness increasing along the $x$-axis), while the duration of each note was mapped to the $y$-axis, (notes being shortest towards the bottom of the screen). Through the re-mapping of these music parameters the original recordings expression was retained but the participants were free to shape the music dramatically.

Two participants would play as a duo, each iPad controlling either the bass (left-hand) or treble (right-hand) parts. The distribution of performance roles allows participants to collaboratively re-interpret the performance.

The piano keyboard interface was used to select different instrument sounds ranging from acoustic grand pianos, forte piano, harpsichord, guitar, harp etc. Care was taken to present high quality realistic sounds each with its own distinct characteristics. The synthesis of each sound was also designed to reward the participant's choices with rich and responsive sonic detail. A great advantage of this system was its ability to distribute separate parts of a multi-part work by Bach or Vivaldi, to up to four participants. This enabled the participants to contribute to a collective sound or ensemble. The additional advantage of exploring repertoire also allowed a piano keyboardist to join the groups and play continued accompaniment.

### *18.5.5 Results from Evaluation*

A six-week trial was conducted with multiple iPads (four) connected with the same participants being involved in the development. Collaborative performance became a focus for a range of investigative work exploring changes in interaction over time. Video and audio data alongside careful observation and staff feedback were gathered for study and analysis. In regard to the four themes from case study one the following was found:

**Ease of Use**. There was little need for explanation, simple demonstration sufficed and participants would engage for a long time.

One participant was a trained pianist and the only one who disliked the piano keyboard repurposed role for switching instruments and not triggering keyboard sounds.

**Engagement and Experiencing Impact**. The quality of the engagement was strongly evident during these trials with participants commenting on the beauty and the quality of the musical experience. "No really thanks for that today that was really great ... I never knew anything about Bach, but this is really good".

A mix of listening and doing also created many opportunities for self-expression and fun. Once a participant was brought to the session in a very subdued state, head down and unable to extend their fingers. Over the course of an hour we noticed a major transformation in posture, mood, energy and communicativeness. At the session end the participant had confessed they had been a singing teacher and sang several impromptu songs together with the musician keyboardist. All of this was witnessed by the participant's granddaughter who had brought to tears by the transformation. "I had never thought about her music … I swear this is how she used to be … I will get her old keyboard out from the garage and set it up in her room as soon as we get home …"

**Increased and Shared Social Interaction**. After a period of exploration, the participants were encouraged to listen to each other's playing and then respond. Each participant could clearly identify their own part and how their actions related to changes in the sound and expression of the music. At this point the interaction would change, as each participant listened and adjusted to one another's dynamics, touch and expression. Often players would adjust their note durations to be in opposite, accentuating their own parts in the mix.

Usually there would be no talking while the music was going. Participants would look over to what each other were doing on their respective iPads. Occasionally a participant would leave or arrive to a respectful silent acknowledgement of nodding smiling and welcoming gestures from the ensemble group. There would often be a range of participants playing together some with very advanced dementia.

The sessions with musicians would start many interesting conversations; A participant said to a bassoon player "I really don't know how you do that, but I really like it" where the bassoonist replied; "Well what did you do?" with a conversation evolving from the question.

There were always conversations at the end of pieces too about music, memories, concerts, bands, loves and hates. Participants were typically chatty and talkative by the end of the 45–60 min play sessions.

**Skill Development**. Interaction studies explored the development of gestural touches (gestalts) over time. Two participant's touch-pad traces demonstrated a change (refinement) in interaction style across a six-week trial period with the Bach interactive activity. This was also confirmed by music and video data. Certain features emerged such as shorter deliberate movements (changing note durations), leaving the 2D controller (x/y control pad) in a specific place for a period, bursts of focused activity within a specific sector and long sideways sweeps changing musical dynamics.

These examples of actions (see examples in Fig. 18.8) contrasted markedly with their initial style of play (see examples Fig. 18.7). The initial style of play typically maintained a slow steady speed, with minimal changes to direction and circuit tracing around the perimeter of the touchpad. This development of style over time signalled a shift from *exploratory* to *performatory* action [33] or development of *adroit* action [34]. The emergence of *performative* data features suggests that these two participants developed and retained performance skill during the six-week trial period.

Findings presented in this case study demonstrate that people with dementia can successfully perform and engage in collaborative music performance activities with little or no scaffolded instruction. Furthermore, the evolution of control gesture over time suggests people with dementia can develop and retain musical performance skills over time.

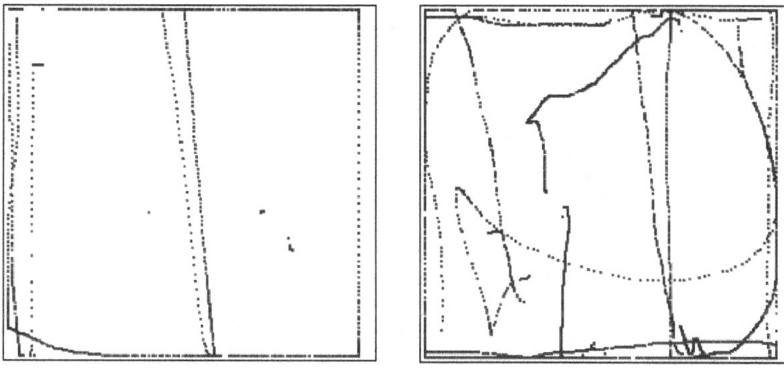

Fig. 18.7 1-min exploratory actions (2 users, week 1)

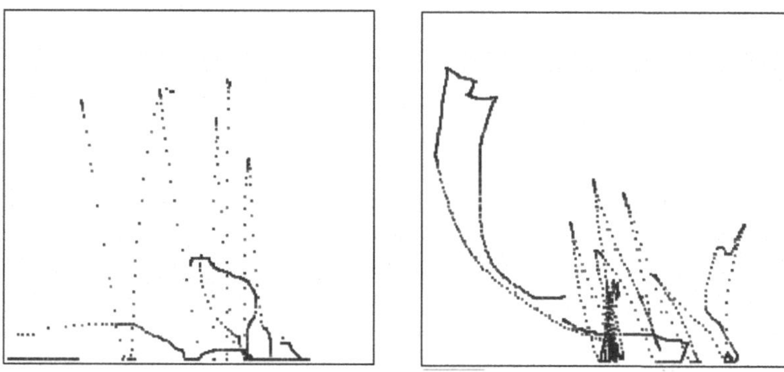

Fig. 18.8 30 s of performative actions (same users, week 6)

## 18.6 Recommendations

Our experience working with older people with dementia and iPads suggests that success is dependent on the *quality* and *content* of the interaction. Importantly, the interactions should not just engage the person with dementia but also their loved ones, care staff and social networks.

To assist other researchers and developers wishing to design for people living with dementia we propose the following six design recommendations taking this complex socio-technical system between technologies, people with individual preferences and environment into account. The recommendations are further elaborated with insights from the two case studies detailing what the interactions supported by touchscreen technologies should try to achieve to promote ongoing personhood for people living with dementia. These recommendations should not only be understood as final outcomes to evaluate against, but importantly as guides to the iterative process of interest-based co-design of socially oriented touch technologies with people living with dementia.

### 18.6.1 Design for Rich and Stimulating Exploration (1)

In both case studies it was relevant to create space for individual exploration within the technology supported activities. The activities should be varied, stimulating and accommodating of different preferences. In the music ensemble we found that some participants were happy to play bizarre abstract electronic sounds while others preferred Bach. Unlike the commercial iPad apps that we also trialled, both the electronic music and Bach interactions provided a large variation of expressiveness with a simple layout affording exploration. Similarly, the A Better Visit app provided a wide range of activities for both participants to choose from. The game mechanics themselves afforded no right or wrong way nor the need to finalise certain games or sections. Participants were able to be engaged together at the same time to have fun.

### 18.6.2 Design for Responsive and Expressive Interaction (2)

Art and music making activities promote active interactions over passive consumption (such as watching television). It was important to offer the opportunities for creative expressions in content manipulation and skill development over time. For the music controller we found the participants tuned in quickly to the expressive attributes of the interaction. They delighted in the discovery of bizarre chortles and beating effects. They enjoyed the detail of the sounds and their agency to change the music's character and hear the result. In the game called Traces we applied a similar visual reward for long tracing gestures providing colourful visual patterns and feedback.

### 18.6.3 Provide Participants with Identifiable, but Equal Active Roles (3)

Successful collaborative play required clearly defined roles for each participant (such as playing the bass for example). Our music players benefited from individual foldback speakers and from being seated closely to best hear the impact of their individual play and instrument. In the case of visitor and resident, the game activities encouraged the visitor to take on a facilitating role through prompts (visuals, sounds) and questions to engage the resident in conversations. The guidance was very subtle as to not disturb the aim for shared experience as equal players. In both interaction environments gently prompting was important, guiding all participants to engage in the interaction on their own terms.

### *18.6.4 Encourage Engagement with Each Other in the Moment (Social Connectedness) (4)*

When participants listened to each other and played together the results were quite surprising. Some of the participants who were socially withdrawn clearly benefited from the inclusive and non-competitive activity. The greatest family member's pleasure was seeing their loved one do something well and enjoying themselves in a group, but also through making contact and creating opportunities for connectedness through talking and body language to make a visit worthwhile. Cognitive warm-ups with music and playful activities can set the scene for unexpected and profound moments experienced together. In addition, there was a flow on effect after the interactions of enhanced social behaviour instigated by the activities.

### *18.6.5 Simple Interaction Design but Non-patronising Content (5)*

The users in both studies responded to simple and easy to use or to control interfaces. However, the participants in our study groups responded well to rich and complex musical situations. The actual *sound* of professional looking music iPad apps such as *Tiny Piano* and *Rhythm Pad* were considered by many to be suited for children. The visual and audio content of the *A Better Visit* app was created to a very high design standard while relating directly to topics and activities of interest. While the interactions in themselves were easy, the visual and sound quality was nuanced and not oversimplified. Hence it is important to understand the difference and strike the right balance between content and its potential within interactions as they unfold. Touch technologies work best when the convey the richness and variety of life and cater for individual interests.

### *18.6.6 Minimal Scaffolding (Short Demonstration) for Skill Development (6)*

The participants in our group were much more interested in doing rather than being shown how to do something. Touchscreen interactions, when varied and based on interests, have the potential to provide extended periods of enjoyment and meaningful engagement. They can also help introduce users, living with dementia, to technology in group settings. Touchscreen games inspire direct social interactions and conversations whereas ensemble music provides rich social experiences. While minimal scaffolding can play an important role, shared screens and integrating instructions directly into the activities themselves encourages doing. Encouraging agency through exploration manifests skill from one week to the next requiring less demonstration.

## 18.7 Concluding Comments

Two apps have been developed using an interest-based and co-design approach with people living with dementia and people of their immediate support network. We have identified several key design challenges for researchers concerning touchscreen technologies for older people with dementia. We suggested to overcome these via embracing the concept of personhood with the aim to engaging people living with dementia in technology supported activities that lead to meaningful experiences of creation and social connectedness. It is relevant to say that the focus on personhood was relevant twofold—firstly in the development process and used approach that was applied—secondly as manifestation in the developed outcomes. We presented novel interactive game touchscreen applications (game activities and a music software) for engaging older people with dementia.

The results from two case studies established that older people with dementia can successfully participate in co-creative activities and the resulting touchscreen activities for extended periods with little or no scaffolded instruction. Results also suggest there is potential for older people with dementia to acquire and retain performative and creative action in a collaborative music setting and a one-to-one game setting.

The project offers six design recommendations that we expect can be adapted to a wide range of iPad applications for creative and shared social interaction. Our work suggests that careful attention to sound, collaborative interaction and touchscreen control is essential for success and promotion of personhood.

**Acknowledgements** We would like to thank Ann Lafferty and all the staff from Lifeview Residential Care for their input in this project and their support throughout. We also wish to thank Tanya Petrovich from Dementia Australia for initiating this collaboration with the Future Self and Design Living Lab. A special thank you to all the participating residents from Lifeview and their visitors co-designing these app with us. We would like to thank the City Council of Wyndham for our friendly and constructive research collaboration over many years.

## References

1. Kitwood, T.: Dementia Reconsidered. Open University Press (1997)
2. Leng, F.Y., Yeo, D., George, S., Barr, C.: Comparison of iPad applications with traditional activities using person-centred care approach: Impact on well-being for persons with dementia. Dementia **13**(2), 265–273 (2013)
3. Ryan, E.B., Byrne, K., Spykerman, H., Orange, J.B.: Evidencing Kitwood's Personhood Strategies: Conversation as Care in Dementia, pp. 18–36. Palgrave Macmillan. Alzheimer Talk, Text and Context, UK (2005)
4. Craig, C.: Journeying Through Dementia: An Occupational Approach to Supported Self Management (2017)
5. Tyack, C., Camic, P.M.: Touchscreen interventions and the well-being of people with dementia and caregivers: a systematic review. Int. Psychogeriatr. **29**(8), 1261–1280 (2017). https://doi.org/10.1017/S1041610217000667

6. Upton, D., Upton, P., Jones, T., Jutlla, K., Brooker, D., Grove, H.: Evaluation of the Impact of Touch Screen Technology on People with Dementia and their Carers within care Home Settings. University of Worcester, UK (2011)
7. Loi, S.M., Mazur, A., Huppert, D., Hoy, B., Swan, J., Lautenschlager, N.T.: A pilot study using “apps” as a novel strategy for the management of challenging behaviors seen in people living in residential care. Int. Psychogeriatr. **29**(4), 637–643 (2016). https://doi.org/10.1017/S1041610216002039
8. Mokhtari, M., Aloulou, H., Tiberghien, T., Biswas, J., Racoceanu, D., Yap, P.: New trends to support independence in persons with mild dementia—a mini review. Gerontology **58**, 554–563 (2012). https://doi.org/10.1159/000337827
9. Leahey, A., Singleton, J.F.: Utilizing therapeutic recreation to empower persons with Alzheimer’s in a day center: a case report. Ther. Recreation J. **45**(2), 135–146 (2011)
10. Cutler, C., Hicks, B., Innes, A.: Does digital gaming enable healthy aging for community-dwelling people with dementia? Games Cult. **11**(1-2), 104–129 (2015). https://doi.org/10.1177/1555412015600580
11. Bogost, I., Wright, W.: Purposes of Persuasion, Chap. 11, pp. 317–340 (2003)
12. Hitch, D., Swan, J., Pattison, R., Stefaniak, R.: Use of touchscreen table technology by people with dementia in homes: a scoping review. J. Rehabil. Assistive Technol. Eng. **1–4**, 1–8 (2017). https://doi.org/10.1177/2055668317733382
13. Riley, P., Alm, N., Newell, A.: An interactive tool to promote musical creativity in people with dementia. Comput. Hum. Behav. 599–608 (2009). https://doi.org/10.1016/j.chb.2008.08.014
14. Yasini, M., Marchand, G.: Adoption and use of a mobile health application in older adults for cognitive stimulation. In: Transforming healthcare with the internet of things. Proceedings of the EFMI special topic conference April 2016, UK https://doi.org/10.3233/978-1-61499-633-0-13 (2016)
15. Chidester, A.S., Sautter, S., Aravich, P., Ord, A.: Transforming Dementia Care and Quality of Life Using Innovative Touch Screen Computer Engagement: A Research Study—The Birdsong Initiative (2016)
16. Joddrell, P., Astell, A.J.: Studies involving people with dementia and touchscreen technology: a literature review. JMIR Rehabil. Assistive Technol. **3**(2), e10 (2016). https://doi.org/10.2196/rehab.5788
17. McCallum, S. Boletsis, C.: Dementia games: a literature review of dementia-related serious games. In; International conference on serious games development and applications, pp. 15–27 (2013)
18. Kong, A.P.H.: Conducting cognitive exercise for early dementia with the use of apps on iPads. Commun. Disord. Q. **36**(2), 102–106 (2015). https://doi.org/10.1177/1525740114544026
19. Zmily, A., Mowafi, Y., Mashal, E.: Study of the usability of spaced retrieval exercise using mobile devices for Alzheimer’s disease rehabilitation (2014). https://doi.org/10.2196/mhealth.3136
20. Groenewoud, H., Schikhof, Y., Astell, A. J., Goumans, M., de Lange, J.: iPad happy games for people with dementia as pleasant and meaningful activities. Dement. Care Pract. E-Interv. Training Care (2014)
21. Beh, J., Pedell, S.: When interest pays off: The relationship between motivation, well-being and learning of technologies by older adults. Des. 4 Health 27–30 (2017)
22. Taffe, S.: The hybrid designer/end-user: revealing paradoxes in co-design. Des. Stud. **40**, 39–59 (2015)
23. Sanders, E., Stappers, P.: Co-creation and the new landscapes of design. CoDesign **4**(1), 5–18 (2008)
24. Astell, A.: Technology and personhood in dementia care. Qual. Ageing Policy Pract. Res. **7**(1), 15–25. https://doi.org/10.1108/14717794200600004
25. Rodgers, P.A.: Co-designing with people living with dementia. Int. J. CoCreation Des. Arts **14**(3), 188–202 (2018). https://doi.org/10.1080/15710882.2017.1282527
26. Treadaway, C., Fennell, J., Prytherch, D., Kenning, G., Walters, A.: Designing for wellbeing in late stage dementia. Context Med. Humanit. (2016)

27. Tsekleves, E., Bingley, A.F., Escalamte, M.A.L., Gradinar, A.: Engaging people with dementia in designing playful and creative practices: co-design or co-creation? Dementia **0**(0), 1–17 (2018). https://doi.org/10.1177/1471301218791692
28. Wintermans, M.C., Brankaert, R.G.A., Lu, Y.: Together we do not forget co-designing with people living with dementia towards a design for social inclusion. Des. Manage. Acad. Conf. (2017)
29. Cuddy, L., Duffin, J.: Music, memory, and Alzheimer's disease: is music recognition spared in dementia, and how can it be assessed? Med. Hypotheses **64**(2), 229–235 (2005)
30. Blaine, T., Fels, S.: Contexts of collaborative musical experiences. In: Proceedings of the 2003 conference on New interfaces for musical expression, pp. 129–134 (2003)
31. Ingram, A.: Critical Review: Does Music Therapy have a Positive Impact on Language Functioning in Adults with Dementia? Reviews Western University Canada (2012)
32. Wigram, T., et al.: A Comprehensive Guide to Music Therapy, pp. 30. Transaction Publishers (2002)
33. Jordà, S.: Multi-user instruments: models, examples and promises. In: Proceedings of the 2005 conference on New interfaces for musical expression, pp. 23–26. National University of Singapore, Singapore (2005)
34. Cahill, S., Macijauskiene, J., Nygård, A.-M., Faulkner, J.-P., Hagen, I.: Technology in dementia care. Technol. Disabil. **19**(2), 55–60 (2007)

**Sonja Pedell** Associate Professor, is Director of Swinburne University's Future Self and Design Living Lab. The FSD Living Lab has core development capabilities in the area of innovative socio-technical systems and design solutions for health and wellbeing with a focus on the ageing population and dementia. Prior to taking up this role at Swinburne, Dr. Pedell completed a Masters of Psychology from the Technical University of Berlin and was employed as an Interaction Designer, Usability Consultant and Product Manager in industry for several years.

**Stu Favilla** is Lecturer at Swinburne University of Technology in Interaction Design in the Communication Design and Digital Media Design Department. A musician and composer, his interests include audio design and interaction design of displays. His project work focuses on immersive interactions including 3D Soundfield capture and presentation, dynamic binaural synthesis of 3D Sound fields (3D Headphone listening with head-tracking), specialised speaker solutions, directional and portable ambisonic arrays and 3D immersive environments in Unity 3D.

**Andrew Murphy** is a Digital Designer and Developer working as part of Swinburne University's Future Self and Design Living Lab. His work focuses on delivering user-centred designs solutions across a gamut of media including application/game design and development, animation, web and general UI/UX interaction design in the digital space.

**Jeanie Beh** is currently working as a Research Assistant in the Future Self and Design Living Lab, maximising her passion for overturning the perception that older adults are unable or disinterested in developing technological skills. Her research focuses on ways to engage older adults with mobile touchscreen technologies through the role of human interest. Currently, she is focusing on dementia-centric projects such as the use of technology to support social engagement and improve quality of life.

**Dr. Tanya Petrovich** is Business Innovation Manager, Centre for Dementia Learning, Dementia Australia. She completed studies in the field of Genetics at the University of Melbourne and Post-Doctoral studies at the Marie Curie Research Institute, UK. She has spent many years teaching in the fields of Science, and Health Science. She joined Dementia Australia in 2008. Tanya leads the team in applying innovations in technology throughout dementia care education. This includes the Virtual Dementia Experience, an immersive sensory environment and EDIE (Educational Dementia Immersive Experience).

# Chapter 19
# Realising the Potential of People Living with Dementia Through Co-designing and Making Interventions

**Euan Winton and Paul A. Rodgers**

**Abstract** The UK has an ageing population where there are now more people aged over 65 than those under the age of 16. The impact of this creates increased pressures on the National Health Service (NHS), and on local and regional health and social care services. Key concerns in regards to this aging population include the prevalence of the five most common chronic conditions among the over 65s—arthritis, heart disease, stroke, diabetes and dementia—with the latter expected to increase 25% by 2020 and more than 50% by 2050. In order to counteract the increasing pressures of aging health and mental healthcare issues current government policy aims to encourage people to remain active, engage in regular exercise and refrain from behaviours that could have a detrimental effect on their health. The research, presented here, focuses on developing and implementing innovative design interventions, that seek to encourage people to remain active, promote dignity, and encourage independence particularly for people living with dementia.

**Keywords** Dementia · Co-design

## 19.1 The Wicked Problem of Dementia

Supporting people as they age, in particular people living with dementia, involves the design and production of various tools, devices, systems and services that are multifaceted, holistic and interdisciplinary in nature. These activities comprise diverse parties and stakeholders and overlapping disciplines and specialists. The degenerative nature of dementia requires that these solutions and interventions must be adaptable to progressive changes. In regards to dementia what makes the situation all the more difficult is that no two individual's journeys are the same and therefore there can be no singular prescribed approach. Instead, the nature of support requires

---

E. Winton · P. A. Rodgers (✉)
Imagination, Lancaster University, Bailrigg, UK
e-mail: p.rodgers@lancaster.ac.uk

E. Winton
e-mail: e.winton@lancaster.ac.uk



A. Woodcock et al. (eds.), *Design of Assistive Technology for Ageing Populations*, Intelligent Systems Reference Library 167,
https://doi.org/10.1007/978-3-030-26292-1_19

insights where parties from various backgrounds converse and, most importantly, act to help to alleviate tensions, strains or blocks and to, where possible, create a valued quality of life [11]. Family, clinicians, link workers, occupational health, financial controllers, and the person living with dementia need to work together to develop tailored responses to each person's situation that can nurture personhood [8]. Increasingly, designers work with all of the stakeholders to intervene proactively in supporting positive lived experiences whilst challenging and changing persistent negative experiences. As Cross [2] suggests design is a human activity derived to respond to human behaviours and requirements and for making novel and informed solutions. As such designers are well placed for working in challenging areas such as supporting people living with dementia. It is here that the designer's ingenuity for producing integrated systems, for making things and for intervening in the hope of improving the status quo is increasingly being incorporated into care planning and delivery [9]. This may take the form of service or product designs, systems, environments, and technologies or through collaborating with those they mean to help.

The work in this chapter looks at how the integrative collaborative actions of Co-design [15] and Design Disruption [12] as theoretical approaches, involve people living with dementia in rethinking and reshaping or circumventing conventional approaches to dementia care. Moreover, this work seeks to change mind-sets and existing prejudiced ideas about what people living with dementia might be capable of. The activity of collaboratively designing with people who are not designers themselves, seeks to challenge and alter preconceived ideas about the capabilities of people living with a diagnosis of dementia. In this work, Co-design is used as a facilitating proposition that supports people to act and respond independently and collectively. Here, the people living with dementia inform activities of designing through personal engagement, knowledge, insight and emotional sensibilities that may have been disregarded as a result of their diagnosis. They are then invited to act upon these insights and responses through further design actions. This work poses possibilities as to how people can be empowered by design, how they can find ways to inform a situation, or be supported in informing conditions where their personal identity, values, knowledge, skills, experiences, perspectives and thoughts are integral to the production of new ideas and ways of thinking and doing. The thinking being particularly important in forming propositions, responding to opportunities and decision making that leads to new realities through self-efficacy [1].

## 19.2 The Key Issues of Dementia

Dementia is the umbrella term for a range of brain diseases that are progressive and chronic in their nature. Symptoms include deterioration in cognitive function, behavioural changes and functional limitations. These degenerative conditions occur over an indeterminate period of time and strip a person of their identity, understanding, abilities and independence and can affect people in very different ways [6].

Common symptoms of dementia can include problems with short term memory where new information is difficult to retain. People with dementia can get lost in seemingly familiar places, may experience confusion with names, and may also experience confusion in environments which are unfamiliar to them. Even, language can become troublesome and methods of communication restricted as words and intentions become confused.

People living with dementia may lose interest in engaging with others socially and may become quieter and more introverted. The needs that require support can therefore be varied but often overlapping, and can result in very different personal experiences even though the same diagnosed conditions are being lived through. What this means is how people live is of paramount importance. Even when a person seems to have lost a significant part of what made them a unique individual, core elements of their identity will remain [6]. These characteristic gestures and ways of doing things are what keep alive the sense of the individual they once were, even if the more sophisticated cognitive levels of that individual have been eroded [3]. This has important implications for the approach to providing support and what people require in addition to the basics of daily living.

In the UK, much of the support and care of people living with dementia is delivered by families and friends who often struggle to fulfil the gamut of needs that lie well beyond basic needs. Their challenge is to deliver the initial care and support but to also ensure love, respect and individual rights. Carers are charged with continuing to promote the value of the lived experience and the right of the individual to engage with the world around them and to engage with society as much as possible. This is not always easy or possible. As such, there is a requirement for charitable organisations to underpin care provision by providing respite and activities that can enhance the lived experience of both people living with dementia and their carers. These services endeavour to promote opportunities that engage people through personally relatable actions. It is here that the design research contained in this chapter is situated.

Working with people living with dementia, carers and charitable organisations, the projects presented have aimed to nurture the latent capabilities [7] of participants and in doing so support the attainment of skills and outcomes that engender esteem and self-actualisation [10]. The results of which convey a sense of value for both the participants' and the people their work engages. The work therefore does not set out to resolve better ways of opening things, provide tools for remotely accessing a doctor, or suggest utensils or gadgets for resolving a difficulty that must be overcome. Alternatively, the design interventions here look to build upon emotional intelligence [5] and the power of being involved in activities that support a person's individual identity. The outcomes of which help to narrate a person living with dementia's interests or engagements in groups, excursions and tasks, along with encouraging participation in and influence upon communities.

The following co-design projects have attempted to engage fully with people living with dementia throughout the design process. The approach of working ***with*** people means that intentions and outcomes may have evolved within projects and that the direction of some outcomes have been removed from the original discussion as new opportunities have manifested. Central to the considered approaches undertaken in all of the projects, is the development of working and collaborative relationships that nurture trust and confidence [13]. From conception to completion the projects have been devised and undertaken from a 'designing with' perspective where the user is not viewed as a 'subject' but rather as an active 'partner' in the project [14]. Here, Donetto et al.'s 'Experience-Based Co-Design (EBCD)' has been developed as a participatory research approach that draws upon design tools and ways of thinking in order to bring stakeholders and participants together to improve experiences within care dynamics. In particular, this has afforded co-design practice to bring about reconfigurations of power relations, the appropriate role of design expertise within such processes and their eventual impact on the care framed experiences [4]. As such more often than not it is challenging to distinguish ownership and authorship of the projects and their outcomes, an admirable recognition of the contribution, collaboration and immersion of all of the parties involved.

## 19.3 Designed with Dementia; A Suite of Design Projects

Designed with Me represents a suite of projects that have been undertaken with people who are living with Dementia and by collaborating with Alzheimer Scotland. The projects cover a number of processes and investigations into how we can use design to unlock and reveal personal and collective capabilities, utilising and stimulating interests and arranging new ways of doing things. Essential to this approach is the idea that design is a social activity and as such forms a space for shared enjoyment and fun in safe experimental conditions [12]. In this situation, each person's input is valued and any failure is not seen as problematic or undermining but as a positive part of the process or something to be learned from and responded to. The authors have undertaken a variety of projects to develop, reveal and present designed solutions by people living with dementia. Where they are helping to "shake things up" through their actions that highlight their capabilities, individually and collectively, and indeed, by directing projects with people who do not have a diagnosis of dementia. The result is a number of thought-provoking outcomes that challenge assumptions but that have also developed people living with dementia's sense of self and value through personal experiences.

Throughout the process kits, tools and devices have been developed to further ongoing conversations and to build the importance of those individuals taking part in the projects. These have been responsive interventions and though they have stimulated or supported the next stage in a creative journey they have not been the catalyst for action. Within this work the most important actions have come from open and fun conversations which will be discussed in relation to what this work has achieved. The importance of this work lies in how people living with dementia have been able to form ideas, develop solutions and propose new designs for public appreciation, consumption or use. The Designed with Dementia projects discussed within this chapter include; Dementia Tartan, Perfect Day Cards, 75BC and a Stained-Glass Window for Glasgow. All of which have used collaborative design processes to create new artefacts where individual and collective actions have been valued and displayed participants' self-efficacy. The projects discussed here share intimate small group focussed working practices along with larger projects which have collectively engaged in excess of 200 people who are living with dementia.

## 19.4 Designed with Dementia: 75BC

The "Designed with Dementia" intervention, 75BC, is a collaborative design project that explores representations of the Glaswegian comedian Billy Connolly. Initiated through conversations about visits to the 75BC murals (Glasgow) five people living with dementia posed that the murals were not befitting of their ideas of Connolly. Inspired by further visits to the 75BC Exhibition at the People's Palace, (Glasgow) and to the American artist Tschabalala Self's exhibition at the Tramway Gallery, (Glasgow) the group developed an idea of what Connolly meant to them and how those thoughts might take form. This resulted in the design of new fabrics and products. What occurred in delivery of their ideas was a co-design approach that became recognisable as a typical design process, at the core of this process were discussions and decisions undertaken with all of the participants in an open forum:

**Stage 1**, introduced the group to undertaking primary research where the participants made use of digital cameras to freely photograph displayed artworks at the aforementioned exhibitions. They focussed on documenting composition, structure and personally appealing attributes. The image rich investigations that the group had taken part in informed the framework and content of the design actions in Stage 2.

**Stage 2**, involved a project kit that repurposed Tschabalala Self's artistic process where different scraps of fabric formed different body parts. Here, the participants created new representations of Billy Connolly. The form of which were based upon more widely recognised images of Billy Connolly from highlights of his career.

**Stage 3**, resulted in Pattern Forming of the new visual representations of Billy Connolly. The reconsideration of, and composition of which, stemmed from the unexpected arrangements that occurred in Stage 2.

**Stage 4**, involved the participants reviewing patterns and production scales of the designs that had been created. Five variant sizes of each pattern were printed on paper in order to select patterns and production sizes where three patterns would be printed as prototype textiles. The group selected four. The selection of four were produced as prototypes in heavyweight cotton using a commercial digital textile printer based in Edinburgh.

**Stage 5**, explored where and how these new prototype products might be used. Through discussion it was agreed that the textiles were most likely going to be used in a range of household objects and furnishings. The group chose to work with a range of interior products which included a lamp, a light, cushion, rug, bed linen, a sofa and a lounge chair. Through a kit devised to make the process of applying textiles to the rudimentary household objects. The group collectively created a range of products that made great use of the textiles they had designed. The processes

**Fig. 19.1** The processes and results of designing and applying the 75BC fabrics

allowed for iterations and supported decisive finishing points where the participants were happy with their designs (Fig. 19.1).

The 75BC designs, along with other participatory resources the group had co-designed formed an exhibition and public engagement event in Lancaster. Here, 200 people explored the group's designs and were invited to creative textiles activities. The event afforded children as young as one through to adults in their nineties to be inspired to get involved and to use the co-designed textile printing kits that had been developed by the co-design group. The impact of which included changing perspectives of those who visited, generating new understanding of capabilities of people living with dementia.

## 19.5 Designed with Dementia: Stained-Glass Window for Glasgow

Developed with another group of five people living with dementia and based upon similar kinds of cultural-visits as the previous project. This Collaborative work engaged with excursions to the Scotland Street Museum, Glasgow designed by Charles Rennie Mackintosh and St. Mungo's Museum of Religious Life and Art again in Glasgow. During both visits, the importance to Glasgow of Mackintosh and his influence on contemporary stained-glass design formed a significant part of the discussion. This became particularly pertinent as the second Glasgow School of Art fire had just occurred. The destruction of the building and the significance of his work in relation to what the group had seen encouraged their idea to further explore the design of stained-glass artefacts through collaborative workshops.

**Stage 1**, primary investigations during these visits were much like the 75BC project where images were made and collected of the designs and artworks that were observed. However, the photographs taken on these occasions were guided by the participants rather than taken by them. During these visits, the groups also took part in designing their own versions of stained-glass at both sites through different techniques one using pre-cut tiles of colour and the other using stained-glass colouring pens.

**Stage 2**, in response to their desire to explore the process of designing stained glass and as a result of the influences they had observed during their visits the

**Fig. 19.2** Prototyping the stained glass window for Glasgow. Framed within a backlit panel the work exists somewhere between an artwork of the city and prototype design stained glass-window

group undertook a workshop that allowed them to explore different media. Here they produced a Mackintosh style pendant light using lighting gels and stained-glass pens to explore some of his pattern designs. This process allowed the group to come together in making objects collectively. Furthermore, it triggered discussions about what their future projects should encompass.

**Stage 3**, developed the kind of themes that might be explored and used kits of options to help inspire new designs. The theme of this session was the Glasgow coat of arms. The participants collaged elements of transparent line drawings and texts in order to generate their own layouts. The stained-glass pens were then used to add colour to these designs.

**Stage 4**, progressed the ideas that the group had about developing a Stained-glass Window for Glasgow. The focus of which was built upon the kinds of scenes and buildings that had been discussed as being iconic for Glasgow. These were produced in the form of black printed transparencies that the group could arrange to form a collective image which became the basis of their new stained-glass design.

**Stage 5**, the pasted-up image was scanned and produced as transparent designs for the participants to colour by means of acrylic paint. The result of which was two versions of the same image that they decided looked even better when overlaid upon each other (Fig. 19.2).

## 19.6 Designed with Dementia: Perfect Day Cards

The "Perfect Day" service was initiated following visits to over 20 Alzheimer Scotland day care centres across Scotland. During these visits, first-hand opportunities to work and chat with people living with dementia, their family members, and the Alzheimer Scotland care support workers was achieved. During this time, it quickly became obvious that the day-to-day work of the care support workers is very challenging and demanding. In particular, the care support workers are continually challenged to provide activities and services that are often conflicting in nature (i.e. highly personalized yet open to all, flexible yet focused, etc.) Moreover, the care support workers need to ensure that any service or activity is appropriate for a wide range of people living with dementia who may have different levels of independence,

**Fig. 19.3** Samples of the Perfect Day cards

physical, cognitive and other abilities. It became abundantly clear that a service design intervention aimed at supporting Alzheimer Scotland care support workers would be very welcome and that these could be achieved by including people living with dementia. Here their input, guidance and desires were revealed and used to inform a range of 'Perfect Day' examples. The idea was to explore what the 'Perfect Day' might look/feel/taste/sound like for people living with dementia. The result of which would be a kit of actions that carers could make use of. Working with people living with dementia a semi-structured interview approach, prompted input and collected insight as to what should be contained in the 'Perfect Day'. As such the person living with dementia was invited to think and talk about their lives and their personal preferences by answering (Fig. 19.3):

- Where have you lived?
- What jobs have you done?
- What would your 'Perfect Day' look like?
- What would you do?
- Where would you do it?
- What would it involve?
- What would you need?
- Why is it perfect?

The process put the individual at the centre of a discussion about what they thought was important and asked of them why that might be. Their answers were then fashioned into a card that gave instruction as to how to create their perfect day. The result was set of 75 double-sided 'Perfect Day' cards that day care support workers can use to help plan and run their daily activities and sessions for and with people living with dementia. The collaborations in these forms were social conversational experiences and occurred between the researcher and people living with dementia and their support workers. Though they did not follow the same kinds of processes as discussed within the 75BC and Stained-Glass Window projects the results were equally unachievable without the buy in and contribution of the collaborators. Their input and propositions resulted in shaping real tangible products to be used by people across Scotland. This product provides care support workers with thoughtful guides for activities that they might never have previously considered. In doing so making unseen connections where one person's 'Perfect Day' in Glasgow might be really enjoyable and relevant for another person's 'Perfect Day' in Orkney and vice versa.

Many of the 'Perfect Day' ideas are appropriate for both male and female and cover a range of abilities encouraging active participation for people living with dementia.

## 19.7 Designed with Dementia: Disrupting Dementia Tartan

The catalyst for much of the work that has been discussed previously in this chapter was the Disrupting Dementia Tartan project, that looked to break the preconceived notions of capabilities, or incapability to be more precise. In challenging why people living with dementia have never before been considered capable of designing a commercial product, new methods and approaches were developed and applied. The project set-out to design and test an inclusive and collaborative set of platforms that would facilitate multiple individual designs. The approach utilised both tactile making kits and online conversion software which facilitated each co-designer to produce a sketch prototype and a digital production ready version. These designs were evaluated by an industry and care sector panel in order to select and produce a commercially viable tartan.

The workshops in this project commenced with a short presentation of the rules associated with the creation of the Disrupting Dementia tartan. The rules of the Disrupting Dementia tartan project are that each participant must use no more than 6 colours in their design and one of those colours must be purple (Alzheimer Scotland's primary colour in their new brand identity). Unlike the previously described projects these collaborative creations worked within these parameters in order to be consistent with tartan industry standards and as such restricted more extreme deviations in form. However, each participant was free to determine and shape the tartan design they created during the stages of the design process.

**Stage 1**, involved the creation of each participant's tartan design starting with an acetate-based version.

**Stage 2**, developed the creation of a physical prototype constructed using ribbon.

**Stage 3**, involved the creation of a digital version using a publically available Internet-based tartan design tool from the physical ribbon prototypes, each person with dementia directed the researcher to co-create his or her digital design one colour at a time (Fig. 19.4).

**Fig. 19.4** Processes involved in the production of the disrupting Dementia Tartan

The stages of development required iterative design approaches by the designers (people living with dementia) where they tested ideas, evaluated them and responded accordingly. Sometimes, this would involve going back and forth between different versions and sometimes they would alter the order of the colours to finally achieve the design that they were satisfied with.

In the Disrupting Dementia Tartan project, some of the co-design participants provided rich knowledge of local and regional tartans, their own skills and understanding of the craft of weaving, and even some experiences of working in the tartan industry which was a major bonus. The creative actions within the collaborative approach stimulated a real sense of re-connecting to their 'past lives' that evoked positive memories of association for many individuals.

This 'live' project brought together many stakeholders and gave access to the design process to a wide demographic through exhibition, discussion and the resultant product. Their appreciation of what people living with dementia are capable of undertaking has been challenged proposing the effectiveness of co-design in developing valued and valuable contributions to society, and which in turn promotes self-efficacy of people living with dementia.

The Designed with Dementia projects have been exhibited at national museums including Verdant Works, Dundee and the Stirling Smith Art Gallery, Stirling. Both exhibitions have attracted thousands of visitors and have included public engagement events such as lectures and workshops. Both exhibitions have also attracted national press coverage in press and social media outlets.

## 19.8 Analysis

A range of positive comments have been made by the co-designers (people living with dementia), their carers, and their family members throughout every stage of the design projects described in this paper. These comments have, at various points, driven key activities in the projects that have been undertaken. A selection of some of the co-designers' comments made during these projects is shown in Fig. 19.5.

These comments clearly articulate positive feelings towards the design activities conducted and they suggest value in the design exploration work. Equally important here is the voice that each participant (co-designer) has had in making decisions that have informed the design outcomes and key influences within each project. The suite of design projects described here, at their outset, were conceived to support and nurture the opinions and decision making of the co-designers (people living with dementia). A significant consideration within all of the projects was that the design sessions must be enjoyable and must offer a safe space for diverse ideas to formulate where differing opinions can occur. The responses and feedback collected from the co-designers during the series of co-design projects described in this paper are rich and informative. During the co-design sessions, the co-designers were happy and felt confident to assert their opinions openly: "Nah it's no' for me …", "It gave me the creeps …", "No. I know how it's been done, and it's been done well, but I

**Fig. 19.5** A selection of co-designers' comments made during the projects

**Fig. 19.6** Comments made by members of the public during the design projects

don't like it", "but that's the one I want" were just some of the phrases collected as the co-designers (people living with dementia) discussed various project tasks and actions.

Furthermore, during the workshops, public engagement events, and exhibitions of co-designed work the value of what has been designed and exhibited has been ascertained through a number of means including feedback commentary cards and "share your thoughts" prompts. This has allowed the authors to collect diverse sets of opinions, perceptions, and comments from the general public about what might be important in terms of how we view people living with dementia and their place in wider society (Fig. 19.6).

## 19.9 Why Are These Interventions Important?

All of the co-design projects presented in this chapter have afforded a large number of people living with dementia opportunities to use design as a social activity. Their involvement in the co-designing processes described here have delivered new ways of looking at what is possible, new creative interactions, the creation of innovative tools and products for many people (not just people living with dementia) to enjoy. To evaluate the impact of these co-design projects much consideration has been given to the conversations, comments and feedback that have accompanied each co-design session. Equally important has been the observed responses by the people living with dementia to what has been going on in the co-design activities. More often than not these observations have revealed a clear willingness on the part of the people living with dementia to be involved, to take control of their own design output, and to lay claim to authorship/ownership of the creative outcomes produced. Through their demonstration of creating their designs, the people living with dementia regularly displayed independence in thinking and acting, and the importance of personal preferences. The actions, comments and feedback received in the co-design projects presented here show a strong desire of people living with dementia to be involved as well as engaging with and driving the design decisions that are needed to be taken. In all of these co-design projects, the people living with dementia have proffered a sense of eagerness and enjoyment but also pride and esteem in their co-design work.

## 19.10 Conclusions

In the co-design projects presented in this chapter, the iterative nature of the design projects leads to extended and interlinking thought processes, which in turn, results in design solutions that display the power of people living with dementia and the meaningful ways in which they can interact with and inform the world. This view presents a situation where design acts as an instigator and mediator in small and large social transactions. But more than that, the work presented in this chapter has allowed the authors to access the insights, ideas and comments of people living with dementia that has resulted in designed outcomes that are directed by the people living with dementia and made accessible for the enjoyment and appreciation of other people. The authors are of the opinion that many of the artefacts designed in these co-design sessions have the capacity to generate financial income for their designers and as such provide alternative ways in which people living with dementia might support themselves. Moreover, the authors believe the co-designed work presented here has the potential to change the opinions and perceptions of many in society who believe that a diagnosis of dementia represents the end of a person's creative and other abilities.

Furthermore, the personalised yet collective participation in these co-design activities supports the person living with dementia to have a better quality of life. Where the activities and contributions they make and undertake reinforces the right to active participation and citizenship, the inclusive nature of the co-design activities and resultant designed outcomes provide meaningful and relevant engagement for each individual. The co-design sessions aim to maintain the fabric of their life for as long as possible. Giving additional time to a person with dementia saves time and other valuable resources in the long run. Supporting a person to do things for themselves, rather than carrying out the task for them, maintains function and skills in their other activities of daily living. Throughout the co-design sessions described in this chapter, every individual was involved in decision making, offering advice and feedback, and tasks that stimulated their fine motor skills. The co-design sessions have helped support the UK Government's agenda of keeping people both mentally and physically active. This approach to support continues to empower the person living with dementia to retain personal agendas within their lived experiences reaffirming capabilities and stopping people from being deemed incapable before their time. What design affords people living with dementia, in this context, is group-led and collaborative social interaction, and ultimately the production of designed outcomes that can be purchased and used by audiences beyond the natural reach of the participants. This chapter clearly shows that the co-design sessions described here build self-efficacy to achieve empowerment that change perceptions and expectations whilst creating a wide range of designed outcomes that are desirable for many.

## References


1. Cervone, D.: Thinking about self-efficacy. Behav. Modif. **24**(1), 30–56 (2000)
2. Cross, N.: Design Thinking: Understanding How Designers Think and Work. Bloomsbury, London (2011)
3. Crutch, S.: Dementia and the Arts: Understanding the Differences Between the Dementias and People Living With Them. Lecture, Edinburgh (2018)
4. Donetto, S., Pierri, P., Tsianakas, V., Robert, G.: Experience-based co-design and healthcare improvement: realizing participatory design in the public sector. Des. J. **18**(2), 227–248 (2015)
5. Evans, D.: Emotion: The Science of Sentiment. Oxford University Press, Oxford (2001)
6. Greenfield, S.: You and Me: The Neuroscience of Identity. Notting Hill Editions, London (2011)
7. Kelley, T., Kelley, D.: Creative confidence: unleashing the creative potential within us all. William Collins, London, UK (2015)
8. Kitwood, T.: Quality of life. In: Katz, J., Peace, S., Spur, S. (eds.) Adult Lives: A Life Course Perspective. Policy Press and Open University Press, London (2011)
9. Muratovski, G.: Research for Designers: Guide to Methods and Practice. Sage, London (2016)
10. Maslow, A.: A theory of human motivation. Psychol. Rev. **50**(4), 370–396 (1943)
11. Phillips, D.: Quality of life. In: Katz, J., Peace, S., Spur, S. (eds.) Adult Lives: A Life Course Perspective. Policy Press and Open University Press, London (2011)
12. Rodgers, P.A., Tennant, A.: Disrupting health and social care by design. In: Proceedings of the 9th International Conference on Design & Emotion, Bogota, Colombia, Oct 2014
13. Rodgers, P.A.: Co-designing with people living with dementia. CoDesign (2017). https://doi.org/10.1080/15710882.2017.1282527
14. Sanders, E.B.-N., Stappers, P.J.: Co-creation and the new landscapes of design. CoDesign **4**(1), 5–18 (2008)
15. Scrivener, S.: Editorial: CoDesign (2005). https://doi.org/10.1080/15710880412331289935


**Euan Winton** is a Lecturer (Edinburgh Napier University) and an AHRC Ph.D. Student (Lancaster University). Euan has developed and delivered undergraduate and post graduate programmes in design for ENU including the undergraduate programme of Design and Digital Arts along with seven pathways in postgraduate studies incorporating; Lighting, Interaction, Graphics,

Motion Graphics, Product, Interior and Exhibition design. Nationally and internationally he has undertaken design research practice in areas of installation design and public engagement. Euan embeds inclusive approaches in his work developing co-design practices with non-designers such as his Ph.D. work with people living with dementia. For the past 15 years Euan's public collaborations have involved creating opportunities and platforms that actively engage people with subjects relevant to them. Here he combines a range of tools, technologies and media to inspire input which reveals insight that informs and shapes their designed futures.

**Paul A. Rodgers** is Professor of Design at Imagination, Lancaster University and the Arts and Humanities Research Council (UK) Design Leadership Fellow (2017–2020). Prior to joining Lancaster University in 2016, he was Professor of Design Issues at Northumbria University, School of Design (2009–2016), Reader in Design at Edinburgh Napier University (1999–2009), and a postdoctoral Research Fellow at the University of Cambridge's Engineering Design Centre (1996–1999). Professor Rodgers has had a distinguished and extensive career in design research with nearly 25 years of experience in product design research and development. He has led several research projects for all of the main Research Councils in the UK and led design projects funded by the Scottish Government and The Lighthouse (Scotland's National Centre for Architecture, Design and the City). He is the author of more than 150 papers and 11 books, including The Routledge Companion to Design Research (2014) and Design School: After Boundaries and Disciplines (2019). His books have been translated into several languages including Spanish, Italian, Taiwanese, and Chinese. His current research explores the discipline of design and how disruptive design interventions can enact positive transformational change in health and social care and elsewhere.

The manufacturer's authorised representative in the EU is Springer Nature Customer Service Centre GmbH, Europaplatz 3, 69115 Heidelberg, Germany. If you have any concerns regarding our products, please contact ProductSafety@springernature.com

Printed and bound by CPI Group (UK) Ltd, Croydon, CR0 4YY
22/09/2026
02221049-0001